Editorial Board Members' Collection Series: Metal Crystal/Polycrystal Plastic Strain Hardening

Editorial Board Members' Collection Series: Metal Crystal/Polycrystal Plastic Strain Hardening

Guest Editors

John D. Clayton
Ronald W. Armstrong

MDPI

Basel • Beijing • Wuhan • Barcelona • Belgrade • Novi Sad • Cluj • Manchester

Guest Editors
John D. Clayton
Terminal Effects Division
DEVCOM Army Research Laboratory
Adelphi, MD
USA

Ronald W. Armstrong
Department of Mechanical Engineering
University of Maryland
College Park, MD
USA

Editorial Office
MDPI AG
Grosspeteranlage 5
4052 Basel, Switzerland

This is a reprint of the Special Issue, published open access by the journal *Metals* (ISSN 2075-4701), freely accessible at: https://www.mdpi.com/journal/metals/special_issues/3OS87W55R4.

For citation purposes, cite each article independently as indicated on the article page online and as indicated below:

Lastname, A.A.; Lastname, B.B. Article Title. *Journal Name* **Year**, *Volume Number*, Page Range.

ISBN 978-3-7258-5615-2 (Hbk)
ISBN 978-3-7258-5616-9 (PDF)
https://doi.org/10.3390/books978-3-7258-5616-9

Contents

metals

MDPI

Editorial

Metal Crystal and Polycrystal Plastic Strain Hardening

John D. Clayton

Terminal Effects Division, DEVCOM ARL, Aberdeen, MD 21005-5066, USA; john.d.clayton1.civ@army.mil

1. Introduction

Crystalline metallic solids are key components of engineering devices and structures whose manufacture and performance often crucially involve, or seek to minimize, plastic deformation phenomena. Inelastic behavior originates at the nanometer scale for individual dislocations, whereas the response of a polycrystalline structure is more complex, spanning scales many orders of magnitude larger, and reflects the complexity of physics, including dislocation and defect interactions, grain and phase boundaries, texture effects, and so on. Vastly different plastic and fracture behaviors emerge among metals and alloys when time scales or loading rates, temperatures, and stress-deformation histories differ [1]. Owing to this complexity, many questions remain unanswered, and physics-based models are neither universal nor satisfactory for all contemporary applications, despite decades of research.

The study of the inelastic mechanical responses of single-crystal and polycrystalline metals has a long history. Pioneering works by G.I. Taylor [2] and R. Hill [3] considered the plastic yield and hardening response of crystals and polycrystals. Size effects (e.g., grain diameter) in plasticity and fracture were addressed in seminal studies by Hall [4], Petch [5], and Armstrong et al. [6] in the mid-twentieth century. Classic texts on dislocations by Nabarro [7] and Hirthe and Lothe [8] provide insight at lower length scales, while many books [9–14] and review articles [15–17] describe constitutive models of the inelastic responses of crystalline metals. In the late twentieth century [18–22], several prominent studies were published regarding strain hardening or polycrystal response from the perspectives of crystal interactions, dislocation mechanics, or micromechanics. Keen interest in these subjects still persists, as evidenced by numerous scholarly publications (for which a literature review is outside the scope of this brief article) in the fields of solid mechanics, materials science, and materials physics.

The aforementioned topics related to the inelastic behavior of metallic crystals and polycrystals are explored in this Editorial, showcasing their historical and ongoing interest. Research contributions toward these topics, specifically those appearing in a recent Special Issue of the journal, are collectively summarized and thereby placed in a broader context.

2. Discussion

A Special Issue of *Metals*, entitled "Metal Crystal/Polycrystal Plastic Strain Hardening", was initiated in 2022 by the late Professor Ronald W. Armstrong to advance the current state of the art. Bibliographic details are listed for each contribution following the Section 3 of this Editorial. In its final form (2025), this issue comprises ten published works: an initial Editorial article (contribution 1), six regular research papers (contributions 2 through 7), one research communication (contribution 8), one review article (contribution 9), and the present closing Editorial (contribution 10, unlisted). The key outcomes from each of the first nine contributions are discussed below.

Armstrong (contribution 1) concisely described advances in our understanding of strength and strain hardening from the historic and recent literature on dislocation mechanics that emerged from G.I. Taylor's pioneering work. The specific topics explored were partial dislocations and stacking faults, cross-slip, dislocation reactions, and grain size effects. Armstrong also reported on the plastic instability experienced during the tensile loading of FCC and BCC polycrystals.

Saffarini and Voyiadjis (contribution 2) modeled plastic flow in gold metallic foams at very high compressive strain rates. Large-scale atomistic simulations of foam samples with different domain sizes were used to provide numerical data for upscaling. A continuum plasticity framework accounting for dislocation evolution, strain hardening, and strain rate and thermal effects was parameterized to successfully capture the stress–strain behaviors witnessed in the simulations. The outcomes were compared with those obtained with other continuum models.

Vinogradov and Estrin (contribution 3) analyzed grain-size effects in the context of tensile plastic deformation instability. The Kocks–Mecking–Estrin hardening model, based on dislocation evolution, and in which the grain size enters indirectly, was used to evaluate necking stress. The Hall–Petch effect, in which strength increases with decreasing grain size, was shown to be better described by the necking stress from the Considère criterion than the usual 0.2% offset yield strength. The results, encompassing data for nickel and titanium, have ramifications for pragmatic parameter determination and materials design.

Zubelewicz and Clayton (contribution 4) modeled the low-temperature behavior of BCC metals, focusing on yield surfaces, plastic potentials, and plastic volume changes. Their new description included terms in the yield stress and plastic potential that were associated with dislocation core spreading and atomic friction, leading to non-Schmid effects. The model's accuracy was demonstrated by comparison with experimental data on molybdenum. In the absence of repulsive core pressure, the analysis further predicted a null correlation between plastic dilatation due to anharmonic effects and dislocation core spreading affecting lattice friction.

Kunda et al. (contribution 5) studied the thermo-mechanical responses of copper single crystals and polycrystals by measuring their stress, strain, and temperature evolution for both static and dynamic compression simultaneously at strain rates spanning 10^{-3}/s to 4800/s. A novel finite-strain crystal plasticity constitutive model was implemented in finite element (FE) simulations, with rigorous thermodynamics based on dislocation theory that enabled a non-constant Taylor–Quinney factor. The FE simulations showed promising agreement with experimental stress and temperature data and provided physical insight into dislocation mechanisms and thermodynamic processes at the slip-system scale.

In a series of two papers, You and Hasebe (contributions 6 and 7) incorporated Hasebe's field theory of multiscale plasticity (FTMP) into FE crystal plasticity simulations of the slip bands responsible for metal fatigue crack initiation. The FTMP accounts for differential-geometric incompatibility, physically associated with strain gradients and dislocation structures. The first paper (contribution 6) showed predictions of laddered band morphologies typically obtained in experiments. The second paper (contribution 7) demonstrated an extended field theoretic model augmented with vacancy sources, showing that vacancy diffusion is affected by dislocation flux and the incompatibility rate. The theory was implemented in multi-physics crystal plasticity simulations with diffusion kinetics, using both indirectly and directly coupled schemes. The results were in agreement with those obtained by transmission electron microscopy and provided new insight into the microstructure-based origins of surface roughening followed by grooving. The latter eventually transitioned to cracks under cyclic straining.

Yan and Fu (contribution 8) studied the ductility of BCC refractory alloys, including high-entropy alloys, using Density Functional Theory (DFT). Their calculations produced a new measure of average bond stiffness for solid (alloy) solutions. Decreasing this stiffness, or alternatively increasing the average static displacement of atoms, was shown to correlate with increasing ductility (e.g., fracture strain measured in experiments). This information provides guidance on composition to optimize the properties of refractory alloys.

Clayton et al. (contribution 9) reviewed classical and contemporary experiments and models, both analytical and numerical, for the spherical indentation of polycrystalline metals. A new dynamic dimensional analysis framework was implemented to extract uniaxial-equivalent stress–strain curves and material properties (e.g., yield and strain hardening) using experimental force–depth data from dynamic spherical indentation tests. The uility and limitations of the coupled experimental–analytical methodology were assessed in the context of dynamic indentation data for the aluminum alloy Al 6061-T6.

3. Conclusions

The research published in this recent Special Issue, primarily focused on the constitutive modeling of crystalline metals' plastic response, has been summarized above. These works collectively enhance our current understanding of the effects of dislocations, microstructures, chemistry, and loading conditions on the mechanics and thermodynamics of metals across a range of length and time scales, which we encourage future studies to further expand upon.

Funding: This research received no external funding.

Acknowledgments: The *Metals* journal staff are thanked for their efforts in supporting and sustaining this Special Issue. The authors of all published manuscripts are thanked for their excellent contributions. Most importantly, the late Ronald W. Armstrong, who conceived and initiated this Special Issue in 2022, is acknowledged for his prominent role, widespread and collegial interactions with the scientific community, and profound technical contributions to materials science and solid mechanics in an astounding research career spanning over six decades.

Conflicts of Interest: The author declares no conflicts of interest.

List of Contributions

1. Armstrong, R. Metal crystal/polycrystal plasticity and strengths. *Metals* **2022**, *12*, 2070. https://doi.org/10.3390/met9020154.
2. Saffarini, M.; Voyiadjis, G. Atomistic-continuum constitutive modeling connection for gold foams under compression at high strain rates: the dislocation density effect. *Metals* **2023**, *13*, 652. https://doi.org/10.3390/met13040652.
3. Vinogradov, A.; Estrin, Y. Hall-Petch description of the necking point stress. *Metals* **2023**, *13*, 690. https://doi.org/10.3390/met13040690.
4. Zubelewicz, A.; Clayton, J. Yield surfaces and plastic potentials for metals, with analysis of plastic dilatation and strength asymmetry in BCC crystals. *Metals* **2023**, *13*, 523. https://doi.org/10.3390/met13030523.
5. Kunda, S.; Schmelzer, N.; Pedgaonkar, A.; Rees, J.; Dunham, S.; Lieou, C.; Langbaum, J.; Bronkhorst, C. Study of the thermomechanical behavior of single-crystal and polycrystal copper. *Metals* **2024**, *14*, 1086. https://doi.org/10.3390/met14091086.
6. You, X.; Hasebe, T. Modeling and simulation of fatigue crack initiation process based on field theory of multiscale plasticity (FTMP). Part I: PSB ladder formation and verification. *Metals* **2024**, *14*, 1392. https://doi.org/10.3390/met14121392.
7. You, X.; Hasebe, T. Modeling and simulation of fatigue crack initiation process based on field theory of multiscale plasticity (FTMP). Part II: Modeling vacancy formation

and coupling with diffusion analysis. *Metals* **2024**, *14*, 1406. https://doi.org/10.3390/met14121406.

8. Yan, J.; Fu, C. Understanding room-temperature ductility of bcc refractory alloys from their atomistic-level features. *Metals* **2025**, *15*, 851. https://doi.org/10.3390/met15080851.
9. Clayton, J.; Casem, D.; Lloyd, J.; Retzlaff, E. Toward material property extraction from dynamic spherical indentation experiments on hardening polycrystalline metals. *Metals* **2023**, *13*, 276. https://doi.org/10.3390/met13020276.

References

1. Armstrong, R. Dislocation mechanics pile-up and thermal activation roles in metal plasticity and fracturing. *Metals* **2019**, *9*, 154. [CrossRef]
2. Taylor, G. The mechanism of plastic deformation of crystals. Part I. Theoretical. *Proc. R. Soc. Lond. A* **1934**, *145*, 362–387.
3. Hill, R. *The Mathematical Theory of Plasticity*; Oxford University Press: New York, NY, USA, 1950.
4. Hall, E. The deformation and ageing of mild steel: Discussion of results. *Proc. Phys. Soc. Lond. B* **1951**, *64*, 747–753. [CrossRef]
5. Petch, N. The cleavage strength of polycrystals. *J. Iron Steel Inst.* **1953**, *174*, 25–28.
6. Armstrong, R.; Codd, I.; Douthwaite, R.; Petch, N. The plastic deformation of polycrystalline aggregates. *Philos. Mag.* **1962**, *7*, 45–58. [CrossRef]
7. Nabarro, F. *Theory of Crystal Dislocations*; Oxford University Press: London, UK, 1967.
8. Hirth, J.; Lothe, J. *Theory of Dislocations*; John Wiley and Sons: New York, NY, USA, 1982.
9. Havner, K. *Finite Plastic Deformation of Crystalline Solids*; Cambridge University Press: Cambridge, UK, 1992.
10. Voyiadjis, G.; Kattan, P. *Advances in Damage Mechanics: Metals and Metal Matrix Composites*; Elsevier: Amsterdam, The Netherlands, 1999.
11. Lubarda, V. *Elastoplasticity Theory*; CRC Press: Boca Raton, FL, USA, 2001.
12. Nemat-Nasser, S. *Plasticity: A Treatise on Finite Deformation of Heterogeneous Inelastic Materials*; Cambridge University Press: Cambridge, UK, 2004.
13. Clayton, J. *Nonlinear Mechanics of Crystals*; Springer: Dordrecht, The Netherlands, 2011.
14. Hasebe, T. *Field Theory of Multiscale Plasticity*; Cambridge University Press: Cambridge, UK, 2024.
15. Kocks, U.; Argon, A.; Ashby, M. Thermodynamics and kinetics of slip. *Prog. Mater. Sci.* **1975**, *19*, 1–291.
16. Estrin, Y. Dislocation-density-related constitutive modeling. In *Unified Constitutive Laws of Plastic Deformation*; Krausz, A., Krausz, K., Eds.; Academic Press: San Diego, CA, USA, 1996; pp. 69–106.
17. McDowell, D. Viscoplasticity of heterogeneous metallic materials. *Mater. Sci. Eng. R Rep.* **2008**, *62*, 67–123. [CrossRef]
18. Mecking, H.; Kocks, U. Kinetics of flow and strain-hardening. *Acta Metall.* **1981**, *29*, 1865–1875. [CrossRef]
19. Zerilli, F.; Armstrong, R. Dislocation-mechanics-based constitutive equations for material dynamics calculations. *J. Appl. Phys.* **1987**, *61*, 1816–1825. [CrossRef]
20. Bronkhorst, C.; Kalidindi, S.; Anand, L. Polycrystalline plasticity and the evolution of crystallographic texture in FCC metals. *Philos. Trans. R. Soc. Lond. A* **1992**, *341*, 443–477.
21. Kalidindi, S.; Bronkhorst, C.; Anand, L. Crystallographic texture evolution in bulk deformation processing of FCC metals. *J. Mech. Phys. Solids* **1992**, *40*, 537–569. [CrossRef]
22. Zubelewicz, A. Micromechanical study of ductile polycrystalline materials. *J. Mech. Phys. Solids* **1993**, *41*, 1711–1722. [CrossRef]

metals MDPI

Editorial

Metal Crystal/Polycrystal Plasticity and Strengths

Ronald W. Armstrong

Department of Mechanical Engineering, University of Maryland, College Park, MD 20742, USA; rona@umd.edu

Abstract: A brief historical sketch is given of Taylor's dislocation density-based model description, leading to the prediction of a parabolic, tensile, stress–strain curve for the plastic deformation of aluminum. The present focus is on additional results or analyses obtained on the subject for crystal/polycrystal strain hardening. Our current understanding of such material behavior is attributed to post-Taylor descriptions of sequential deformation stages in stress–strain measurements that are closely tied to specific dislocation interaction and reaction mechanisms. A schematic comparison is given for individual face-centered cubic (fcc), body-centered cubic (bcc), and hexagonal close-packed (hcp) crystal curves and to related strength properties determined for individual crystals and polycrystalline material. For the fcc case, an example sessile dislocation reaction is described based on a stereographic projection. Then, quantitative constitutive-relation-based assessments are presented for the tensile strain hardening leading to the plastic instability behaviors of copper and tantalum materials.

Keywords: Taylor-based strain hardening; crystal plastic deformation stages; deformation-induced dislocation interactions/reactions; polycrystal tensile plastic instability

1. Introduction

Soon after crystal dislocations were invented on an atomic lattice basis in 1934, G.I. Taylor produced a dislocation model calculation that allowed him to obtain parabolic strain hardening in a tensile stress (σ_ε)–strain (ε) curve for the 20th century face-centered cubic (fcc) metal: aluminum [1]. Not much changed on the subject until 1951, when another dislocation inventor, Orowan, stated that "Since the publication of Taylor's (1934) theory there has been no significant progress in the understanding of strain hardening" [2,3]. The sentiment was repeated years later by Cottrell [4], as recently quoted [5]. The present report aims to refute this claim by providing an update on the enhanced knowledge gained over these many years about the dislocation mechanics of plastic deformation, particularly of plastic strain hardening.

The 1950s signaled the beginning of an explosion in dislocation observations, particularly by Hirsch and colleagues [6] with the transmission electron microscope (TEM) and through dislocation associations made with a broad range of mechanical properties being measured comprehensively for fcc, body-centered cubic (bcc) and hexagonal close-packed (hcp) metals. An early sign of progress on strain hardening was reported in the 1960 TMS Institute of Metals Lecture given by Mott [7]. Additional mechanical property measurements/analyses that have been achieved until the present time are included in the present report. Not least are advances made in the development of computationally based metal constitutive relations for determining the full σ_ε–ε behavior and of experimental and simulated crystal/polycrystal deformations.

2. Stress–Strain Characterizations of Fcc, Bcc and Hcp Metal Deformations

Post-Taylor description of the several topics, such as dislocation line intersections; partial dislocations and associated stacking faults; slip plane changes through cross-slip; solute pinning of individual dislocations; and grouped dislocation pile-up mechanisms, were soon associated with the three-crystal-structure-type deformations. Figure 1 shows schematic shear stress (τ_ε)–shear strain (γ) curves for the crystal structure types in which particular regions of the deformation curves are marked either with subscript athermal or thermal dependencies, so also being dependent on the imposed metal strain rate in shear, $(d\gamma/dt)_\varepsilon$.

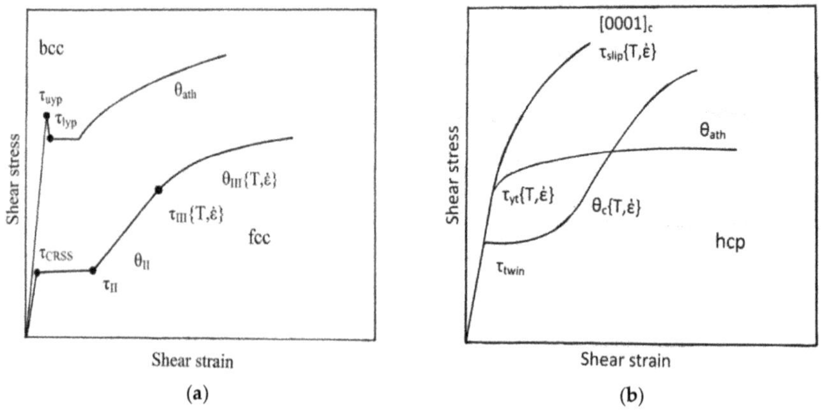

Figure 1. Schematic single-crystal resolved shear stress (τ) versus resolved shear strain (γ) for bcc, fcc (**a**) and hcp (**b**) crystal structure types.

The interstitial solute pinning of dislocations produces the indicated upper and lower yield point behavior for the bcc curve in Figure 1a, whereas the subsequent plastic strain hardening, $(d\gamma/dt)_\varepsilon = \theta_{ath}$, is essentially athermal. The bcc yield stress, with or without a yield point, exhibits a strong thermal dependence. In contrast, multiple stages are also shown in Figure 1a for the fcc curve corresponding to the initial yielding on a single-slip system in stage I, then moving to the activation of double-slip systems in stage II, followed by the advent of cross-slipping in stage III. Stage I for aluminum is normally mostly absent and the combination of θ_{II} and θ_{III} stages approximate to the parabolic strain hardening produced in the Taylor model. The latter two stages are more clearly identified for copper, as explained on the basis of a higher stacking fault energy for aluminum providing smaller partial dislocation separations and, hence, easier cross-slip [8].

The schematic curves for the hcp crystal case in Figure 1b show a significant variation because of the restricted distribution of available slip (or deformation twinning) systems and the significantly different shear stress levels required for their operation, whether in tension or compression. In the case of magnesium, for example, the highest critical resolved shear stress, τ_{CRSS}, is obtained for the compression of a single crystal along the [0001] direction for pyramidal slip, and the lowest shear stress for the initial deformation twinning of a polycrystal that transitions at larger strain to an analogous steeper strain hardening for pyramidal (and prism) slip is also obtained [9]. An intermediate thermally dependent shear stress at yield, $\tau_{yt}\{T, [d\varepsilon/dt]\}$, and athermal strain hardening, θ_{ath}, is shown to be bcc-like for an AZ31B magnesium alloy (0001)-plane textured sheet material subjected to in-plane tension tests [10]. Xie, Zhu, Kang and Yu have presented model constitutive relations for extruded AZ31B material under tensile (θ_{ath}) and compressive (θ_c) loading conditions [11].

3. Dislocation Interactions/Reactions

An early post-Taylor investigation of the thermal dependence of stage I, "easy glide", in (hcp) magnesium single crystals led to the interpretation of dislocation 'forest' intersections controlling the rate dependence of plastic flow [12]. Dislocation 'jogs' and point defects are generated for the further passage of the otherwise mobile dislocations. The same mechanism carries over for the fcc single-crystal stage I case to stage II, in which exceptional hardening is produced by dislocation interactions at slip plane intersections. An important element of the hardening is attributed to dislocation reactions that occur along combined line length segments of the intersecting dislocations. The basis for the reacted dislocations is their relatively lower self-energies compared to the paired dislocation reactants [13]. Hirth provided a detailed geometric description of the possible dislocation reactions for the fcc case [14]. Figure 2 is a stereographic projection that has been constructed to show the particular case described by Hirth for $[\bar{1}10](111)$ and $[011](1\bar{1}1)$ slip system intersections to produce hardening via a sessile "stair rod" dislocation obstacle. For the bcc case, reacted sessile [100] Burgers vector dislocations are produced and are made even more complicated for multiple sessile dislocation reactions produced in hcp crystal deformations involving the basal, pyramidal and prism slip systems.

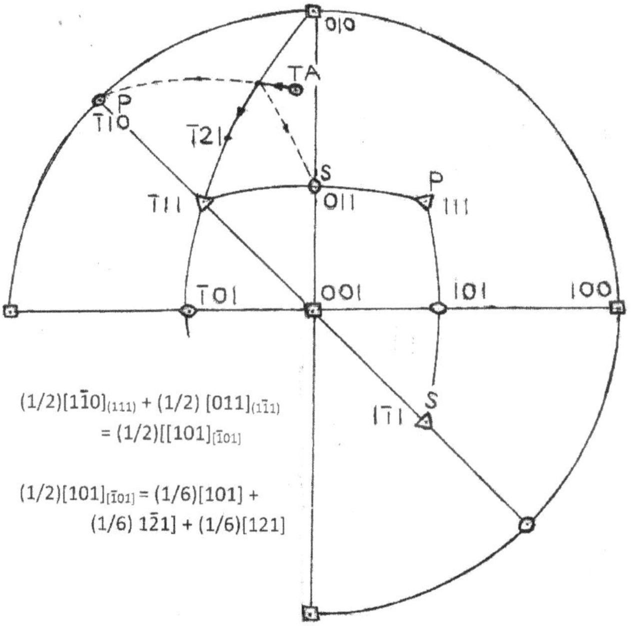

Figure 2. Stereographic projection for tensile axis (TA) rotation associated with the start of stage I type $[\bar{1}10]$ slip on the (111) primary slip plane (P), then leading to stage II rotation with addition of [011] slip on the secondary $(1\bar{1}1)$ slip plane (S), for which a sessile "stair rod" dislocation obstacle, among other partial dislocations, is produced with Burgers vector, (1/6)a[101], and line direction along $[\bar{1}01]$.

4. Strain Hardening and (Tensile) Plastic Instability

A seemingly minor influence of the polycrystal grain size on the tensile plastic deformation of aluminum, an effect that had been missed by Taylor, was established in the pioneering work by Hansen [15]. A later report by Hansen and Ralph established a stronger grain size effect for copper [16]. In both cases, the grain size effect was explained in terms of an analogous and much stronger grain size dependence reported in the early 1950s for α-iron and steel materials by Hall [17] and by Petch [18], respectively. The grain size effect was attributed to dislocation pile-ups in slip bands being needed to overcome the grain boundary's resistance to the transmission of plastic flow between the grains within a polycrystal. A constitutive relation incorporating the grain size effect for polycrystalline copper was originally reported by Zerilli and Armstrong [19] and was updated by Zerilli [20] as follows:

$$\sigma_\varepsilon = \sigma_{0G\varepsilon} + B_0[\varepsilon_r(1 - \mathbf{exp}\{\varepsilon/\varepsilon_r\})]^{1/2}(\mathbf{exp}[-\alpha T]) + k_\varepsilon \ell^{-1/2} \quad (1)$$

In Equation (1), $\sigma_{0G\varepsilon}$ is an athermal stress; B_0 is the thermal stress at absolute temperature, T = 0; and $k_\varepsilon \ell^{-1/2}$ is the Hall–Petch term for an inverse square root of the grain diameter, $\ell^{-1/2}$, multiplied by the stress intensity parameter, k_ε; ε_r is a recovery strain; and

$\alpha = \alpha_0 - \alpha_1\mathbf{ln}(d\varepsilon/dt)$ for a coupled T and strain rate dependence. The first two terms on the right side of Equation (1) are normally combined in the single Hall–Petch intercept stress, $\sigma_{0\varepsilon}$. Under small strain, a Taylor-type parabolic stress–strain dependence is obtained.

Figure 3 provides an assessment of the strain-hardening behavior able to be derived from the Hansen and Ralph investigation of copper $\sigma_\varepsilon - \varepsilon$ behavior in tension and leading to a grain size dependence of the plastic instability condition based on the Considère relationship: $\sigma_\varepsilon = (d\sigma_\varepsilon/d\varepsilon)$ at the tensile maximum load point. For the copper material (and many other metals), the tensile strain hardening was found to be contained in $(d\sigma_{0\varepsilon}/d\varepsilon)$, while k_ε was constant at ~5 MPa.mm$^{1/2}$. The product $B\mathbf{exp}(-\alpha T) = B' = 500$ Mpa was determined for $\sigma_{0\varepsilon}$ from the experimental strain dependence of $\sigma_{0\varepsilon}$, thus providing for the assumption of $\sigma_{0G\varepsilon} = 0$. The value of ε_r was estimated as 1.2, and its influence on the deviation of the $\sigma_\varepsilon - \varepsilon$ curve from the Taylor prediction is shown to begin at ε = ~0.15 for material with a 40 µm grain size. Otherwise, a comparison of the computed curves for the 4.0 and 40 µm grain size curves shows that plastic instability occurs sooner for material with a smaller grain size. The computations are consistent with a limiting value of $\varepsilon = 0.3$ reported by Hansen and Ralph for their highest uniform tensile strain measurements.

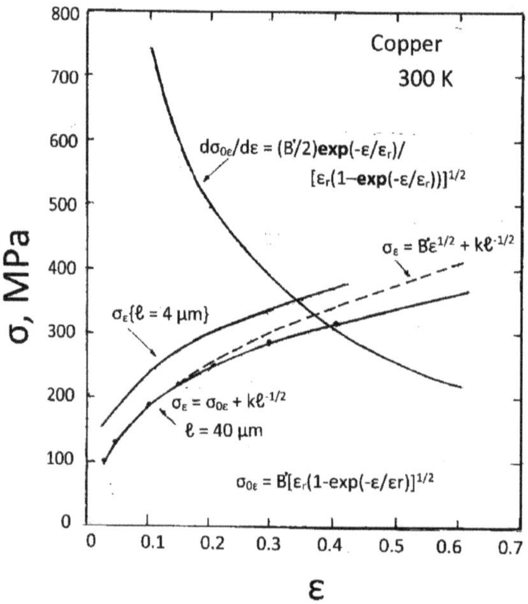

Figure 3. Computations specifying the maximum uniform strain achieved in tension for copper material measurements reported on a Hall–Petch basis by Hansen and Ralph [16]; in the figure, $B' = B\mathbf{exp}[-\alpha T]$.

5. Plastic Instability for the Bcc Case

The following constitutive relation has been validated for bcc ARMCO iron material [19]:

$$\sigma_\varepsilon = \sigma_{0G\varepsilon} + B_0\mathbf{exp}(-\beta T) + A\varepsilon^n + k_\varepsilon \ell^{-1/2} \quad (2)$$

In Equation (2), $\sigma_{0G\varepsilon}$ is an analogous athermal stress component to that described for Equation (1); B_0 is the thermal stress at T = 0; $\beta = \beta_0 - \beta_1\mathbf{ln}(d\varepsilon/dt)$ as for the α parameter in Equation (1); and A and n are constants measuring the material strain hardening. Thus, again, the first three terms in Equation (2) are an expanded interpretation of $\sigma_{0\varepsilon}$ as for Equation (1). It is of particular note that the bcc $(d\varepsilon/dt)$ and T dependencies are in the yield stress, while the strain hardening is essentially athermal.

Beyond the evaluation of the constants in Equation (2) for ARMCO iron material, an extensive investigation was conducted for a wide range of experimental measurements reported for tantalum materials [21]. In particular, Hoge and Mukherjee had reported complete $\sigma_\varepsilon - \varepsilon$ measurements over a wide range of T and strain rates [22]. Figure 4 provides a summary description for the application of Equation (2) to describe the accumulated measurements. In the figure, $c_0 = \sigma_{0G\varepsilon}$ and K = A. One might note that the n = 0.44 value in the figure is close to the Taylor specification. Otherwise, the locus for the maximum load point for the Hoge and Mukherjee measurements is seen to follow an essentially athermal strain hardening behavior, in line with the designation in the Figure 1 schematic.

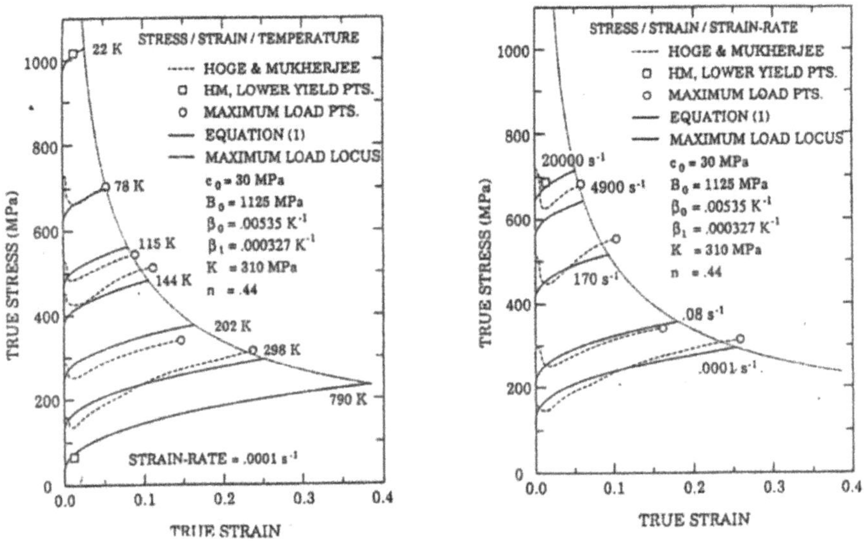

Figure 4. Stress–strain measurements obtained over a wide range of temperatures and strain rates for tantalum by Hoge and Mukherjee [22], as shown to be describable on a bcc constitutive relation basis [21], shown in combined form with permission from AIP Publishing, 2022.

6. Discussion

The preceding description of strength and strain hardening measurements and their analyses on a dislocation mechanics basis provides evidence of progress made in understanding crystal/polycrystal plasticity carrying on from the outstanding work performed by Taylor. Here, additional references are added to provide further evidence for the increased understanding being brought to bear on the subject.

Peeters, Bacroix, Teodosiu, Van Houtte and Aernoudt have monitored via TEM the dislocation densities within slip bands of individual grains in an interstitial-free (IF) polycrystalline iron material and correlated the measurements with an extension of the Taylor model [23]. Messemaeker, Verlinden and Van Humbeeck presented measurements influenced by the complementary grain boundary, for which the generated dislocation densities were contained in $\sigma_{0\varepsilon}$ and thus implicitly contain the influence of strain hardening [24]. Madec and Kubin reported on both bcc and fcc dislocation reactions occurring during straining [25], while Messner, Rhee, Arsenlis and Barton attributed strain hardening in the hcp lattice to dislocation reaction (junctions) occurring in accordance with the Frank and Nicholas dislocation self-energies [13,26]. Most recently, Mishra and Alankar have applied discrete dislocation dynamics for plastic deformation of copper being controlled by the formation of dislocation reacted junctions and cross-slip [27].

An interesting report made by Li, Cui, Yan, Zhang et al. put forth the suggestion for additive manufacturing that the dislocation cell walls formed at larger deformation are anchored by reacted sessile dislocations and thereby require small dislocation pile-ups for breaching [28]. Such model consideration relates to the subject of deformation structures induced by severe plastic deformation, for example, as reported by Muñoz, Higuera and Cabrera for ARMCO iron material deformed to additional stages IV and V by equal-channel angular pressing (ECAP) [29]. The measurements were analyzed in relation to accumulated dislocation measurements and to a Hall–Petch grain size description.

Lastly, there is mention of a relationship between $\sigma_\varepsilon - \varepsilon$ behavior and other properties, for example, fatigue and hardness behaviors. Brown has given emphasis to the importance of the thermally activated annihilation of screw dislocation dipoles at the endurance limit measured in fatigue tests of copper, thus establishing a relationship with strain hardening in stages III and IV of tensile tests [30]. In further work connecting the strain hardening and fatigue behavior of fcc crystals, emphasis was given to understanding the role of dislocation mechanisms at the slip band level [31]. Alhafez, Ruestes, Bringa and Urbassek reported quantitative counts of (a/2) <111> slip and a<100>-reacted Burgers-vector dislocations produced within the plastic deformation zones of simulated nano-indentations made in a (001) α-iron crystal surface [32]. Goel, Cross, Stukowski, Gamsjäger, Beake and Agrawal demonstrated that a[100] reacted dislocation line lengths formed at the earliest stages of deformation in simulations of nano-indentations made in tungsten crystals [33]. Armstrong and Elban have reported comparative plastic strain hardening behaviors in post-pop-in plastic deformation at nano-indents and in drawn wire and micro-pillar deformation tests [34].

7. Summary

A brief description has been given, first, of the important impetus given by Taylor of understanding the nature of crystal plasticity, particularly connecting with an understanding of crystal strain hardening. Subsequent developments of the dislocation mechanics of the subject have been reviewed, with focus on further refinements in dislocation model developments, for example: the role of crystal deformation for partial dislocations and stacking faults; dislocation interactions/reactions and cross-slips; metal grain size; and crystal/polycrystal stress–strain applications, including their tensile plastic instability properties and grouped dislocation behaviors in slip band stress concentrations.

Funding: This research received no external funding.

Acknowledgments: Appreciation is expressed for the correspondence with Yuri Estrin, Alexei Vinogradov and Dierk Raabe on the subject of the present report.

Conflicts of Interest: The author declares no conflict of interest.

References

1. Taylor, G.I. The Mechanism of Plastic Deformation of Crystals. *Part I. Theoretical. Proc. R. Soc. Lond.* **1934**, *145*, 362–387.
2. Nabarro, F.R.N.; Basinski, Z.S.; Holt, D.B. The plasticity of pure single crystals. *Adv. Phys.* **1964**, *13*, 193–323. [CrossRef]
3. Pande, C.S. Strain hardening in FCC metals and alloys. *Mater. Sci. Eng.* **2001**, *A309–A310*, 328–330. [CrossRef]
4. Cottrell, A.H. Commentary: A brief review on work hardening. *Dislocations Solids* **2002**, *11*, vii–xvii.
5. Zepeda-Ruiz, L.A.; Stukowski, A.; Oppelstrup, T.; Bertin, N.; Barton, N.A.; Freitas, R.; Bulatov, V.V. Atomic insights into metal hardening. *Nat. Mater.* **2021**, *5*, 315–320. [CrossRef]
6. Hirsch, P.B.; Horne, R.W.; Whelan, M.J. Direct observations of the arrangement and motion of dislocations in aluminum. *Philos. Mag.* **1956**, *1*, 677–684, reprinted in Special Issue: 50 years of TEM of dislocations, *Philos. Mag.* **2006**, *86*, 4553–4572.
7. Mott, N.F. The Work Hardening of Metals. *Trans. TMS-AIME* **1960**, *218*, 962–968.
8. Armstrong, R.W. Plasticity: Grain Size Effects III. In *Reference Module in Materials Science and Materials Engineering*; Hashmi, S., Ed.; Elsevier: Oxford, UK, 2016; pp. 1–23.
9. Armstrong, R.W.; Li, Q.-Z. Dislocation mechanisms of high-rate deformations. *Metall. Mater. Trans. A* **2014**, *46A*, 4438–4453.
10. Kurukuri, S.; Worswick, M.J.; Tari, D.G.; Misra, R.K.; Carter, J.T. Rate sensitivity and tension-compression asymmetry in AZ31B magnesium alloy sheet. *Philos. Trans. R. Soc. A* **2014**, *372*, 20130216. [CrossRef]
11. Xie, Q.; Zhu, Z.; Kang, G.; Yu, C. Crystal plasticity-based impact dynamic constitutive model of magnesium alloy. *Int. J. Mech. Sci.* **2016**, *119*, 107–113. [CrossRef]
12. Conrad, H.; Armstrong, R.W.; Wiedersich, H.; Schoeck, G. Thermally-activated Glide in Magnesium Crystals from 4.2 to 120 K. *Philos. Mag.* **1961**, *6*, 177–188. [CrossRef]
13. Frank, F.C.; Nicholas, J.F. Stable dislocations in the common crystal lattices. *Philos. Mag.* **1953**, *44*, 1213–1235. [CrossRef]

14. Hirth, J.P. On dislocation reactions in the fcc lattice. *J. Appl. Phys.* **1961**, *32*, 700–706. [CrossRef]
15. Hansen, N. Effect of grain size and strain on the tensile flow stress of aluminum at room temperature. *Acta Metall.* **1977**, *25*, 863–869. [CrossRef]
16. Hansen, N.; Ralph, B. The Strain and Grain Size Dependence of the Flow Stress of Copper. *Acta Metall.* **1982**, *30*, 411–417. [CrossRef]
17. Hall, E.O. The deformation and ageing of mild steel: Discussion of results. *Proc. Phys. Soc. Lond. B* **1951**, *64*, 747–753. [CrossRef]
18. Petch, N.J. The cleavage strength of polycrystals. *J. Iron Steel Inst.* **1953**, *174*, 25–28.
19. Zerilli, F.J.; Armstrong, R.W. Dislocation-mechanics–based constitutive equations for material dynamics calculations. *J. Appl. Phys.* **1987**, *61*, 1816–1825. [CrossRef]
20. Zerilli, F.J. Dislocation mechanics based constitutive relations. *Metall. Mater. Trans. A.* **2004**, *35A*, 2547–2555. [CrossRef]
21. Zerilli, F.J.; Armstrong, R.W. Description of tantalum deformation behavior by dislocation mechanics based constitutive relations. *J. Appl. Phys.* **1990**, *68*, 1580–1591. [CrossRef]
22. Hoge, K.G.; Mukherjee, A.K. Temperature and strain rate dependence of flow stress of tantalum. *J. Mater. Sci.* **1977**, *12*, 1666–1672. [CrossRef]
23. Peeters, B.; Bacroix, B.; Teodosiu, C.; Van Houtte, P.; Aernoudt, E. Work-Hardening/Softening Behavior of BCC Polycrystals during Changing Strain Paths: II. TEM Observations of Dislocation Sheets in an IF Steel during Two-Stage Strain Paths and Their Representation in Terms of Dislocation Densities. *Acta Mater.* **2001**, *49*, 1621–1632. [CrossRef]
24. De Messemaeker, J.; Verlinden, B.; Van Humbeeck, J. On the strength of boundaries in submicron IF steel. *Mater. Lett.* **2004**, *58*, 3782–3786. [CrossRef]
25. Madec, R.; Kubin, L.P. Dislocation strengthening in FCC metals and in BCC metals at high temperatures. *Acta Mater.* **2017**, *126*, 166–173. [CrossRef]
26. Messner, M.C.; Rhee, M.; Arsenlis, A.; Barton, N.R. A crystal plasticity model for slip in hexagonal close packed metals based on discrete dislocation simulations. *Model. Simul. Mater. Sci. Eng.* **2017**, *25*, 044001. [CrossRef]
27. Mishra, A.; Alankar, A. Revisiting dislocation reactions and their role in uniaxial deformation of copper single crystal micro-pillars. *Model. Simul. Mater. Sci. Eng.* **2019**, *27*, 055010. [CrossRef]
28. Li, Z.; Cui, Y.; Yan, W.; Zhang, D.; Fang, Y.; Chen, Y.; Yu, Q.; Wang, G.; Ouyang, H.; Fan, C.; et al. Enhanced strengthening and hardening via self-stabilized dislocation network in additively manufactured metals. *Mater. Today* **2021**, *50*, 79–88. [CrossRef]
29. Muñoz, J.A.; Higuera, O.F.; Cabrera, J.M. Microstructural and mechanical study in the plastic zone of ARMCO iron processed by ECAP. *Mater. Sci. Eng. A* **2017**, *697*, 24–26. [CrossRef]
30. Brown, L.M. Unifying concepts in dislocation plasticity. *Philos. Mag.* **2005**, *85*, 2989–3001. [CrossRef]
31. Brown, L.M. Constant intermittent flow of dislocations: Central problems in plasticity. *Mater. Sci. Tech.* **2012**, *28*, 1209–1232. [CrossRef]
32. Alhafez, I.A.; Ruestes, C.J.; Bringa, E.M.; Urbassek, H.M. Influence of pre-existing plasticity on nano-indentation—An atomistic analysis of the dislocation fields produced. *J. Mech. Phys. Sol.* **2019**, *132*, 103674.
33. Goel, S.; Cross, G.; Stukowski, A.; Gamsjäger, E.; Beake, B.; Agrawal, A. Designing nanoindentation simulation studies by appropriate indenter choices: Case study on single crystal tungsten. *Comput. Mater. Sci.* **2018**, *152*, 196–210. [CrossRef]
34. Armstrong, R.W.; Elban, W.L. Dislocation Reaction Mechanism for Enhanced Strain Hardening in Crystal Nano-indentations. *Crystals* **2020**, *10*, 9. [CrossRef]

metals

MDPI

Article

Atomistic-Continuum Constitutive Modeling Connection for Gold Foams under Compression at High Strain Rates: The Dislocation Density Effect

Mohammed H. Saffarini [1] and George Z. Voyiadjis [2,*]

[1] Department of Civil and Environmental Engineering, University of Missouri, Columbia, MO 65201, USA
[2] Computational Solid Mechanics Laboratory, Department of Civil and Environmental Engineering, Louisiana State University, Baton Rouge, LA 70803, USA
* Correspondence: voyiadjis@eng.lsu.edu

Abstract: Constitutive description of the plastic flow in metallic foams has been rarely explored in the literature. Even though the material is of great interest to researchers, its plasticity remains a topic that has a much room for exploration. With the help of the rich literature that explored the material deformation mechanism, it is possible to introduce a connection between the results of the atomistic simulations and the well-established continuum constitutive models that were developed for various loading scenarios. In this work, we perform large-scale atomistic simulations of metallic gold foams of two different sizes at a wide range of strain rates ($10^7 - 10^9$ s^{-1}) under uniaxial compression. By utilizing the results of those simulations, as well as the results we reported in our previous works, a physical atomistic-continuum dislocations-based constitutive modeling connection is proposed to capture the compressive plastic flow in gold foams for a wide range of sizes, strain rates, temperatures, and porosities. The results reported in this work present curated datasets that can be of extreme usefulness for the data-driven AI design of metallic foams with tunable nanoscale properties. Eventually, we aim to produce an optimal physical description to improve integrated physics-based and AI-enabled design, manufacture, and validation of hierarchical architected metallic foams that deliver tailored mechanical responses and precision failure patterns at different scales.

Keywords: size effect; strain rate; strain hardening; atomistic-continuum; multiscale; large-scale simulations

1. Introduction

Cellular solids have always attracted attention because they have one of the most desirable material characteristics that a researcher could ask for, and that is a high stiffness-to-weight ratio [1]. This promotes the material capacity to be used in a wide range of applications; it is excellent in terms of shock absorbency [1,2], energy absorbency as the core of sandwich metal tubes [3,4], catalysts [5–7], sensors and super capacitors [8–12], actuators [13], and radiation-tolerant materials [14–16]. Bicontinuous metallic nanofoam, which is a class of cellular solid, is made of a network of interconnected ligaments and pores. This material, which is the subject of this work, has been researched for some time by using various approaches, including pure experimental work (continuum scale) [17–23] and pure computer simulations (mainly atomistic scale) [24–31]. Despite the fact that there were trials to directly compare the results of molecular dynamics (MD) simulations with the results of experiments [32–35], no adequate connection has been made between the two scales in terms of the constitutive description of the material plastic flow. In fact, a constitutive description of the plastic flow of metallic nanofoams has rarely been explored in the literature. That can be attributed first to the fact that constitutive description of the plastic flow of any full density (nonporous) solid is a very challenging task, given that there are a lot of variables that play important roles in the material response (size, strain rate, temperature, crystal structure, etc.). Additionally, for the case of bicontinuous metallic nanofoam, there is the added complexity of having more variables to take into account, such as porosity, network connectivity, structural disorder, and free surface effects. All of this makes atomistic simulations of the plasticity of nanofoams computationally expensive and structurally complex.

However, this should not hold researchers back from trying to provide the required constitutive models that capture the material response. The need for either physical or phenomenological constitutive models is a fundamental necessity. For example, it is required to ensure the applicability of the second law of thermodynamics in the continuum description of a system in equilibrium in addition to the basic laws of mass, energy, and momentum conservation. Without the constitutive models, we will end up with a number of equations less than the number of unknowns, and in turn, produce a nonsolvable scheme. In other words, unlike atomistic simulations, which are dependent on the interatomic potential description rather than the constitutive description, continuum-based simulations are extremely dependent on constitutive description. For example, constitutive models that account for material deformation mechanisms as well as the effects of temperature, strain rate, size, and density are required to perform continuum-based simulations such as finite element (FE) and/or meshless-based simulations. Therefore, the challenge becomes more about what constitutive model to use or to develop rather than discussing the need for those constitutive descriptions.

To avoid repeating the works of others, and because we would not provide any better extensive comparisons between different existing constitutive models than those provided by references [36–40], we would refer the readers to those references. In general, to choose a suitable continuum-based constitutive model to describe the plastic flow of foams, especially that those constitutive models were originally developed for full density (nonporous) solids, a physical-based constitutive model should be the first type to consider. Additionally, because the deformation mechanism of gold metallic foams is controlled by dislocation mechanics, as explained extensively in the literature and in our previous works [25,26,29–31,41–43], the physical-based constitutive model needs to be dislocation-controlled. As we will show later, the only available constitutive description for gold foams is based on dislocation dynamics. Moreover, the flexibility of determining the material constants from a limited set of experimental data and capturing the static and/or dynamic behavior should be considered in the selection process. For those reasons, among the well-known available constitutive models in the literature [38,44–50], we choose to use the Voyiadjis–Abed (VA) constitutive model [49] which is a modified/extension version of the well-known Zerilli–Armstrong (ZA) constitutive model [47]. In later sections, we provide more details on the theory that promotes the constitutive model efficiency, aside from it being flexible in determining its parameters.

There were few attempts in the literature to explore the constitutive modeling of the plastic flow in metallic foams [51–56]. However, to the best of authors' knowledge, there were no trials in developing a physical-based and dislocation-based constitutive model for that matter. Moreover, there is no known effort in establishing a connection between any physical-based continuum constitutive model and the atomistic observations of the material behavior. The proposed work utilizes a continuum, physical-based, and dislocation-based constitutive model to provide a physical meaning and to modify the authors' previously published atomistic constitutive description in metallic foams. The newly proposed constitutive model will establish a connection between the two scales (atomistic and continuum). Accordingly, we report the first trials to propose a connection between atomistic and continuum constitutive description of the plastic flow of face center cubic (FCC) nanofoams after performing large-scale MD simulations of different sizes and strain rates of gold foam. In Section 2, the authors recall the dislocation-based atomistic constitutive description of gold foams proposed in their previous publications. In Section 3, the authors review the dislocation-based continuum constitutive description of plasticity in nonporous FCC metals as proposed by Voyiadjis and Abed [57]. In Section 4, the authors introduce the proposed connection between the atomistic and the continuum constitutive models. In Section 5, the computational modeling part is presented, in which large-scale MD simulations are performed over a wide range of strain rates for two different average ligament sizes. The results of those simulations in addition to the results of the simulations that were reported by the authors in their previous works are used to test the proposed atomistic-continuum connection. In Section 6, the results are discussed in light of the simulations.

2. Atomistic Dislocation-Based Constitutive Model for Gold Foams

In this section, the authors recall the atomistic constitutive model that was developed by the same authors in their previous work [29–31,58]. Saffarini et al. [29] proposed an expression to describe the plastic flow in gold foams under compression as a function of the total dislocation density, as well as other phenomenological parameters. The parameters were determined based on large-scale simulations that were conducted for a range of ligament sizes, temperatures, and strain rates. However, no coupling effect was considered during the development of the formulation. In other words, every series of simulations was done by varying one aspect and fixing all others. The proposed expression is shown in Equation (1),

$$\sigma^{\text{Compression}} = \underbrace{C\sigma_y}_{\substack{\text{Yield} - \text{Controlled Stress} \\ \text{Component}}} + \underbrace{De^{n\rho}}_{\substack{\text{Dislocation} - \text{Controlled Hardening Stress} \\ \text{Component}}}, \quad (1)$$

where σ_y is the yield strength, C is a size dependent constant, D is a stress parameter that is size- and temperature-dependent, n (mm^2) is an exponent that is related to the ligament area and is size- and strain rate-dependent, and ρ (mm^{-2}) is the total dislocation density. Table 1 shows the values of Equation (1) parameters as provided by Saffarini et al. [29].

Table 1. Equation (1) parameters values according to Saffarini et al. [29].

Avg. Ligament Diameter (nm)		**6.4**									**13.1**	**25.6**
Strain Rate (s^{-1})		10^6	10^7	10^8					5×10^8	10^9	10^8	
Temperature (K)		300		300	400	500	600	700	300			
Compression	C	1.10									1.05	0.95
	D (MPa)	11			12	13	17.5	23	11		5.5	1.0
	n ($\times10^{-12}$) (mm^2)	9.5	8.2	7.5					7.0		10.5	23.5

The constitutive model suggests that the plastic flow in gold foam can be decomposed into two stress components: (1) a yield-controlled stress component, and (2) dislocation-controlled hardening (densification) stress component. The second component corresponds to the amount of stress needed to continuously ensure the nucleation of dislocations in gold foams to plastically deform the material. Due to the small ligament size and the presence of porosity that typically comprise the microstructure of the foam, dislocations annihilate almost instantly upon nucleation by escaping at the ligament's free surfaces. This type of dislocation interaction and accumulation mechanism prior to densification requires a continuous supply of stress to plastically deform the material. For more details about the dislocation dynamics controlling the deformation mechanism, the reader is referred to the authors' previous works [29–31].

The hardening term in Equation (1) takes the exponential form to capture the densification regime that is dominant in all cellular materials under compression. As shown by Saffarini et al. [31], this regime is initiated once the material transforms from being open-cell foam into closed-cell foam, losing its bicontinuous microstructure and allowing dislocation to accumulate and interact without being annihilated [29].

The total dislocation density ρ (mm^{-2}) variable that appears in Equation (1) can be determined by using the formulation that Saffarini et al. [29] previously proposed. The developed exponential expression describes the evolution of total dislocation density as a function of relative density under compression.

Aside from the effect of relative density, the effect of ligament size, strain rate, and temperature are captured by the remaining phenomenological parameters. The model is described by Equation (2),

$$\rho = Ae^{m\varphi} + B, \tag{2}$$

where ρ is the total dislocation density, A (m^{-2}) and B (m^{-2}) are parameters that capture the coupling effect of size, temperature, and strain rate, and m (m^2) is an exponential decay factor that is size-dependent only (i.e., strain rate- and temperature-independent). Table 2 shows the values of the parameters as reported in their work.

Table 2. Equation (2) parameters values according to Saffarini et al. [29].

Avg. Ligament Diameter (nm)		6.4									13.1	25.6
Strain Rate (s^{-1})		10^6	10^7	10^8					5×10^8	10^9	10^8	
Temperature (K)		300		300	400	500	600	700	300			
Comp	A ($\times10^{17}$)	−27.5	−33	−42	−38.5	−35	−31.5	−23.5	−45		−22.62	−16.56
	B ($\times10^{17}$)	4	4.7	5.9	5.4	5	4.5	3.5	6.4		8.6	8.2
	m	−4									−2	−1.42

3. Continuum Dislocation-Based Constitutive Model for Nonporous FCC Metals

Plastic deformation in FCC metals is controlled by dislocations' motion and their respective interactions, which are well established by the dislocation theory [59]. Accordingly, constitutive models describing metal plasticity were developed to accommodate the microscopic interaction of dislocations and how they control the deformation and the stress evolution upon initiation of plasticity. In this section, a brief overview is presented of the continuum dislocation-based constitutive model that is utilized to establish the connection with the atomistic constitutive description described in the previous section. Specifically, we refer to the dislocation dynamics theory in full density (nonporous) FCC metals as the basis of this proposed connection.

Voyiadjis and Abed [49] developed a physical- and dislocation-based constitutive model that describes the plastic flow in FCC metals over a wide range of strain rates and temperatures based on the concepts of thermal activation processes that were proven to control dislocation dynamics at low strain rates. They modified the ZA constitutive model [47], to overcome two major shortcomings. The first is that the explicit definition of one of its parameters which assumes that the thermal activation area is constant instead of being temperature-dependent. The second shortcoming is that the model assumes a simple mathematical expansion that can be only applicable for very low strain rates and temperature ranges. Such an assumption limits the usage of the model with regard to the reference strain rate. Accordingly, Voyiadjis and Abed [47] proposed significant modifications to those assumptions. Here, we present the final form of the VA constitutive model which will be used later to provide the physical basis for our proposed continuum-atomistic connection.

In FCC metals in particular, the plastic deformation is dominated by the evolution of a "heterogeneous microstructure of dislocations (mobile)" and the long-range intersections between dislocations (forest), especially at strain rates less than 10^4. For that, the thermal activation analysis depends on plastic strain [49]. Moreover, dislocations during plastic deformation can be cumulatively trapped to form a forest of dislocations. Forest dislocations act as a barrier that hinders the motion of mobile dislocations. Overcoming such a barrier requires introducing thermal energy that in turn can provide thermal hardening. For that reason, the plastic flow stress must contain a thermal component that is dependent on the plastic strain. Moreover, because "slip-crystal flow stress" is controlled by dislocation density and intersections (activation area), this thermal component of the stress is proportional to the inverse square root of the plastic strain. Accordingly, the VA constitutive model decomposes the stress response of the material into athermal stress (σ_{ath}) and thermal stress (σ_{th}) components as shown by Equation (3):

$$\sigma = \sigma_{ath} + \sigma_{th}. \quad (3)$$

The athermal stress component is plastic strain-independent and is completely related to the initial yield stress Y_a. Since there is no strain rate and temperature dependency on the initial yield stress, Y_a becomes constant. The thermal stress component describes the coupling effect of temperature, strain rate, and plastic strain. Based on that, Voyiadjis and Abed [47] presented the final form of the total flow stress expression as shown by Equation (4),

$$\sigma = \underbrace{Y_a}_{\text{athermal stress}} \underbrace{+ \tilde{B}\varepsilon_p^{0.5}\left(1-(\beta T)^{\frac{1}{q}}\right)^{\frac{1}{p}} + Y_d\left(1-(\beta T)^{\frac{1}{q}}\right)^{\frac{1}{p}}}_{\text{thermal stress}} \quad (4)$$

where $\tilde{B}$ is a hardening parameter and is defined as $\tilde{B} = \frac{m\alpha_o\mu_o b^2}{A_o''}$, parameter Y_d is the resultant drag-stress at the reference velocity or zero absolute temperature and is defined as $Y_d = \frac{m\alpha_o\mu_o b^2}{A_o'}$, constants p and q define the shape of the obstacle where p ranges from 0 to 1, and q ranges from 1 to 2. Moreover, the thermal activation parameter β is defined as $\beta = \beta_1 - \beta_2 \ln\frac{\dot{\varepsilon}_p}{\dot{\varepsilon}_{po}}$ with parameters β_1 and β_2 defined as $\beta_1 = \frac{k}{G_o}\ln\left(\bar{m}bv_o\rho\right)$ and $\beta_2 = \frac{k}{G_o}$, where k is the Boltzmann's constant, b is Burger's vector, ρ is the dislocation density, v_o is the dislocation velocity, and G_o is the reference Gibbs free energy.

4. Proposed Atomistic-Continuum Constitutive Connection

Except for a few constitutive models [45,60], most of the existing continuum constitutive models that describe the plastic flow in FCC metals, including the one presented earlier, were well-established for strain rates less than 10^4 s^{-1} because they were developed by using the concept that thermal activation processes are the driving force for the motion of dislocation. For the case of higher strain rates than 10^4 s^{-1}, and specifically at certain threshold values, there will be an "upturn" in the flow stress as proven experimentally by Follansbee and Kocks [50]. In more detailed terminology, there appears to be a rise in the strain rate sensitivity at such level of strain rates. For that reason, those constitutive models need to be revisited to include a description that considers some of the physics that are driving this strain rate sensitivity. That is, the constitutive model's parameters need to either be reevaluated, expanded, or combined with additional parameters.

Despite the upturn in flow stress being an experimental fact, the main driving force behind it is still debatable [37]. Follansbee and Kocks [50] stated that it is still controlled by thermal activation processes instead of the dislocation drag. Rusinek et al. [61] emphasized that it is controlled by both dislocation drag and thermal activation. Zerilli and Armstrong [62] showed that dislocation drag is not the driving force because strain under tension does not increase as it should if dislocation drag is active. This means that there is an unclear thermal-activation-guided process that is playing a key role. However, as shown by Saffarini et al. [30], the dislocation evolution becomes significantly high at very high strain rates due to the rapid nucleation and annihilation of dislocations, as well as the speed at which they move. Such phenomena are difficult to attribute to thermal activation processes and are believed to be controlled by dislocation drag [62,63].

Following this debate, several trials ranging from attributing the strain rate sensitivity to structure evolution, redefining the dislocation spacing and density, to introducing a "characteristic length evolution" parameter, were made to explain the upturn in flow stress [50,64,65]. Despite the fact that those trials do not exactly match the case we are presenting in this work, they fall under the same umbrella. It is shown later that the main goal of this proposed connection and modification is to combine terms that attribute dislocation behavior under very high strain rates in metallic foams to both the thermal activation processes (especially yielding) and dislocation drag (especially hardening and densification).

All of the abovementioned studies concluded that the rate of dislocations' evolution at very high strain rates is significantly higher, and in turn, this increases the dislocation density and decreases the dislocation spacing, activation area, and other characteristic lengths. This is believed to induce the upturn in the stress [37]. For that reason, there needs to be a dedicated term in our proposed constitutive model that describes the unique, abrupt change in the dislocation interactions at such high strain rate, which in turn captures the said significant increase in dislocation density and the corresponding material hardening. Moreover, because the dislocation spacing and interaction in metallic foams is defined mainly by the material porosity [30,31], the effect of porosity (relative density) needs to be a fundamental aspect of the constitutive model (see Equations (2), (8) and (10)).

As explained earlier, at very high strain rates, it is debatable whether the upturn in stress is controlled solely by thermal activation processes or by dislocation drag. We show here that adopting a constitutive description that combines a derivation based on the concept of thermal activation and a phenomenological description of the other processes that were shown to control the deformation mechanisms in metallic foams can accurately capture the stress–strain response of gold foams. By trying to describe the unique dislocation density evolution mechanism in metallic foams, we will be able to capture the stress–strain response of gold foams under compression for different sizes and at strain rates in the range $10^7 - 10^9\ s^{-1}$.

Several constitutive models were proposed to describe dislocation density evolution [44,66–69]. However, it is a slightly different case in this work due to the different nature of the microstructure encountered in metallic foams that are uniquely characterized by the presence of porosity. In other words, the fact that dislocations in the case of porous medium evolve in a different manner than that in nonporous medium plays an important role in the behavior of dislocation evolution. As shown by our previous works [29–31], dislocations in porous media travel significantly shorter distances than those in nonporous media because they get annihilated after escaping the solids at the ligament's surfaces. This leads to the need of a continuous stress increase to renucleate dislocations and plastically deform the material. This process remains active until material compaction reaches a point where the bicontinuous microstructure starts to disappear, and densification is initiated. At that stage, we start to observe some of the dislocation dynamics that occur in full density (nonporous) solids to take place, and thus, significant hardening starts to appear in the material response. However, that does not preclude the fact that dislocation dynamics during the instances when dislocations are traveling within ligaments prior to annihilation (especially before densification initiation), are the same as that in nonporous media. The fact is that the stress response that is controlling the plastic deformation of metallic foams is dominated by dislocation annihilation. Other than that, dislocation dynamics within the solid will still abide by the same dislocation dynamics observed in nonporous media at very high strain rates. Therefore, the following continuum formulation is used to physically justify Equation (2), which determines the dislocation density as a function of relative density in gold foams. The first connection is that the continuum formulation to predict the dislocation density in FCC metals takes the same form as the atomistic formulation as will be shown next. This is because, as stated earlier, the dynamics of dislocations remain the same for porous and nonporous media at very high strain rates.

Armstrong et al. [57,70] pointed out that the activation area decreases significantly and approaches an atomistic dimension when the upturn in flow stress occurs at high strain rates. This means that the dislocation distance will decrease, leading to a dislocation density increase. Accordingly, the hardening parameter controlled by dislocation density can take the form shown by Equation (5),

$$\rho_f = \rho_{fi} + \overline{f}\frac{M}{k_a}(1 - e^{-k_a \varepsilon_p}), \tag{5}$$

where ρ_{fi} is the initial forest dislocation, ρ_f is the forest dislocations, $\overline{f}$ the forest dislocation density fraction of the total dislocation density, M is the Schmidt factor, and k_a is the dislocation annihilation factor. The term shown by Equation (5) can be rewritten to take the form shown by Equation (6),

$$\rho_f = \rho_{fi} + \overline{f}\frac{M}{k_a} - \overline{f}\frac{M}{k_a}e^{-k_a \varepsilon_p}, \tag{6}$$

where the first two terms can be assumed to be constant in a statistical averaging sense while the third term is a varying term as a function of plastic strain. This can help us in rewriting Equation (6) to take a form similar to Equation (2), such that:

$$\rho = C_1 + C_2 e^{r\varepsilon_p}, \tag{7}$$

where $C_1 = \rho_{fi} + \overline{f}\frac{M}{k_a}$, $C_2 = -\overline{f}\frac{M}{k_a}$, and $r = -k_a$. For the case of metallic foams, as shown previously by Saffarini and coworkers [29–31], relative density in our case is directly proportional to the engineering plastic strain, and the dislocation distance. The intersection spacing in metallic foams is in the order of atomic diameter (a factor of the average ligament size ~5 × Atomic Diameter) as Armstrong et al. [57,70] pointed out (see above).

Since the relative density is directly proportional to the engineering plastic strain, as stated earlier, it is used as an alternative representation that describes the average activation area as well as the dislocation distance in the material. This allows one to reasonably and directly replace the plastic strain by the relative density, leading one to rewrite Equation (7) to follow the same form presented in Equation (2) as by Equation (8):

$$\rho = C_1 + C_2 e^{r\varphi}. \tag{8}$$

By comparison, one has the parameters of Equation (8) follow the same form presented in Equation (2) such that $C_1 = B$, $C_2 = A$, and $r = m$. By using Equation (8), one can evaluate the second term (hardening stress term) in Equation (1).

Despite the fact that Saffarini et al.'s constitutive model (Equation (1)) has been shown to significantly capture the MD simulation results as displayed in their previous studies [71] and the fact that the second term of the model (dislocation-controlled stress component) is physically justified by the dislocation dynamics reported in their work and well-established by the literature, the first term (yield-controlled stress term) still lacks the required physical reasoning. For that, here one introduces the second continuum-atomistic connection, in which it is proposed that the first part of the atomistic constitutive model (yield-controlled stress component) is directly related to the continuum microscopic description that the VA constitutive model provides. In other words, one proposes the equality shown by Equation (8):

$$\underbrace{C\sigma_y}_{\text{Yield − Controlled stress component}} = \underbrace{Y_a}_{\text{athermalstress component}} + \underbrace{\widetilde{B}\varepsilon_p^{0.5}\left(1 - (\beta T)^{\frac{1}{q}}\right)^{\frac{1}{p}} + Y_d\left(1 - (\beta T)^{\frac{1}{q}}\right)^{\frac{1}{p}}}_{\text{thermal stress component}}. \tag{9}$$

Substituting Equation (9) into Equation (1) gives the proposed model as shown by Equation (10):

$$\sigma = \underbrace{\underbrace{Y_a}_{\substack{\text{athermal stress}\\ \text{component}}} + \underbrace{\tilde{B}\varepsilon_p^{0.5}\left(1-(\beta T)^{\frac{1}{q}}\right)^{\frac{1}{p}} + Y_d\left(1-(\beta T)^{\frac{1}{q}}\right)^{\frac{1}{p}}}_{\text{thermal stress component}}}_{\text{VA Model}} + \underbrace{\underbrace{De^{n\rho}}_{\substack{\text{"upturn" stress}\\ \text{component}}}}_{\text{Hardening Term From Equation (1)}}, \quad (10)$$

where ρ is the dislocation density as determined by Equation (8). That being said, the proposed constitutive model now consists of three components: (1) the stress component that controls the material yielding (athermal stress component), (2) the stress component that controls the thermal activation processes (thermal stress component), and (3) the stress component that controls both the densification (hardening) and the upturn in flow stress due to the effect of very high strain rate (dislocation-controlled hardening stress component). The third stress component is dislocation-controlled and should capture the exponential increase in stress in metallic foams due to the material densification.

5. Computational Modeling

The microstructures of the nanoporous gold samples used in this work simulations were produced by using the phase field method through spinodal decomposition [72–74]. More details about the process can be found in our previous work [31] which has the details of generating the samples used in this work. Two Au single crystals were generated; one has the dimensions of 100 × 100 × 100 lattice spacing (a_o = 4.08 Å), and the other is scaled by a factor of two in each direction. The smaller sample microstructure and morphology is shown in Figure 1 after being postprocessed by using the OVITO surface reconstruction modifier [59,60]. This sample template was applied as is in the finite element solver, and then scaled by a factor of 2 as mentioned earlier to generate the second larger sample. This protocol guarantees self-similarity, the same porosity, and the same network connectivity for both samples. Such conditions will reduce the variability between the two samples, limiting the difference between them to the difference in size and difference in response to strain rate effect. At the end of the process, the first sample ended up containing ~2 million atoms, whereas the second one has ~16 million atoms.

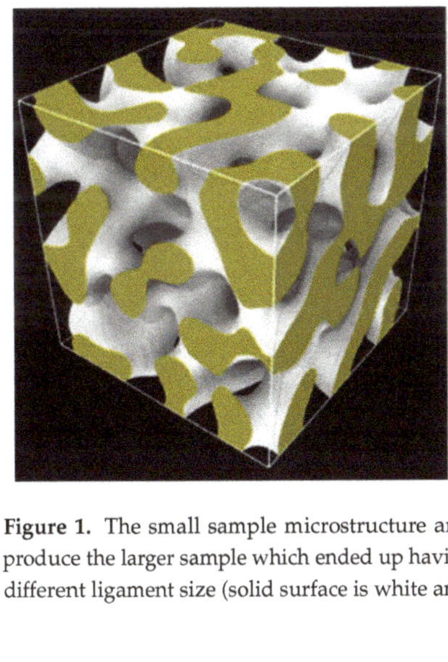

Figure 1. The small sample microstructure and morphology. This template was scaled as is to produce the larger sample which ended up having same porosity and network connectivity, but with different ligament size (solid surface is white and interior of solid is gold).

The samples morphologies were characterized using AQUAMI [75] software. This analysis yields that the average ligament diameter (L) for first sample is 6.4 nm and 13.1 nm for the second one. Both generated samples have 50% porosity.

The large-scale atomic molecular massively parallel simulator (LAMMPS) open-source code [76] was used in this work to perform the MD simulations. The Foiles et al. embedded atomic method (EAM) interatomic potential for gold [77] was used for the interatomic interaction. Before performing any production simulations, the statistical minimization of the atomic potential energy was performed by using the conjugate gradient method with a final condition of zero stress in each direction. Upon finalization of energy minimization, a thermal relaxation was performed for each sample to achieve a final target temperature of 300 K and final target pressure of zero over the span of 0.5 ns in NPT ensemble.

For the production simulations, a uniaxial strain through scaling of the simulation box was applied along the [001] direction at four different strain rates of 10^7 s^{-1}, 10^8 s^{-1}, 5×10^8 s^{-1}, and 10^9 s^{-1} and room temperature of 300 K while maintaining zero stress condition in the other two perpendicular directions in an NPT ensemble. The 3D periodicity was maintained during all simulation stages (minimization, equilibration, and production). Finally, the visualization and postprocessing was performed by using the OVITO software and crystal analysis tool [78–82].

6. Results and Discussion

In this section, the authors present and discuss the results of the proposed constitutive model compared with the MD simulation results. In addition, they provide a comparison between the proposed constitutive model, the VA constitutive model, and the hardening term in Equation (1) to further show the usefulness of the newly proposed connection.

The stress–strain curves obtained from the simulations were determined based on the global stress tensor of the full sample as per the virial theorem in LAMMPS [83]. The global stress is computed by the formula $\sigma = \frac{Nk_BT}{V} + \frac{1}{Vd}\sum_{i=1}^{N'} \vec{r_i}.\vec{f_i}$, where N is the number of atoms in the system, k_B is the Boltzmann constant, T is the temperature, d is the dimensionality of the system (2 for 2D, 3 for 3D), and V is the system volume. The second term is the virial, equal to $-\mathrm{d}U/\mathrm{d}V$ (U is the potential energy), computed for all pairwise as well as 2-body, 3-body, 4-body, many-body, and long-range interactions, where $\vec{r_i}$ and $\vec{f_i}$ are the position and force vector of atom i, and the dot indicates the dot product (scalar product). As for the engineering strain, it is the simple change of length along the load direction since the loading was applied by scaling the simulation box at a predefined rate (refer to Section 5). Accordingly, the global von Mises stress (J_2) is computed by using Equation (11) to produce the stress-strain plots:

$$J_2 = \sqrt{\frac{(\sigma_{xx} - \sigma_{yy})^2 + (\sigma_{yy} - \sigma_{zz})^2 + (\sigma_{zz} - \sigma_{xx})^2 + 6\left(\sigma_{xy}^2 + \sigma_{xz}^2 + \sigma_{yz}^2\right)}{2}} \tag{11}$$

Tables 3 and 4 show the parameters' values of Equations (8) and (10), respectively. The parameters of Equation (8) were obtained from Saffarini et al. [29] for all strain rates in the case of the small sample and for the strain rate of 10^8 s^{-1} in the case of the medium sample. As for the remaining strain rates of the medium sample, the parameters have been evaluated by using the same approach of Saffarini et al. [29] by using the MD simulations performed for this work and as per the definitions of Equation (6).

Table 3. Equation (8) parameter values.

L (nm)	**6.4**				**13.1**			
$\dot{\varepsilon}$ (s^{-1})	**10^7**	**10^8**	**5×10^8**	**10^9**	**10^7**	**10^8**	**5×10^8**	**10^9**
C_1 ($\times 10^{17}$) (m^{-2})	4.7	5.9	6.4	6.4	8.85	8.6	8	7.75
C_2($\times 10^{17}$) (m^{-2})	−33	−42	−45	−45	−24.2	−22.62	−20	−18
r(m^2)	−4				−2			

Table 4. Equation (10) parameter values.

L	**6.4**				**13.1**			
$\dot{\varepsilon}$ (s^{-1})	**10^7**	**10^8**	**5×10^8**	**10^9**	**10^7**	**10^8**	**5×10^8**	**10^9**
Y_a (MPa)	180	200	220	240	170	200	250	290
$\widetilde{B}$ (MPa)	1000							
β	0.003315	0.003323	0.003330	0.003333	0.003315	0.003323	0.003330	0.003333
Y_d (MPa)	50							
p	0.5							
q	1.5							
D (MPa)	11				5.5			
n ($\times 10^{-12}$) (mm^2)	7.5				9.5			

The parameters of the VA constitutive model part of Equation (10) were determined by using the definition of each parameter as listed in Section 3 and by performing the simple regression analysis explained by Voyiadjis and Abed [49,57] using the results of the small sample. As the table shows, the parameters of the VA constitutive model, except for the yield stress (Y_a), are size-independent. The strain rate dependency is captured by the β parameter which allows the remaining VA parameters, except for the yield stress, to be strain rate-independent. The yield stress term is both size and strain rate-dependent. Those variations in the parameters' values, along with their dependencies, are in line with the physical definition of the VA constitutive model described in Section 3, and with the physical meaning intended from combining the VA constitutive model with the hardening term from the Saffarini et al. constitutive model.

The upturn flow stress component, as represented by the hardening stress term in Equation (10), is strain rate-independent because the strain rate contribution is taken care of by C_1 and C_2 parameters of Equation (8), as well as the β parameter in the VA part. Additionally, the size dependency is captured by both the parameters of the upturn stress component and Equation (8) parameters.

Figure 2 shows the proposed constitutive model predictions of the plastic flow in gold foam compared to the MD simulation results, the prediction of the VA constitutive model, and the predictions of the hardening term from the Saffarini et al. model (dislocation-controlled stress component) at different strain rates of $\dot{\varepsilon} = 10^7$ s^{-1}, $\dot{\varepsilon} = 10^8$ s^{-1}, $\dot{\varepsilon} = 5 \times 10^8$ s^{-1}, and $\dot{\varepsilon} = 10^9$ s^{-1} for the smaller sample (L = 6.4 nm). Figure 3 shows the same but for the larger sample (L = 13.1 nm). Figure 4 shows a comparison of the stress component corresponding to the VA model contribution to the newly proposed model due to thermal activation processes at the different strain rates of $\dot{\varepsilon} = 10^7$ s^{-1}, $\dot{\varepsilon} = 10^8$ s^{-1}, $\dot{\varepsilon} = 5 \times 10^8$ s^{-1}, and $\dot{\varepsilon} = 10^9$ s^{-1} for both sizes simulated in this work (L = 6.4 nm and L = 13.1 nm)

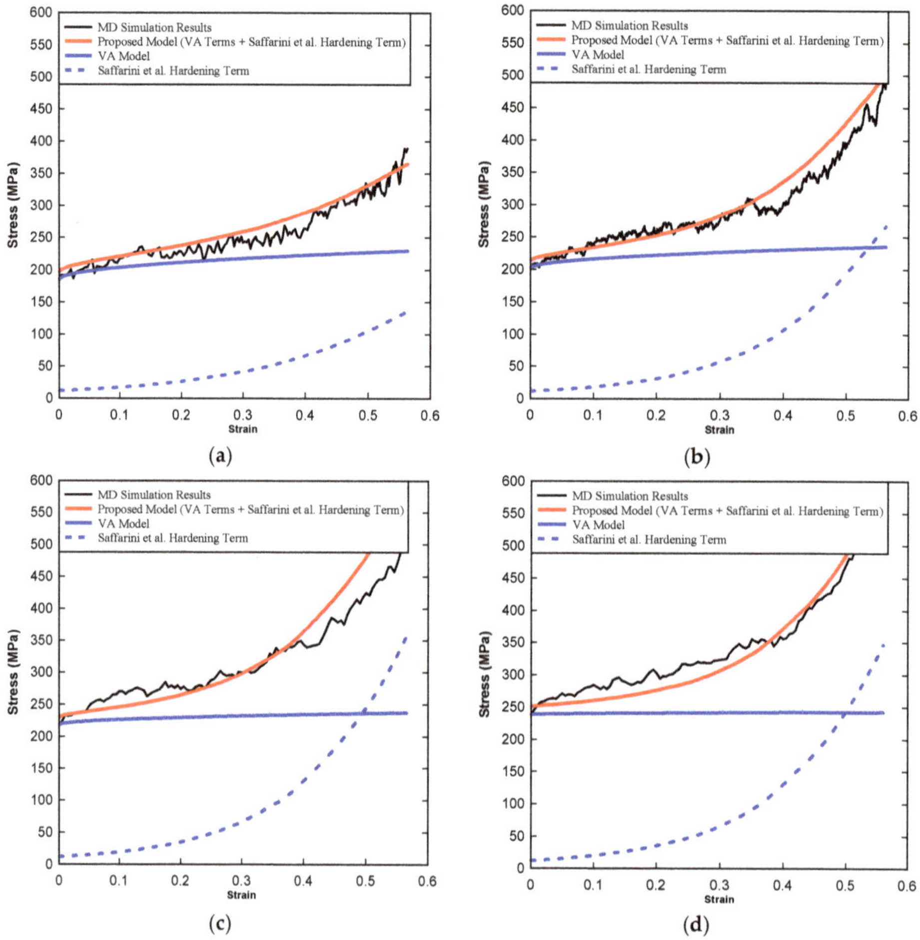

Figure 2. Model predictions of the plastic flow in gold foam compared to the MD simulation results performed in this work, the VA model predictions, and the hardening term from the Saffarini et al. model (dislocation-controlled stress component) in Equation (1) at (**a**) $\dot{\varepsilon} = 10^7\ s^{-1}$, (**b**) $\dot{\varepsilon} = 10^8\ s^{-1}$, (**c**) $\dot{\varepsilon} = 5 \times 10^8\ s^{-1}$, (**d**) $\dot{\varepsilon} = 10^9\ s^{-1}$. All for the smaller sized sample with $L = 6.4$ nm.

Several observations regarding the constitutive model behavior can be noted from the three figures. The first observation is that the proposed equation results are in good agreement with the MD simulation results. The proposed equation captures both the initial plateau-like plasticity and the later stages of densification, which are controlled mainly by the rapid increase in dislocation evolution upon porosity annihilation. As shown by Equation (10), the proposed equation is simply the sum of the VA part and the Saffarini et al. hardening part. For this, it is obvious that each component fails to capture the full stress–strain response of the material. Here appears the superiority of the atomistic-continuum constitutive model over the standalone continuum constitutive model. It is also obvious that the hardening term is not supposed to capture the MD simulation stress–strain response because it is only half the original model shown in Equation (1). However, it is important to highlight that the hardening stress component of the model comprises less than 15% of the total stress throughout the deformation process. The remaining 85% contribution comes from the VA stress component. In the previous atomistic formulation, the VA constitutive model was replaced by a simple multiplication between a constant and the yield stress. Although it phenomenologically captured the VA constitutive model contribution, that term did not have a physical reasoning behind it. Using the VA constitutive model, one can simply show that this term is physically justified.

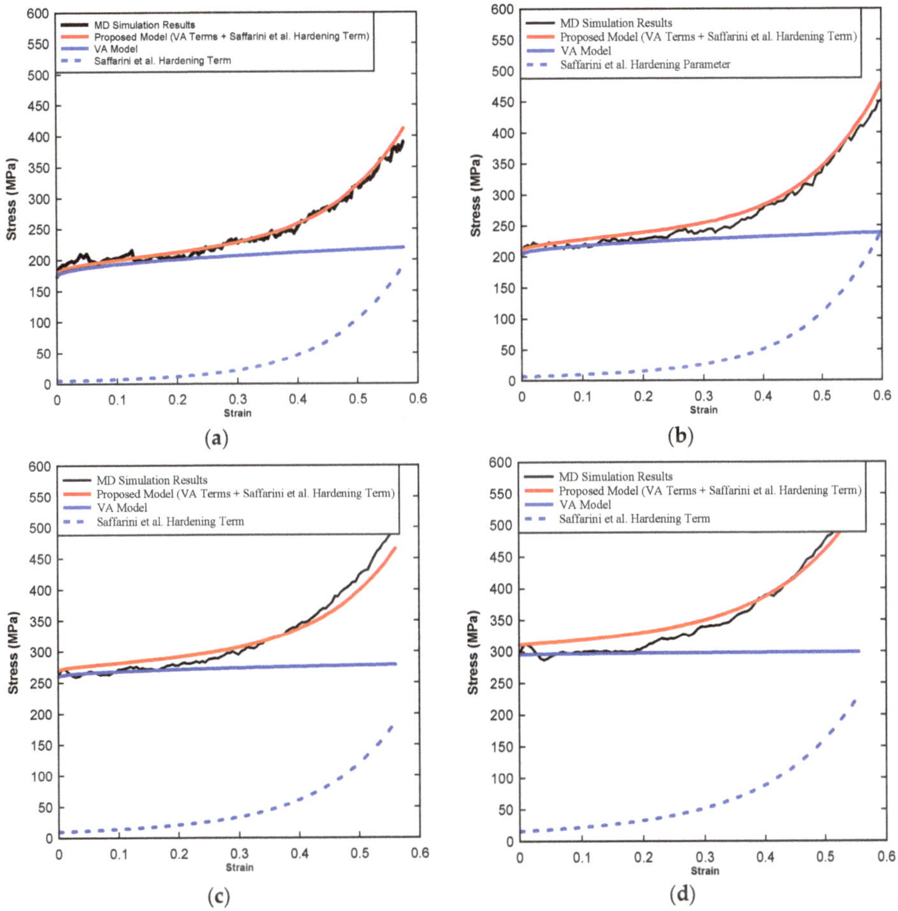

Figure 3. Model predictions of the plastic flow in gold foam compared to the MD simulation results performed in this work, the VA model predictions, and the hardening term from the Saffarini et al. model (dislocation-controlled stress component) in Equation (1) at (**a**) $\dot{\varepsilon} = 10^7\ s^{-1}$, (**b**) $\dot{\varepsilon} = 10^8\ s^{-1}$, (**c**) $\dot{\varepsilon} = 5 \times 10^8\ s^{-1}$, (**d**) $\dot{\varepsilon} = 10^9\ s^{-1}$. All for the medium-sized sample with $L = 13.1$ nm.

Moreover, the previous atomistic formulation assumes that the yield-controlled stress term (refer to Equation (1)) is a constant value throughout the deformation process at each strain rate. That is simply because the constant parameter (C) is independent of the plastic strain or relative density. This means that for any strain rate, the yield-controlled stress component is a fixed value from plasticity initiation until complete densification. Despite it being true for a very high strain rate, it is not quite accurate for lower values of strain rate. As shown by Figure 4, the VA stress component shows that there is slight hardening due to thermal activation energy. This is more accurate constitutive description in light of the discussion in Section 4 where it shows that thermal activation processes still contribute to the material hardening even at high strain rates [47,50]. This shows that the updated model proposed in Equation (10) is better than that in Equation (1).

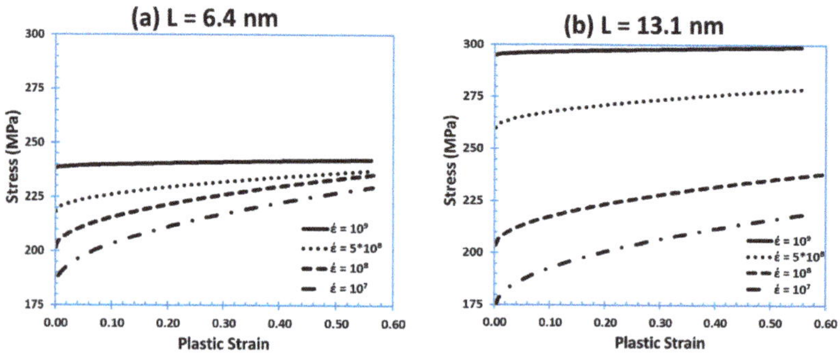

Figure 4. The stress component value corresponding to the VA model contribution of the newly proposed model due to thermal activation processes at $\dot{\varepsilon} = 10^7\ \text{s}^{-1}$, $\dot{\varepsilon} = 10^8\ \text{s}^{-1}$, $\dot{\varepsilon} = 5 \times 10^8\ \text{s}^{-1}$, and $\dot{\varepsilon} = 10^9\ \text{s}^{-1}$ for both sizes simulated in this work.

Another observation is that at very high strain rate, the VA component of Equation (10) reaches a value equal to the yield stress and maintains this value to be constant throughout the full deformation process. That behavior, which reaches the value of the yield stress or the athermal stress component of the VA constitutive model, can be noticed for both sizes at strain rate of $10^9\ \text{s}^{-1}$ in Figure 4a,b. This means that at very high strain rates, the contribution of the thermal activation part of the model towards the plastic deformation of the foam is becoming completely ineffective, and that the plastic flow is completely controlled by the hardening stress component (especially after densification initiation at $\varepsilon \approx 0.22$). This follows the discussion presented earlier in Section 4, in which it was shown that the contribution of the thermal activation processes decay and that the control will be mainly due to dislocation drag at very high strain rate, during which the dislocation evolution and speed become significantly high. In fact, those physical phenomena go hand to hand with the dislocation dynamics presented in the authors' previous work about the effect of strain rate in gold foams at a wide range of strain rates [30].

The last observation is that in the case of the small sample, the hardening term contribution in the regions of material densification ($\varepsilon \geq 0.22$) increases significantly until it reaches or surpasses the value predicted by the VA stress component. This happens at later stages of the deformation when the material completely compresses and the porosity vanishes, transforming it from an open-cell bicontinuous foam microstructure into a full-density nonporous-like microstructure. At this stage the contribution of the upturn stress component matches the contribution of the VA stress component because the material is now fully populated by forest dislocations, and there is no dislocation annihilation at free surface anymore.

7. Conclusions

In this work, the authors performed large-scale simulations at wide range of strain rates ($10^7\ s^{-1}$ to $10^9\ \text{s}^{-1}$) and different ligament sizes ($L = 6.4$ nm and 13.1 nm) to propose a new constitutive description of the plastic flow in gold foams. The description is based on an atomistic-continuum connection that physically captures both thermal activation processes at low strain rates as well as the upturn in flow stress controlled by dislocation drag and the rapid increase in their evolution at very high strain rates. The paper presents an overview of the theory behind the two combined models (atomistic constitutive model and the continuum constitutive model) to later introduce the connection between the two scales, and in turn, present the proposed model results. The proposed constitutive model shows good agreement with the simulation results and captures the physical description discussed throughout the paper. The model shows promising results in capturing the physics of the deformation mechanism in gold foams at a wide range of strain rates and for different sizes. The proposed equation captures both the initial plateau-like plasticity and the later stages of densification, which is controlled mainly by the rapid increase in dislocation evolution upon porosity annihilation. Moreover, it is observed that the proposed constitutive description is more accurate than the standalone VA model or the hardening term from the Saffarini et al. model. This comes from the fact that the model combines the best of the two models to cover all the processes involved in controlling the dislocation density effect (thermal activation energy and dislocation drag). Such a promising constitutive description can help improving integrated physics-based and AI-enabled design of architected metallic foams that deliver tailored mechanical responses at different scales.

Author Contributions: M.H.S., conceptualization, methodology, software, validation, formal analysis, investigation, data curation, writing—original draft, writing—review and editing, visualization, and funding acquisition; G.Z.V., conceptualization, methodology, software, validation, formal analysis, investigation, data curation, writing—original draft, writing—review and editing, visualization, supervision, project administration and funding acquisition. All authors have read and agreed to the published version of the manuscript.

Funding: This research received no external funding.

Data Availability Statement: The raw/processed data required to reproduce these findings is available from the authors upon reasonable request.

Acknowledgments: The authors wish to acknowledge that portions of this research were conducted with high performance computational resources provided by the Louisiana Optical Network Infrastructure (http://www.loni.org). Additionally, the authors would like to acknowledge the contribution of Carlos Ruestes for his input and feedback on the preparation of the samples and the simulation's scripts.

Conflicts of Interest: The authors declare that they have no known competing financial interests or personal relationships that could have appeared to influence the work reported in this paper.

References

1. Ashby, M.F.; Evans, T.; Fleck, N.A.; Gibson, L.J.; Hutchinson, J.W.; Wadley, H.N.G. Metal foams: A design guide. *Mater. Des.* **2002**, *23*, 119. [CrossRef]
2. Gibson, L.J. Cellular Solids. *MRS Bull.* **2003**, *28*, 270–274. [CrossRef]
3. Zhang, J.; Guo, H. Large deflection of rectangular sandwich tubes with metal foam core. *Compos. Struct.* **2022**, *293*, 115745. [CrossRef]
4. Zhang, J.; Ye, Y.; Zhu, Y.; Yuan, H.; Qin, Q.; Wang, T. On axial splitting and curling behaviour of circular sandwich metal tubes with metal foam core. *Int. J. Solids Struct.* **2020**, *202*, 111–125. [CrossRef]
5. Wittstock, A.; Biener, J.; Bäumer, M. Nanoporous gold: A new material for catalytic and sensor applications. *Phys. Chem. Chem. Phys.* **2010**, *12*, 12919–12930. [CrossRef]
6. Biener, J.; Biener, M.M.; Madix, R.J.; Friend, C.M. Nanoporous Gold: Understanding the Origin of the Reactivity of a 21st Century Catalyst Made by Pre-Columbian Technology. *ACS Catal.* **2015**, *5*, 6263–6270. [CrossRef]
7. Zielasek, V.; Jürgens, B.; Schulz, C.; Biener, J.; Biener, M.M.; Hamza, A.V.; Baeumer, M. Gold Catalysts: Nanoporous Gold Foams. *Angew. Chem. Int. Ed.* **2006**, *45*, 8241–8244. [CrossRef]
8. Yogeswaran, U.; Chen, S.-M. A Review on the Electrochemical Sensors and Biosensors Composed of Nanowires as Sensing Material. *Sensors* **2008**, *8*, 290–313. [CrossRef]
9. Xiao, X.; Si, P.; Magner, E. An overview of dealloyed nanoporous gold in bioelectrochemistry. *Bioelectrochemistry* **2016**, *109*, 117–126. [CrossRef]
10. Lang, X.; Hirata, A.; Fujita, T.; Chen, M. Nanoporous metal/oxide hybrid electrodes for electrochemical supercapacitors. *Nat. Nanotechnol.* **2011**, *6*, 232–236. [CrossRef]
11. Hou, Y.; Chen, L.; Hirata, A.; Fujita, T.; Chen, M. Non-aqueous nanoporous gold based supercapacitors with high specific energy. *Scr. Mater.* **2016**, *116*, 76–81. [CrossRef]
12. Quynh, B.T.P.; Byun, J.Y.; Kim, S.H. Non-enzymatic amperometric detection of phenol and catechol using nanoporous gold. *Sens. Actuators B Chem.* **2015**, *221*, 191–200. [CrossRef]
13. Biener, J.; Wittstock, A.; Zepeda-Ruiz, L.A.; Biener, M.M.; Zielasek, V.; Kramer, D.; Viswanath, R.N.; Weissmüller, J.; Bäumer, M.; Hamza, A.V. Surface-chemistry-driven actuation in nanoporous gold. *Nat. Mater.* **2009**, *8*, 47–51. [CrossRef] [PubMed]
14. Bringa, E.; Monk, J.D.; Caro, A.; Misra, A.; Zepeda-Ruiz, L.; Duchaineau, M.; Abraham, F.; Nastasi, M.; Picraux, S.T.; Wang, Y.Q.; et al. Are Nanoporous Materials Radiation Resistant? *Nano Lett.* **2012**, *12*, 3351–3355. [CrossRef] [PubMed]

15. Caro, M.; Mook, W.M.; Fu, E.G.; Wang, Y.Q.; Sheehan, C.; Martinez, E.; Baldwin, J.K.; Caro, A. Radiation induced effects on mechanical properties of nanoporous gold foams. *Appl. Phys. Lett.* **2014**, *104*, 233109. [CrossRef]
16. Ruestes, C.J.; Anders, C.; Bringa, E.M.; Urbassek, H.M. Nanoindentation tests of heavy-ion-irradiated Au foams—Molecular dynamics simulation. *J. Appl. Phys.* **2018**, *123*, 225903. [CrossRef]
17. Hodge, A.M.; Hayes, J.R.; Caro, J.A.; Biener, J.; Hamza, A.V. Characterization and Mechanical Behavior of Nanoporous Gold. *Adv. Eng. Mater.* **2006**, *8*, 853–857. [CrossRef]
18. Hodge, A.M.; Biener, J.; Hayes, J.R.; Bythrow, P.M.; Volkert, C.A.; Hamza, A.V. Scaling equation for yield strength of nanoporous open-cell foams. *Acta Mater.* **2007**, *55*, 1343–1349. [CrossRef]
19. Mathur, A.; Erlebacher, J. Size dependence of effective Young's modulus of nanoporous gold. *Appl. Phys. Lett.* **2007**, *90*, 061910. [CrossRef]
20. Hakamada, M.; Mabuchi, M. Mechanical strength of nanoporous gold fabricated by dealloying. *Scr. Mater.* **2007**, *56*, 1003–1006. [CrossRef]
21. Briot, N.J.; Kennerknecht, T.; Eberl, C.; Balk, T.J. Mechanical properties of bulk single crystalline nanoporous gold investigated by millimetre-scale tension and compression testing. *Philos. Mag.* **2014**, *94*, 847–866. [CrossRef]
22. Bürckert, M.; Briot, N.J.; Balk, T.J. Uniaxial compression testing of bulk nanoporous gold. *Philos. Mag.* **2017**, *97*, 1157–1178. [CrossRef]
23. Pia, G.; Carta, M.; Delogu, F. Nanoporous Au foams: Variation of effective Young's modulus with ligament size. *Scr. Mater.* **2018**, *144*, 22–26. [CrossRef]
24. Sun, X.-Y.; Xu, G.-K.; Li, X.; Feng, X.-Q.; Gao, H. Mechanical properties and scaling laws of nanoporous gold. *J. Appl. Phys.* **2013**, *113*, 023505. [CrossRef]
25. Farkas, D.; Caro, A.; Bringa, E.; Crowson, D. Mechanical response of nanoporous gold. *Acta Mater.* **2013**, *61*, 3249–3256. [CrossRef]
26. Beets, N.; Farkas, D.; Corcoran, S. Deformation mechanisms and scaling relations in the mechanical response of nano-porous Au. *Acta Mater.* **2019**, *165*, 626–637. [CrossRef]
27. Giri, A.; Tao, J.; Wang, L.; Kirca, M.; To, A.C. Compressive Behavior and Deformation Mechanism of Nanoporous Open-Cell Foam with Ultrathin Ligaments. *J. Nanomech. Micromech.* **2014**, *4*, A4013012. [CrossRef]
28. Jiao, J.; Huber, N. Deformation mechanisms in nanoporous metals: Effect of ligament shape and disorder. *Comput. Mater. Sci.* **2017**, *127*, 194–203. [CrossRef]
29. Saffarini, M.H.; Voyiadjis, G.Z.; Ruestes, C.J. Temperature effect on nanoporous gold under uniaxial tension and compression. *Comput. Mater. Sci.* **2021**, *200*, 110766. [CrossRef]
30. Voyiadjis, G.Z.; Saffarini, M.H.; Ruestes, C.J. Characterization of the Strain Rate Effect under Uniaxial Loading for Nanoporous Gold George. *Comput. Mater. Sci.* **2021**, *194*, 110425. [CrossRef]
31. Saffarini, M.H.; Voyiadjis, G.Z.; Ruestes, C.J.; Yaghoobi, M. Ligament size dependency of strain hardening and ductility in nanoporous gold. *Comput. Mater. Sci.* **2021**, *186*, 109920. [CrossRef]
32. Biener, J.; Hodge, A.M.; Hayes, J.R.; Volkert, C.A.; Zepeda-Ruiz, L.A.; Hamza, A.V.; Abraham, F.F. Size Effects on the Mechanical Behavior of Nanoporous Au. *Nano Lett.* **2006**, *6*, 2379–2382. [CrossRef]
33. Lührs, L.; Soyarslan, C.; Markmann, J.; Bargmann, S.; Weissmüller, J. Elastic and plastic Poisson's ratios of nanoporous gold. *Scr. Mater.* **2016**, *110*, 65–69. [CrossRef]
34. Badwe, N.; Chen, X.; Sieradzki, K. Mechanical properties of nanoporous gold in tension. *Acta Mater.* **2017**, *129*, 251–258. [CrossRef]
35. Lührs, L.; Zandersons, B.; Huber, N.; Weissmüller, J. Plastic Poisson's Ratio of Nanoporous Metals: A Macroscopic Signature of Tension–Compression Asymmetry at the Nanoscale. *Nano Lett.* **2017**, *17*, 6258–6266. [CrossRef]
36. Chaboche, J.L. A review of some plasticity and viscoplasticity constitutive theories. *Int. J. Plast.* **2008**, *24*, 1642–1693. [CrossRef]
37. Gao, C.Y.; Zhang, L.C. Constitutive modelling of plasticity of fcc metals under extremely high strain rates. *Int. J. Plast.* **2012**, *32*, 121–133. [CrossRef]
38. Huang, M.; Rivera-Díaz-Del-Castillo, P.E.; Bouaziz, O.; Van der Zwaag, S. A constitutive model for high strain rate deformation in FCC metals based on irreversible thermodynamics. *Mech. Mater.* **2009**, *41*, 982–988. [CrossRef]
39. Xu, Z.; Huang, F. Comparison of constitutive models for FCC metals over wide temperature and strain rate ranges with application to pure copper. *Int. J. Impact Eng.* **2015**, *79*, 65–74. [CrossRef]
40. Liang, R.; Khan, A.S. A critical review of experimental results and constitutive models for BCC and FCC metals over a wide range of strain rates and temperatures. *Int. J. Plast.* **1999**, *15*, 963–980. [CrossRef]
41. Jin, H.-J.; Weissmüller, J.; Farkas, D. Mechanical response of nanoporous metals: A story of size, surface stress, and severed struts. *MRS Bull.* **2018**, *43*, 35–42. [CrossRef]
42. Ruestes, C.J.; Farkas, D.; Caro, A.; Bringa, E.M. Hardening under compression in Au foams. *Acta Mater.* **2016**, *108*, 1–7. [CrossRef]
43. Ruestes, C.J.; Schwen, D.; Millán, E.N.; Aparicio, E.; Bringa, E.M. Mechanical properties of Au foams under nanoindentation. *Comput. Mater. Sci.* **2018**, *147*, 154–167. [CrossRef]
44. Austin, R.A.; McDowell, D.L. A dislocation-based constitutive model for viscoplastic deformation of fcc metals at very high strain rates. *Int. J. Plast.* **2011**, *27*, 1–24. [CrossRef]
45. Preston, D.L.; Tonks, D.L.; Wallace, D.C. Model of plastic deformation for extreme loading conditions. *J. Appl. Phys.* **2003**, *93*, 211–220. [CrossRef]

46. Hansen, B.L.; Beyerlein, I.J.; Bronkhorst, C.A.; Cerreta, E.K.; Dennis-Koller, D. A dislocation-based multi-rate single crystal plasticity model. *Int. J. Plast.* **2013**, *44*, 129–146. [CrossRef]
47. Zerilli, F.J.; Armstrong, R.W. Dislocation-mechanics-based constitutive relations for material dynamics calculations. *J. Appl. Phys.* **1987**, *61*, 1816–1825. [CrossRef]
48. Rodríguez-Martínez, J.; Rodríguez-Millán, M.; Rusinek, A.; Arias, A. A dislocation-based constitutive description for modeling the behavior of FCC metals within wide ranges of strain rate and temperature. *Mech. Mater.* **2011**, *43*, 901–912. [CrossRef]
49. Voyiadjis, G.Z.; Abed, F.H. Microstructural based models for bcc and fcc metals with temperature and strain rate dependency. *Mech. Mater.* **2005**, *37*, 355–378. [CrossRef]
50. Follansbee, P.S.; Kocks, U.F. A constitutive description of the deformation of copper based on the use of the mechanical threshold stress as an internal state variable. *Acta Metall.* **1988**, *36*, 81–93. [CrossRef]
51. Fang, H.; Bi, J.; Zhang, C.; Gutowski, M.; Palta, E.; Wang, Q. A constitutive model of aluminum foam for crash simulations. *Int. J. Non-Linear Mech.* **2017**, *90*, 124–136. [CrossRef]
52. Wang, Z.-H.; Jing, L.; Zhao, L.-M. Elasto-plastic constitutive model of aluminum alloy foam subjected to impact loading. *Trans. Nonferrous Met. Soc. China* **2011**, *21*, 449–454. [CrossRef]
53. Reyes, A.; Hopperstad, O.S.; Berstad, T.; Hanssen, A.; Langseth, M. Constitutive modeling of aluminum foam including fracture and statistical variation of density. *Eur. J. Mech.—A/Solids* **2003**, *22*, 815–835. [CrossRef]
54. Chang, F.S.; Song, Y.; Lu, D.X.; DeSilva, C.N. Unified Constitutive Equations of Foam Materials. *J. Eng. Mater. Technol.* **1998**, *120*, 212–217. [CrossRef]
55. Af Segerstad, P.H.; Larsson, R.; Toll, S. A constitutive equation for open-cell cellular solids, including viscoplasticity, damage and deformation induced anisotropy. *Int. J. Plast.* **2008**, *24*, 896–914. [CrossRef]
56. Af Segerstad, P.H.; Toll, S. Open-cell cellular solids: A constitutive equation for hyperelasticity with deformation induced anisotropy. *Int. J. Solids Struct.* **2008**, *45*, 1978–1992. [CrossRef]
57. Voyiadjis, G.Z.; Abed, F.H. Effect of dislocation density evolution on the thermomechanical response of metals with different crystal structures at low and high strain rates and temperatures. *Arch. Mech.* **2005**, *57*, 299–343. [CrossRef]
58. Saffarini, M.H.; Voyiadjis, G.Z.; Ruestes, C.J. Scaling laws for nanoporous metals under uniaxial loading. *J. Mater. Res.* **2021**, *36*, 2729–2741. [CrossRef]
59. Hull, D.; Bacon, D. *Introduction to Dislocations*; Butterworth-Heinemann: Oxford, UK, 2011.
60. Steinberg, D.J.; Lund, C.M. A constitutive model for strain rates from 10^{-4} to 10^{6} s^{-1}. *J. Appl. Phys.* **1988**, *49*, C3-433. [CrossRef]
61. Rusinek, A.; Rodríguez-Martínez, J.; Arias, A. A thermo-viscoplastic constitutive model for FCC metals with application to OFHC copper. *Int. J. Mech. Sci.* **2010**, *52*, 120–135. [CrossRef]
62. Zerilli, F.J.; Armstrong, R.W. The effect of dislocation drag on the stress-strain behavior of F.C.C. metals. *Acta Met. Mater.* **1992**, *40*, 1803–1808. [CrossRef]
63. Blaschke, D.N.; Burakovsky, L.; Preston, D.L. On the temperature and density dependence of dislocation drag from phonon wind. *J. Appl. Phys.* **2021**, *130*, 015901. [CrossRef]
64. Nemat-Nasser, S.; Li, Y. Flow stress of f.c.c. polycrystals with application to OFHC Cu. *Acta Mater.* **1998**, *46*, 565–577. [CrossRef]
65. Molinari, A.; Ravichandran, G. Constitutive modeling of high-strain-rate deformation in metals based on the evolution of an effective microstructural length. *Mech. Mater.* **2005**, *37*, 737–752. [CrossRef]
66. Klepaczko, J.R.; Chiem, C.Y. On rate sensitivity of f.c.c. metals, instantaneous rate sensitivity and rate sensitivity of strain hardening. *J. Mech. Phys. Solids* **1986**, *34*, 29–54. [CrossRef]
67. Klepaczko, J.R. Physical-state variables—The key to constitutive modeling in dynamic plasticity. *Nucl. Eng. Des.* **1991**, *127*, 103–115. [CrossRef]
68. Kubin, L.P.; Estrin, Y. Evolution of dislocation densities and the critical conditions for the Portevin-Le Châtelier effect. *Acta Met. Mater.* **1990**, *38*, 697–708. [CrossRef]
69. Voyiadjis, G.Z.; Almasri, A.H. A physically based constitutive model for fcc metals with applications to dynamic hardness. *Mech. Mater.* **2008**, *40*, 549–563. [CrossRef]
70. Armstrong, R.W.; Ramachandran, V.; Zerilli, F.J. Materials for Advanced Technology System. In *Indian Institute of Metals Symposium*; Indian Institute of Metals: New Delhi, India, 1993.
71. Saffarini, M.H.Y. Atomistic Thermo-Mechanical Description of the Deformation Behavior, Scaling Laws, and Constitutive Modeling of Nanoporous Gold. 2021. Available online: https://digitalcommons.lsu.edu/gradschool_dissertations/5568 (accessed on 5 January 2023).
72. Crowson, D.A.; Farkas, D.; Corcoran, S.G. Mechanical stability of nanoporous metals with small ligament sizes. *Scr. Mater.* **2009**, *61*, 497–499. [CrossRef]
73. Crowson, D.A.; Farkas, D.; Corcoran, S.G. Geometric relaxation of nanoporous metals: The role of surface relaxation. *Scr. Mater.* **2007**, *56*, 919–922. [CrossRef]
74. Newman, R.C.; Corcoran, S.G.; Erlebacher, J.; Aziz, M.J.; Sieradzki, K. Alloy Corrosion. *MRS Bull.* **1999**, *24*, 24–28. [CrossRef]
75. Stuckner, J.; Frei, K.; McCue, I.; Demkowicz, M.J.; Murayama, M. AQUAMI: An open source Python package and GUI for the automatic quantitative analysis of morphologically complex multiphase materials. *Comput. Mater. Sci.* **2017**, *139*, 320–329. [CrossRef]
76. Plimpton, S. Fast Parallel Algorithms for Short-Range Molecular Dynamics. *J. Comput. Phys.* **1995**, *117*, 1–19. [CrossRef]

77. Foiles, S.M.; Baskes, M.I.; Daw, M.S. Embedded-atom-method functions for the fcc metals Cu, Ag, Au, Ni, Pd, Pt, and their alloys. *Phys. Rev. B* **1986**, *33*, 7983–7991. [CrossRef]
78. Stukowski, A.; Bulatov, V.V.; Arsenlis, A. Automated identification and indexing of dislocations in crystal interfaces. *Model. Simul. Mater. Sci. Eng.* **2012**, *20*, 085007. [CrossRef]
79. Stukowski, A. Computational Analysis Methods in Atomistic Modeling of Crystals. *Jom* **2014**, *66*, 399–407. [CrossRef]
80. Stukowski, A.; Albe, K. Extracting dislocations and non-dislocation crystal defects from atomistic simulation data. *Model. Simul. Mater. Sci. Eng.* **2010**, *18*, 085001. [CrossRef]
81. Stukowski, A. Structure identification methods for atomistic simulations of crystalline materials. *Model. Simul. Mater. Sci. Eng.* **2012**, *20*, 045021. [CrossRef]
82. Stukowski, A. Visualization and analysis of atomistic simulation data with OVITO—The Open Visualization Tool. *Model. Simul. Mater. Sci. Eng.* **2009**, *18*, 015012. [CrossRef]
83. Thompson, A.P.; Plimpton, S.J.; Mattson, W. General formulation of pressure and stress tensor for arbitrary many-body interaction potentials under periodic boundary conditions. *J. Chem. Phys.* **2009**, *131*, 154107. [CrossRef]

metals

MDPI

Article

Hall–Petch Description of the Necking Point Stress

Alexey Vinogradov [1,*] and Yuri Estrin [2,3]

[1] Magnesium Research Center, Kumamoto University, Kumamoto 860-8555, Japan
[2] Department of Materials Science & Engineering, Monash University, Clayton, Melbourne, VIC 3800, Australia
[3] School of Mechanical & Chemical Engineering, University of Western Australia, Crawley, Perth, WA 6009, Australia
* Correspondence: vinogradov@kumamoto-u.c.jp

Abstract: We posited that the grain size dependence of the tensile necking stress, as determined by the Considère criterion for plastic instability, is a more meaningful characteristic of the Hall–Petch (H–P) effect than that of the yield stress or the 0.2% proof stress. An inverse square-root dependence of the necking stress on the grain size was derived from a dislocation dynamics-based constitutive model. In this model, the grain size effect enters the stress indirectly via the evolution of the dislocation density. Model predictions were confirmed by the experimental data for nickel and titanium.

Keywords: strain hardening; grain size; necking instability; dislocation kinetics modeling

1. Introduction

The Hall–Petch (H–P) relation between the yield strength and the average grain size of polycrystalline materials is one of the pillars of physical metallurgy. It has been quite some time since an H–P description was established for the entire stress–strain curve, i.e., for any strain ε—initially for the polycrystalline mild steel tested in tension [1]. The H–P relation is commonly presented in form of the equation

$$\sigma_\varepsilon = \sigma_{0\varepsilon} + K_\varepsilon^{HP} d^{-1/2} \tag{1}$$

relating the flow stress σ_ε at a given strain ε to the average grain size (or average intercept length in micrographs) d. In Equation (1), the 'friction stress', $\sigma_{0\varepsilon}$, represents the flow stress in a single crystal limit of 'infinitely large' d. The friction stress absorbs the contributions to flow stress from mechanisms not related to the dislocation interaction effects, such as solute drag, the Peierls stress, etc. For pure fcc metals, $\sigma_{0\varepsilon}$ can be assumed to be negligible, at least after a sufficiently large strain. The Hall–Petch coefficient K_ε^{HP} measures the stress intensity required for the transmission of plastic flow across the grain boundaries [2]. It is material-specific and is determined empirically, as reviewed in [3,4].

In recent years, the research focus has been on incorporating the H–P description into constitutive models that include temperature and strain rate dependences, as encompassed, for example, in the Zerilli-Armstrong (Z-A) relations that are different for face-centered cubic (fcc) and body-centered cubic (bcc) metals [5]:

$$\sigma_\varepsilon = \sigma_{G\varepsilon} + B_0[\varepsilon_r(1\ \exp(-\varepsilon/\varepsilon_r))]^{1/2} \exp\{-\alpha^* T\} + K_\varepsilon^{HP} d^{-1/2}\ \text{(fcc)} \tag{2}$$

$$\sigma_\varepsilon = \sigma_{G\varepsilon} + B\exp(-\beta^* T) + K\varepsilon^n + K_\varepsilon^{HP} d^{-1/2} \qquad \text{(bcc)} \tag{3}$$

These phenomenological equations can be seen as special cases of Equation (1). The specific form of the friction stress $\sigma_{0\varepsilon}$ involves an athermal stress component $\sigma_{G\varepsilon}$, which is dependent on the solute content and the dislocation density. The remainder accounts for temperature-dependent strain hardening, with the strain- and temperature-dependent terms taken as a product for the fcc case and in an additive form for the bcc case. In

Equation (2), the strain hardening that is measured in terms of B_0 and a recovery strain, ε_r, is coupled with an exponential temperature-dependent factor, where $\alpha^* = \alpha_0 - \alpha_1 \ln(d\gamma/dt)$ introduces a dependence on the shear strain rate $d\gamma/dt$. Here α_0 and α_1 are material parameters. In Equation (3), B is the athermal stress reflecting the resistance to dislocation glide at zero absolute temperature, $T = 0$. The rate dependence enters via $\beta^* = \beta_0 - \beta_1 \ln(d\gamma/dt)$, in the same manner as in α^*. The parameters K and n in the Ludwik term $K\varepsilon^n$ [6] govern the strain hardening of the material. The hexagonal close-packed (hcp) metals α-titanium, zirconium, and hafnium behave like bcc metals, while magnesium, zinc, and cadmium follow an fcc-type behavior. Accordingly, Equations (2) and (3) can be applied for these two respective groups of hcp metals.

An excellent descriptive capability of the Z-A model has been demonstrated in many examples, cf. [7,8]. A deficiency of the Ludwik ansatz, however, is that a trend of the flow stress to saturate at large strains is not reflected by Equation (3), and that the constants K and n do not have any microstructurally-based meaning, even when the approximation of the stress–strain data by the Ludwik-type models is seemingly acceptable. A viable alternative is a physically based modeling approach going back to Kocks and Mecking [9], in which the dislocation density plays the role of an internal state variable evolving in the process of straining. The model was later extended by Estrin and Mecking [10] to account for the grain size effect (see also [11]). In the next section, an analysis of the H–P effect in terms of the Kocks–Mecking–Estrin (KME) model [10,11] will be presented. This will be followed by gauging the KME model against the experimental data for nickel and titanium as representative fcc and hcp materials.

2. Dislocation Density Based Constitutive Modeling of the Hall–Petch Behavior

Virtually all dislocation-based strain hardening models involving a single internal variable comprise two constitutive equations: a semi-empirical equation relating the flow stress σ to the total dislocation density ρ, and an equation describing the evolution of ρ with plastic strain ε. The first equation is the Taylor relation:

$$\sigma = \sigma_f + M\alpha Gb\sqrt{\rho} = \sigma_f + \beta\sqrt{\rho} \quad (4)$$

where b is the magnitude of the dislocation Burgers vector, G is the shear modulus, and α is a microstructure-sensitive factor, typically ranging between 0.1 and 0.4. Its magnitude is governed by the dislocation arrangement and includes the strain-rate and temperature dependence of the flow stress; M is the texture-dependent orientation factor converting the resolved shear stress to the axial stress. The factor $\beta = M\alpha Gb$ is introduced on the right-hand side of Equation (4) for brevity. The evolution equation for the total dislocation density may take different forms, of which the simplest one reads as [10–13]

$$\frac{d\rho}{d\varepsilon} = M(k_0 - k_2\rho) \quad (5)$$

The dislocation density evolution described by Equation (5) is a result of the competing processes of dislocation storage and annihilation (dynamic recovery). If one assumes that dislocations are stored quasi-homogenously at a rate inversely proportional to their mean free path $\langle\Lambda\rangle$, k_0 can be expressed as $k_0 = \frac{\widetilde{k_0}}{b\langle\Lambda\rangle}$, with $\widetilde{k_0}$ being a non-dimensional constant. The term $k_2\rho$ represents the average rate of dynamic dislocation recovery, which is a thermally activated process governed by the cross-slip of screw dislocations or the diffusion-controlled climb of edge dislocations. This is reflected in the temperature and strain rate dependence of the dynamic recovery coefficient $k_2 = k_2(\dot{\varepsilon}, T)$. If the dislocation travel is geometrically constrained by grain boundaries, i.e., for sufficiently small grain size, the mean free path can be identified with the average grain size d. In this case, the evolution equation assumes the form

$$\frac{d\rho}{d\varepsilon} = M\left(\frac{\tilde{k}_0}{b\,d} - k_2\rho\right) \tag{6}$$

where the storage term is explicitly related to d.

In a more general formulation, assuming that both grain boundaries and dislocations within the grains (distributed randomly or organized in dislocation cells) contribute to dislocation storage concurrently, Equation (5) can be extended to a generalized evolution equation of the Kocks–Mecking–Estrin (KME) model [10,11]:

$$\frac{d\rho}{d\varepsilon} = M(k_0 + k_1\sqrt{\rho} - k_2\,\rho) \tag{7}$$

This equation reduces to the evolution equation of the Kocks–Mecking (KM) model in the limit of large d when $k_0 = \frac{\tilde{k}_0}{b\,d}$ is significantly smaller than $k_1\sqrt{\rho}$, and the dislocation production is controlled by dislocation reactions in the grain interior.

The constitutive model expressed by Equations (4), (5), and (7) recovers an H–P type relation between the flow stress and the grain size. Specifically, it delivers the classical H–P formula for the conventional yield stress at the 0.2% proof strain:

$$\sigma_{0.2} = \sigma_f + \frac{K_{0.2}^{HP}}{\sqrt{d}} \tag{8}$$

Here, the H–P coefficient is given by

$$K_{0.2}^{HP} = M^{3/2}\alpha G b\sqrt{\left(\tilde{k}_0/b + k_1\sqrt{\rho_0}d\right)\varepsilon_{0.2}} \tag{9}$$

which, as opposed to the predictions made in [14], may be microstructure-sensitive due to the influence of microstructure on the parameters $\tilde{k}_0$ and k_1.

In the limit of $k_1 \to 0$ (or sufficiently small initial dislocation density ρ_0 and/or grain size d) the yield stress dependence on the grain size reads as

$$\sigma_{0.2} = \sigma_f + M^{3/2}\alpha G\sqrt{\frac{\tilde{k}_0 b\,\varepsilon_{0.2}}{d}} \tag{10}$$

and Equation (9) reduces to

$$K_{0.2}^{HP} = M^{3/2}\alpha G\sqrt{2\cdot 10^{-3}\,\tilde{k}_0 b} \tag{11}$$

We recall that as early as in 1967, Conrad et al. [15] arrived at the following semi-empirical relation between the flow stress and the grain size at the notional onset of plastic flow (i.e., at the 0.2% proof strain):

$$\sigma_{0.2} = \sigma_f + C'\sqrt{\frac{b\,\varepsilon_{0.2}}{d}} \tag{12}$$

Here, C' is a material dependent parameter. The functional dependence on the grain size given by this equation is obviously the same as that in Equation (10).

To derive this equation, Conrad et al. assumed that grain boundaries affect the total dislocation density, which determines the flow stress through the Taylor Equation (4). The experimental data they used showed a linear dependence of the dislocation density on plastic strain, $\rho \sim \varepsilon$, at the onset of yielding. Note that this experimental observation resulting in the well-known 'parabolic' hardening, $\sigma \sim \sqrt{\varepsilon}$, is naturally predicted by

Equation (5) when the first term on the right-hand side describing the dislocation production prevails over the annihilation term, e.g., for well-annealed polycrystals. Shortly after the publication by Conrad et al., Ashby [16] used the slip distance concept that he developed for the description of plasticity in heterogeneous materials and came up with essentially the same form of the $\sigma(d)$ relation. This work prompted the emergence of a family of strain hardening based models of the H–P effect as reviewed, for example, by Cordero et al. [17]. In this context one should mention the work by Hazzledine [18,19] and the companion theoretical analysis [20] on the basis of the KME kinetic Equation (7), which has led to the H–P expression resembling Equation (12). Godon et al. [21] expressed the Hall–Petch slope for a group of strain-hardening based models in a general form similar to Equation (11): $K_{0.2}^{HP} = M^{3/2}\alpha G\sqrt{2 \cdot 10^{-3}\, b}\; \delta$, albeit without referring to the dislocation multiplication coefficient $\widetilde{k}_0$ directly. Here, δ is a model-specific microstructure-sensitive parameter depending on the dislocation multiplication process, in general, and the number of activated slip systems [22], in particular; it may also depend on the grain boundary state through the density of ledges [23]. Furthermore, using discrete dislocation dynamics simulations of the deformation behavior of Cu polycrystals with grain sizes ranging from 0.5 to 2 μm, Lefebvre et al. [22,24] demonstrated that regardless of the details of the dislocation multiplication process, the H–P effect can be interpreted as Taylor hardening owing to a heterogenous dislocation storage process controlled by the accumulation of dislocations at the grain boundaries.

3. The Necking Stress Locus as a Basis for an H–P Analysis

Historically, the H–P relation was first established experimentally for the yield strength. The latter quantity can be represented, e.g., by the lower yield point in bcc α-iron [2,25]. However, for fcc materials, the true yield strength is difficult to determine experimentally, which is why the generally accepted convention of using the stress $\sigma_{0.2}$ determined at the 0.2% irreversible strain is adopted as a measure of the yield strength, as mentioned above (see Equation (8)).

In their early work, dated 1962, Armstrong et al. [2] pointed out that the K_ε^{HP} value can vary, depending on the plastic strain ε at which it was measured. In fact, virtually all strain hardening models predict the square-root strain dependence of the H–P factor in the form akin to Equations (10) and (12). This parabolic strain dependence of K_ε^{HP} has long been a matter of controversy in experimental measurements performed on various materials, including fcc Cu [26–28], Al [29,30], Ni [31,32], and 30/70 brass [2]; bcc Fe [2,25,33]; and hcp Zn [2], Ti [34], Zr [35], and Hf [36]. Most studies converged on the conclusion that K_ε^{HP} increases with plastic strain, yet with some exceptions, as reviewed in [17,37]. It was found that K_ε^{HP} may exhibit an opposite trend, decreasing with strain due to deformation twinning or a strong crystallographic texture. The most remarkable discrepancy seen from these assessments of the strain hardening models lies in the fact that none of the experimental measurements confirm the parabolic K_ε^{HP} behavior predicted by Equation (10). Cordero et al. [17] suggested that one possible reason for the observed inconsistency is that the K_ε^{HP} values have historically been measured at stresses that are so large that Ashby's assumption that the density of geometrically necessary dislocations is much larger than that of statistically stored dislocations is no longer valid. We share this viewpoint and take it a step further. In our opinion, the contribution from the dynamic recovery term in the strain hardening models including the KM, Equation (5), and the KME, Equation (7), models cannot be neglected even at the beginning of plastic flow.

Thus, the KME strain hardening models predict an H–P type dependence of the flow stress on the grain size for *any* given strain, despite an ambiguity with the choice of the reference strain discussed in the previous section. However, the strain is not a state variable [38], and it is therefore desirable to compare the flow stresses for different grain sizes in a condition free of any arbitrariness. We propose to use the locus of the maximum loads, corresponding to the onset of tensile necking under various deformation conditions, as representing a well-defined state that offers itself for an H–P analysis. The uniqueness of this locus defined by the necking instability condition eliminates the arbitrariness of other approaches in which the strain is considered (explicitly or tacitly) as a state variable. The necking condition is obtained in terms of extrinsic variables, the true stress and true strain, for a given plastic strain rate $\dot{\varepsilon}$ according to the Considère criterion [39]:

$$\theta \equiv \left.\frac{\partial\sigma}{\partial\varepsilon}\right|_{\dot{\varepsilon}} = \sigma \tag{13}$$

In this criterion, the strain rate sensitivity of the flow stress is neglected, which is acceptable for many fcc and hcp metals and alloys. In principle, it has been demonstrated that both the Considère condition and the more general Hart instability condition [40], which does account for the strain rate sensitivity, follow from the evolution laws for the principal internal variable—the total dislocation density [41,42]. Undoubtedly, the advantage of utilizing the necking point to characterize the H–P grain size effects is that unlike the conventional yield point (or the flow stress at a given strain), which is not a special point on the stress–strain curve, the point of onset of necking instability is actually a special point. It is controlled entirely by a combination of the parameters governing the strain hardening of a material [41], and the slope of the H–P plot is uniquely determined by these parameters, and not by any convention with regard to strain.

Combining the solution of the constitutive equations of the KME model for constant plastic strain rate with the Considère condition, Equation (9), yields the plastic strain and the flow stress at the necking point:

$$\varepsilon_N^{KME} = -\frac{1}{k_2 M}\left[\left(1+\frac{k_1}{K}\right)\ln\left|\frac{\frac{\sigma_N^{KME}}{\beta}-\frac{k_1+K}{2k_2}}{\frac{\sigma_0}{\beta}-\frac{k_1+K}{2k_2}}\right| + \left(1-\frac{k_1}{K}\right)\ln\left|\frac{\frac{\sigma_N^{KME}}{\beta}-\frac{k_1-K}{2k_2}}{\frac{\sigma_0}{\beta}-\frac{k_1-K}{2k_2}}\right|\right] \tag{14}$$

and

$$\sigma_N^{KME} = \frac{\frac{\beta k_1 M}{2} + \beta\sqrt{\left(\frac{KM}{2}\right)^2 + \frac{M}{2k_2}\cdot\left[K^2 - {k_1}^2\right]}}{(2+k_2 M)} \tag{15}$$

with $K = \sqrt{k_1^2 + 4k_0 k_2}$.

Turning now to the analysis of the grain size dependence of the stress at maximum load, or the necking stress, we obtain

$$\sigma_N^{KME} = \beta\sqrt{\frac{\tilde{k}_0 M}{d\,b}\frac{(1+k_2 M)}{(2+k_2 M)^2}} \Rightarrow (k_2 M \gg 1) \Rightarrow M\alpha G b\sqrt{\frac{1}{d}\frac{\tilde{k}_0}{b k_2}} = \frac{K_N^{HP}}{\sqrt{d}} \tag{16}$$

which yields the Hall–Petch type behavior of the necking stress. Here, the H–P factor is redefined at the Considère point as

$$K_N^{HP} = M\alpha G\sqrt{b}\sqrt{\frac{\tilde{k}_0}{k_2}} \tag{17}$$

We used the condition $k_2 M \gg 1$, which is fulfilled particularly well for fine-grained and ultrafine-grained materials [41–45].

The simplicity of both relations, Equations (16) and (17), makes them very attractive for practical use.

4. An H–P Dependence for σ_N of Nickel and Titanium—A Compilation of Experimental Results

The tenet of the Z-A approach that strain hardening is contained chiefly in $\sigma_{0\varepsilon}$, while K_ε^{HP} is essentially unchanged, has been taken a step further in the comparison of compiled H–P measurements of $\sigma_{0.2}$ and σ_N covering a large range of grain sizes for pure nickel and commercial purity (CP) titanium (Grade 2), as shown in Figures 1 and 2, respectively. In these figures, the subplots (a) and (b) refer to the conventional yield stress $\sigma_{0.2}$ and the necking stress σ_N, respectively. For the sake of generality, both datasets for Ni and Ti represent the results of independent investigations by different researchers. For nickel, the compilation by Di Leo et al. [14] (see the references therein) is used, with an addition of data from several other sources. The shown measurements of Yasnikov et al. [45] comprise two groupings of conventional and ultrafine-grained materials, in the latter case, having been produced by the annealing of the material that had been severely pre-strained by equal-channel angular pressing (ECAP). The ultrafine grain size measurements of Bui et al. [46,47] (see also [48] and Krasilnikov et al. [49]), adopted here, were obtained by various techniques of severe plastic deformation (SPD). The solid line corresponds to the least square regression line, with the intercept value of σ_f = 66 MPa and the slope of $K_{0.2}^{HP}$ = 204 MPa × μm$^{1/2}$, which is in fair agreement with what is typically reported for Ni [14]; the Pearson's r value for the entire dataset approximation is 0.95.

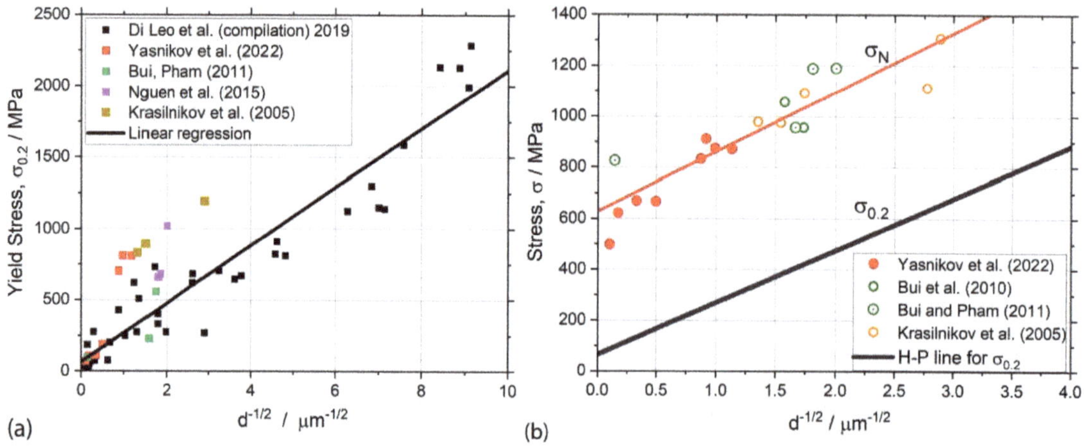

Figure 1. Hall–Petch plots showing the grain size dependence of (**a**) the yield stress $\sigma_{0.2}$ and (**b**) the true stress at maximum load σ_N for nickel polycrystals. The linear regression line for $\sigma_{0.2}$ presented in (**a**) is reintroduced in (**b**) for easier comparison. The references for the datapoints collected from different studies by Di Leo et al. (2019) are given in [14]; other points are adapted from [45–49].

Figure 1b compares the slope $K_{0.2}^{HP}$ represented by the regression line of Figure 1a with K_N^{HP} obtained from the data collected at the true necking stress. The latter quantity, 231 MPa × μm$^{1/2}$, is only slightly higher than the value of $K_{0.2}^{HP}$ corresponding to the 0.2% proof strain. Considering the significant scatter of experimental data, it is fair to say that both values are practically indistinguishable, and as a first order approximation, a near equality $K_{0.2}^{HP} \approx K_N^{HP}$ holds.

The same trends are observed for commercially pure titanium (Grade 2) with a wide range of grain sizes produced by various severe plastic deformation routes and annealing procedures [50–65], as shown in Figure 2.

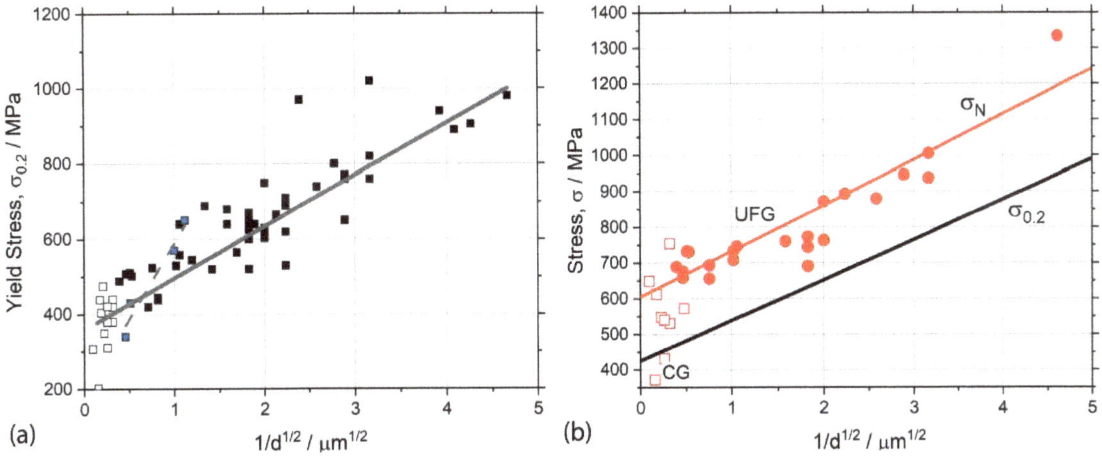

Figure 2. Hall–Petch plots showing the grain size dependence of (**a**) the yield stress $\sigma_{0.2}$ and (**b**) the true stress at maximum load σ_N for CP titanium (Grade 2); the data were collected from [50–64].

5. Discussion

The present measurements and the analysis of the data add to a previously reported compilation of results assembled for the tensile test characteristics of bcc, fcc, and hcp metals [66]. Special emphasis is put on the H–P analysis of the stress at the maximum load locus (necking stress) as an important convention-free measure of the material state. Both the phenomenological Z-A model and the KME model, with its greater microstructural underpinning, were revealed to provide an adequate description of the H–P behavior, the latter being capable, in principle, of predicting different slopes of the H–P curves for the yield stress and the necking stress. Even though no significant difference between these two slopes was seen in the present work for either Ni or Ti, this is not what is often observed. For example, Tsuji et al. [67] found that the $K_{0.2}^{HP}$ value for fine grained CP aluminum alloy Al1100 was considerably larger than that of K_N^{HP} measured at the ultimate tensile strength. The microstructural sensitivity of the H–P effect is reflected in Figures 1 and 2 in several aspects. The significant scatter seen in large experimental datasets assembled from different sources for materials of the same type (see also recent comprehensive compilations of H–P data for Ti by Takebe and Ushioda [68], Ti and Al by Figueiredo and Langdon [69], and for Ti, Al, Mg, Cu, and Fe by Dangwal et al. [70]) suggests that while the general $1/\sqrt{d}$ scaling law for the flow stress does hold, the microstructural factors other than grain size (e.g., crystallographic texture, dislocation storage in the substructure, the grain boundary state, etc.) strongly influence the mechanical response. Thus, K_ε^{HP} is not a universal material constant that can be used to quantitatively predict the flow stress dependence on the grain size. The slopes of the H–P plots generated by different investigators for essentially the same material can differ quite appreciably. As an example, in Figure 2, we highlighted the data by Luo et al. [64] (grey squares), showing that the H–P slope of 404 MPa × $\mu m^{1/2}$ for a specific sub-set (plotted as a dashed linear regression line) is almost triple the slope of the linear regression line for the whole dataset (137 MPa × $\mu m^{1/2}$). Furthermore, both these values are remarkably different from the H–P coefficient of 173 MPa × $\mu m^{1/2}$ derived from the data presented by Figueiredo and Langdon [69] for ultrafine grained Grade 2 Ti. The results reported by Khamsuk et al. [71] for fine grain Al1100, with a different processing history, corroborate the above statement that the H–P coefficient is sensitive to the microstructure. These authors showed that within approximately the same range of grain sizes, the slope of the H–P diagrams varied by a factor of five—from 28 MPa × $\mu m^{1/2}$ for cold-rolled and annealed samples to 58 MPa × $\mu m^{1/2}$ for those produced by torsion, and 139 MPa × $\mu m^{1/2}$ for the material fabricated by accumulated roll bonding.

We should thus contend that the slope of the H–P line is sensitive to the microstructure of a material and may depend appreciably on its processing history. The data by Tian et al. [72], Bai et al. [73], and Dangwal et al. [70] provide further support for this posit. These authors observed a two-stage H–P regime with remarkably different slopes in coarse- and fine-grain domains in differently processed pure Cu, Al, Mg, Ti, and ultrafine grained Fe-31Mn-3Al-3Si alloy specimens, as schematically illustrated in Figure 3. Similar results indicating the existence of a critical grain size d_c, where a sharp transition in the H–P behavior is observed with grain refinement to the sub-micron scale, were reported by Fu et al. [74] for IF steels with the grain size ranging from 0.5 to 500 µm. A similar bi-linearity in the H–P behavior for strains up to 5% has been reported earlier by Kashyap and Tangri [75] for 316 stainless steel (note that the single H–P relation was, however, observed at larger strains up to necking) by Thompson [76] for Ni, by Begrström and Hallen [77] for Fe, and by Lloyd [78] and Armstrong [79] for aluminum. This microstructure sensitivity can, in principle, be readily accounted for in a phenomenological way by allowing a dependence of the dislocation production coefficients $\widetilde{k}_0$ and k_1 in Equation (9) on the grain microstructure, texture, and grain boundary state resulting from a specific processing route. For example, the puzzling two-stage H–P behavior reported in the publications cited above can be seamlessly explained by the KME model we use. The authors of [70] have heuristically related the observed change-over in the slope of the H–P plot, with grain reduction to the sub-micron range, by the increased number of dislocations stored during SPD processing. We concur with this view and expand on that. Recalling the general form of the H–P coefficient proceeding from the KME model, Equation (9), one can notice that the initial dislocation density ρ_0 enters it as an addition to the 'classical' H–P term $\widetilde{k}_0/bd$ under the square root sign. An upward change in the slope of the H–P diagram occurs when the $k_1\sqrt{\rho_0}d$ term in Equation (9) comes into play and becomes comparable with or greater than $\widetilde{k}_0/b$. Then Equation (8) for the yield stress is transformed to the more general form

$$\sigma_{0.2} \sim \sqrt{\frac{1}{d} + \frac{k_1 b}{\widetilde{k}_0}\sqrt{\rho_0}} \tag{18}$$

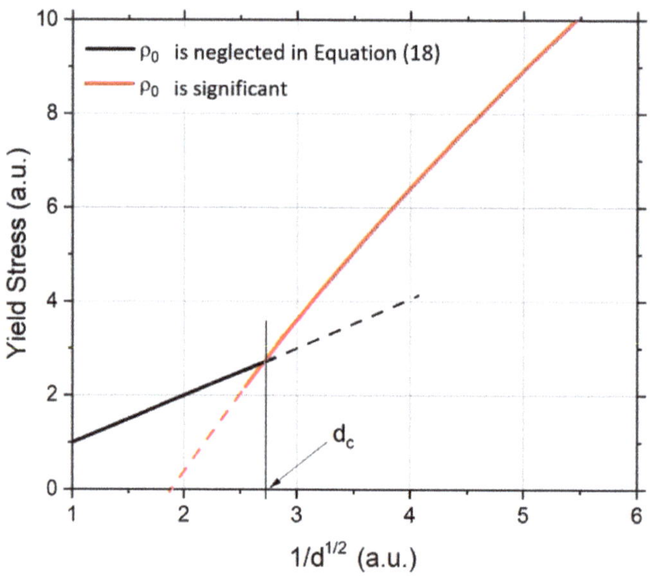

Figure 3. Schematic illustration of the two-stage Hall–Petch behavior that has been frequently reported for pure metals and alloys [70,72–74].

Assuming that the initial dislocation density scales with the inverse grain/cell size is $\rho_0 \sim \frac{1}{d}$ [80] (cf. Figure 4 confirming this relation for ultrafine grained materials, according to the X-ray profile analysis by Zhilyaev et al. [81] for Ni, Gubicza et al. [82] for Ti, and Dalla Torre et al. [43] for Cu), the last expression reads simply as $\sigma_{0.2} \sim \sqrt{\frac{1+A\sqrt{d}}{d}}$. Here A is a microstructure-sensitive parameter integrating the coefficients of the KME model. To highlight the versatility of the proposed approach, we note that, generally speaking, A can also be grain size-dependent through the k_1 factor tending to increase with decreasing grain size. A discussion about these minutiae of the model is beyond the scope of the present paper and will be extended elsewhere.

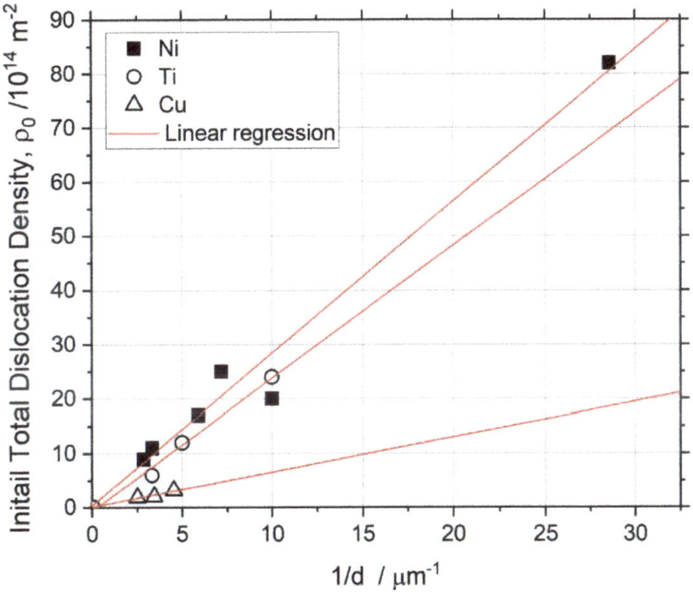

Figure 4. Initial dislocation density scaling with the inverse grain size in ultrafine-grained Ni [81], Ti [82] (cf. also similar data in [83]), and Cu [43].

We would like to emphasize that we do not associate the observed change in the slope of the H–P diagram with a break of the H–P relation. In our opinion, this relation holds universally true for both coarse-grained and fine-grained materials, but with different H–P coefficients governed by the phenomenological parameters in the dislocation kinetics laws. We should reiterate that for grain sizes above the nano scale, the KME model provides a coherent view for both observed H–P regimes, without a need to invoke a new deformation mechanism. Rather, the differences observed in the H–P behavior in coarse- and fine-grained materials or differently manufactured materials are associated with an interplay between the contributions of the omnipresent grain size-dependent and grain size-independent dislocation storage processes.

We should also note that the results reported in [67,74], for the H–P effect measured at 0.2% proof strain and at the maximum tensile load compare favorably with our findings and model predictions. It is fair to say that the predictions for the H–P behavior of the ultimate tensile strength (i.e., the necking stress), which follow from Equation (18), are largely fulfilled. This refers particularly to a decrease in the slope of the H–P diagram with the grain size reduction to the nano-scale [70], which can be attributed to the experimentally established strong inverse dependence of the dynamic recovery rate coefficient k_2 on the grain size. This dependence is bound to level off as the grain size is reduced to the nano scale [41–45]. Indeed, at this scale, the diffusional processes at the grain boundaries prevail and control the plastic flow [84] (see also [70] and references therein). This domain, which is very interesting scientifically, is rarely reached with common materials processing techniques, however.

In conclusion, the simplicity and generality of the present analytical model, coupled with its sensitivity to the microstructure, provides it with a high predictive capability. This refers in particular to the Hall–Petch relation derived on the basis of the model, thus making it a valuable practical tool for materials design. Although we tested it only against Ni and Ti as representatives of fcc and hcp materials, the model can easily be applied to other metals and alloys from these classes. Moreover, without loss of generality, it can be adapted to bcc systems where the Peierls stress can no longer be neglected. The validation of the proposed approach to the Hall–Petch relation for bcc metals is thus a future target of research in this field.

6. Summary

Using experimental data for two archetypal metals—fcc nickel and hcp titanium—the grain size dependence of the stress at maximum load, which is determined by the Considère criterion for necking, was shown to obey a Hall–Petch type relation. Calculations based on the Kocks–Mecking–Estrin model confirmed a great predictive capability of the proposed modeling method. A distinctive difference between the Z-A and the KME approaches is that the grain size effect enters through an additive term in stress in the Z-A model and appears indirectly, through its effect on the strain hardening rate controlled by dislocation density evolution, in the KME model. We contend that the Hall–Petch behavior of the necking stress is a more meaningful measure of the grain size dependence than the classical H–P relation of the yield stress represented by the 0.2% proof stress. The greatest advantage of using the former measure is that it does not rely on any arbitrary conventions. Still, even this measure we favor is not free of problems, as the slope K_N^{HP} is sensitive to the microstructure of the material and cannot be regarded as a universal material characteristic. The concluding judgement is that while the general validity of the H–P relation is not put in question by our results, its universality over the entire grain size range can no longer be claimed.

Author Contributions: Conceptualization, Y.E. and A.V.; investigation, A.V.; writing—original draft preparation, A.V.; writing—review and editing, Y.E. and A.V.; visualization, A.V. All authors have read and agreed to the published version of the manuscript.

Funding: This research received no external funding.

Data Availability Statement: No new experimental data were created in this study. Data sharing is not applicable to this article.

Acknowledgments: We are indebted to R.W. Armstrong, who initiated a discourse on the origins of the Hall–Petch relation that led to this article. Fruitful discussions with I.S. Yasnikov are appreciated. A.V. appreciates the support from MRC, Kumamoto University, for titanium research.

Conflicts of Interest: The authors declare no conflict of interest.

References

1. Armstrong, R.W. The influence of polycrystal grain size on several mechanical properties of materials. *Metall. Mater. Trans. B* **1970**, *1*, 1169–1176. [CrossRef]
2. Armstrong, R.; Codd, I.; Douthwaite, R.M.; Petch, N.J. The plastic deformation of polycrystalline aggregates. *Philos. Mag. A J. Theor. Exp. Appl. Phys.* **1962**, *7*, 45–58. [CrossRef]
3. Armstrong, R.W. 60 years of Hall-Petch: Past to present nano-scale connections. *Mater. Trans.* **2014**, *55*, 2–12. [CrossRef]
4. Armstrong, R.W.; Balasubramanian, N. Unified Hall-Petch description of nano-grain nickel hardness, flow stress and strain rate sensitivity measurements. *AIP Adv.* **2017**, *7*, 085010. [CrossRef]
5. Zerilli, F.J.; Armstrong, R.W. Dislocation-mechanics-based constitutive relations for material dynamics calculations. *J. Appl. Phys.* **1987**, *61*, 1816–1825. [CrossRef]
6. Ludwik, P. *Elemente der Technologischen Mechanik*; Springer: Berlin/Heidelberg, Germany, 1909; p. 57S.
7. Armstrong, R.W. Engineering science aspects of the Hall–Petch relation. *Acta Mech.* **2014**, *225*, 1013–1028. [CrossRef]

8. Armstrong, R.W. The Dislocation Mechanics of Crystal/Polycrystal Plasticity. *Crystals* **2022**, *12*, 1199. [CrossRef]
9. Kocks, U.F. Laws for work-hardening and low-temperature creep. *J. Eng. Mater. Technol.-Trans. Asme* **1976**, *98*, 76–85. [CrossRef]
10. Estrin, Y.; Mecking, H. A unified phenomenological description of work hardening and creep based on one-parameter models. *Acta Metall.* **1984**, *32*, 57–70. [CrossRef]
11. Estrin, Y. *Unified Constitutive Laws of Plastic Deformation*; Academic Press: San Diego, CA, USA; London, UK, 1996.
12. Bergström, Y. A dislocation model for the stress-strain behaviour of polycrystalline α-Fe with special emphasis on the variation of the densities of mobile and immobile dislocations. *Mater. Sci. Eng.* **1970**, *5*, 193–200. [CrossRef]
13. Klepaczko, J. Thermally activated flow and strain rate history effects for some polycrystalline f.c.c. metals. *Mater. Sci. Eng.* **1975**, *18*, 121–135. [CrossRef]
14. Di Leo, C.V.; Rimoli, J.J. New perspectives on the grain-size dependent yield strength of polycrystalline metals. *Scr. Mater.* **2019**, *166*, 149–153. [CrossRef]
15. Conrad, H.; Feuerstein, S.; Rice, L. Effects of grain size on the dislocation density and flow stress of niobium. *Mater. Sci. Eng.* **1967**, *2*, 157–168. [CrossRef]
16. Ashby, M.F. The deformation of plastically non-homogeneous materials. *Philos. Mag.* **1970**, *21*, 399–424. [CrossRef]
17. Cordero, Z.C.; Knight, B.E.; Schuh, C.A. Six decades of the Hall–Petch effect—A survey of grain-size strengthening studies on pure metals. *Int. Mater. Rev.* **2016**, *61*, 495–512. [CrossRef]
18. Hazzledine, P.M. Direct versus indirect dispersion hardening. *Scr. Metall. Mater.* **1992**, *26*, 57–58. [CrossRef]
19. Hazzledine, P.M.; Louat, N.P. Direct as against indirect dispersion work hardening. *Philos. Mag. Lett.* **1994**, *70*, 129–133. [CrossRef]
20. Estrin, Y.; Mecking, H. A remark in connection with 'direct versus indirect dispersion hardening'. *Scr. Metall. Mater.* **1992**, *27*, 647–648. [CrossRef]
21. Godon, A.; Creus, J.; Cohendoz, S.; Conforto, E.; Feaugas, X.; Girault, P.; Savall, C. Effects of grain orientation on the Hall–Petch relationship in electrodeposited nickel with nanocrystalline grains. *Scr. Mater.* **2010**, *62*, 403–406. [CrossRef]
22. Lefebvre, S.; Devincre, B.; Hoc, T. Yield stress strengthening in ultrafine-grained metals: A two-dimensional simulation of dislocation dynamics. *J. Mech. Phys. Solids* **2007**, *55*, 788–802. [CrossRef]
23. Malygin, G.A. Plasticity and strength of micro- and nanocrystalline materials. *Phys. Solid State* **2007**, *49*, 1013–1033. [CrossRef]
24. Lefebvre, S.; Devincre, B.; Hoc, T. Simulation of the Hall–Petch effect in ultra-fine grained copper. *Mater. Sci. Eng. A* **2005**, *400–401*, 150–153. [CrossRef]
25. Tjerkstra, H.H. The effect of grain size on the stress-strain curve of α-iron and the connection with the plastic deformation of the grain boundaries. *Acta Metall.* **1961**, *9*, 259–263. [CrossRef]
26. Ono, N.; Karashima, S. Grain size dependence of flow stress in copper polycrystals. *Scr. Metall.* **1982**, *16*, 381–384. [CrossRef]
27. Thompson, A.W.; Baskes, M.I. The influence of grain size on the work hardening of face-center cubic polycrystals. *Philos. Mag. A J. Theor. Exp. Appl. Phys.* **1973**, *28*, 301–308. [CrossRef]
28. Thompson, A.W.; Baskes, M.I.; Flanagan, W.F. The dependence of polycrystal work hardening on grain size. *Acta Metall.* **1973**, *21*, 1017–1028. [CrossRef]
29. Hansen, N. The effect of grain size and strain on the tensile flow stress of aluminium at room temperature. *Acta Metall.* **1977**, *25*, 863–869. [CrossRef]
30. Al-haidary, J.T.; Petch, N.J.; de los Rios, E.R. The plastic deformation of polycrystals I. Aluminium between room temperature and 400 °C. *Philos. Mag. A* **1983**, *47*, 869–890. [CrossRef]
31. Sonon, D.; Smith, G. Effect of grain size and temperature on the strengthening of nickel and a nickel-cobalt alloy by carbon. *Trans. Metall. Soc. AIME* **1968**, *242*, 1527–1533.
32. Feaugas, X.; Haddou, H. Grain-size effects on tensile behavior of nickel and AISI 316L stainless steel. *Met. Mater. Trans. A* **2003**, *34*, 2329–2340. [CrossRef]
33. Jago, R.A.; Hansen, N. Grain size effects in the deformation of polycrystalline iron. *Acta Metall.* **1986**, *34*, 1711–1720. [CrossRef]
34. Lederich, R.J.; Sastry, S.M.L.; O'Neal, J.E.; Rath, B.B. The effect of grain size on yield stress and work hardening of polycrystalline titanium at 295 K and 575 K. *Mater. Sci. Eng.* **1978**, *33*, 183–188. [CrossRef]
35. Ramani, S.V.; Rodriguez, P. Grain size dependence of the deformation behaviour of alpha zirconium. *Can. Met. Q* **1972**, *11*, 61–67. [CrossRef]
36. Cerreta, E.; Yablinsky, C.A.; Gray, G.T.; Vogel, S.C.; Brown, D.W. The influence of grain size and texture on the mechanical response of high purity hafnium. *Mater. Sci. Eng. A* **2007**, *456*, 243–251. [CrossRef]
37. Lasalmonie, A.; Strudel, J.L. Influence of grain size on the mechanical behaviour of some high strength materials. *J. Mater. Sci.* **1986**, *21*, 1837–1852. [CrossRef]
38. Kocks, U.F.; Mecking, H. Physics and phenomenology of strain hardening: The FCC case. *Prog. Mater. Sci.* **2003**, *48*, 171–273. [CrossRef]
39. Considère, A. *Mémoire sur l'emploi du fer et de l'acier dans les Constructions*; Ch. Dunod, Editeur: Paris, France, 1885; Volume 9, pp. 574–775.
40. Hart, E.W. Theory of the tensile test. *Acta Metall.* **1967**, *15*, 351–355. [CrossRef]
41. Yasnikov, I.S.; Vinogradov, A.; Estrin, Y. Revisiting the Considère criterion from the viewpoint of dislocation theory fundamentals. *Scr. Mater.* **2014**, *76*, 37–40. [CrossRef]

42. Yasnikov, I.S.; Estrin, Y.; Vinogradov, A. What governs ductility of ultrafine-grained metals? A microstructure based approach to necking instability. *Acta Mater.* **2017**, *141*, 18–28. [CrossRef]
43. Dalla Torre, F.; Lapovok, R.; Sandlin, J.; Thomson, P.F.; Davies, C.H.J.; Pereloma, E.V. Microstructures and properties of copper processed by equal channel angular extrusion for 1–16 passes. *Acta Mater.* **2004**, *52*, 4819–4832. [CrossRef]
44. Vinogradov, A.; Yasnikov, I.S.; Matsuyama, H.; Uchida, M.; Kaneko, Y.; Estrin, Y. Controlling strength and ductility: Dislocation-based model of necking instability and its verification for ultrafine grain 316L steel. *Acta Mater.* **2016**, *106*, 295–303. [CrossRef]
45. Yasnikov, I.S.; Kaneko, Y.; Uchida, M.; Vinogradov, A. The grain size effect on strain hardening and necking instability revisited from the dislocation density evolution approach. *Mater. Sci. Eng. A* **2022**, *831*, 142330. [CrossRef]
46. Bui, Q.H.; Dirras, G.; Ramtani, S.; Gubicza, J. On the strengthening behavior of ultrafine-grained nickel processed from nanopowders. *Mater. Sci. Eng. A* **2010**, *527*, 3227–3235. [CrossRef]
47. Bui, Q.H.; Pham, X.T. Modeling of microstructure effects on the mechanical behavior of ultrafine-grained nickels processed by hot isostatic pressing. *Int. J. Mech. Sci.* **2011**, *53*, 812–826. [CrossRef]
48. Nguyen, T.-D.; Phan, V.-T.; Bui, Q.-H. Modeling of Microstructure Effects on the Mechanical Behavior of Ultrafine-Grained Nickels Processed by Severe Plastic Deformation by Crystal Plasticity Finite Element Model. *J. Eng. Mater. Technol.* **2015**, *137*, 021010. [CrossRef]
49. Krasilnikov, N.; Lojkowski, W.; Pakiela, Z.; Valiev, R. Tensile strength and ductility of ultra-fine-grained nickel processed by severe plastic deformation. *Mater. Sci. Eng. A* **2005**, *397*, 330–337. [CrossRef]
50. Beygelzimer, Y.; Orlov, D. Metal plasticity during the twist extrusion. *Defect Diffus. Forum* **2002**, *208–209*, 311–314. [CrossRef]
51. Pachla, W.; Kulczyk, M.; Sus-Ryszkowska, M.; Mazur, A.; Kurzydlowski, K.J. Nanocrystalline titanium produced by hydrostatic extrusion. *J. Mater. Process. Technol.* **2008**, *205*, 173–182. [CrossRef]
52. Kim, W.-J.; Hyun, C.-Y.; Kim, H.-K. Fatigue strength of ultrafine-grained pure Ti after severe plastic deformation. *Scr. Mater.* **2006**, *54*, 1745–1750. [CrossRef]
53. Stolyarov, V.V.; Zhu, Y.T.; Lowe, T.C.; Islamgaliev, R.K.; Valiev, R.Z. A two step SPD processing of ultrafine-grained titanium. *Nanostruct. Mater.* **1999**, *11*, 947–954. [CrossRef]
54. Stolyarov, V.V.; Beigel'zimer, Y.E.; Orlov, D.V.; Valiev, R.Z. Refinement of microstructure and mechanical properties of titanium processed by twist extrusion and subsequent rolling. *Phys. Met. Met.* **2005**, *99*, 204–211.
55. Yapici, G.G.; Karaman, I.; Maier, H.J. Mechanical flow anisotropy in severely deformed pure titanium. *Mater. Sci. Eng. A* **2006**, *434*, 294–302. [CrossRef]
56. Anibal, M.; Andrea, M.K.; Maurizio, F.; Vitor, L.S. How severe plastic deformation at cryogenic temperature affects strength, fatigue, and impact behaviour of grade 2 titanium. *IOP Conf. Ser. Mater. Sci. Eng.* **2014**, *63*, 012161.
57. Rao, M.S.; Chakkingal, U.; Raghu, T. Mechanical behavior of commercial purity titanium processed by equal channel angular pressing followed by cold rolling. *Trans. Indian Inst. Met.* **2013**, *66*, 357–362. [CrossRef]
58. Salishchev, G.A.; Galeev, R.M.; Malysheva, S.P.; Zherebtsov, S.V.; Mironov, S.Y.; Valiakhmetov, O.R.; Ivanisenko, É.I. Formation of submicrocrystalline structure in titanium and titanium alloys and their mechanical properties. *Met. Sci. Heat Treat.* **2006**, *48*, 63–69. [CrossRef]
59. Miura, H.; Kobayashi, M.; Aoba, T.; Aoyama, H.; Benjanarasuth, T. An approach for room-temperature multi-directional forging of pure titanium for strengthening. *Mater. Sci. Eng. A* **2018**, *731*, 603–608. [CrossRef]
60. Kang, D.-H.; Kim, T.-W. Mechanical behavior and microstructural evolution of commercially pure titanium in enhanced multi-pass equal channel angular pressing and cold extrusion. *Mater. Des.* **2010**, *31*, S54–S60. [CrossRef]
61. Terada, D.; Inoue, S.; Tsuji, N. Microstructure and mechanical properties of commercial purity titanium severely deformed by ARB process. *J. Mater. Sci.* **2007**, *42*, 1673–1681. [CrossRef]
62. Milner, J.L.; Abu-Farha, F.; Bunget, C.; Kurfess, T.; Hammond, V.H. Grain refinement and mechanical properties of CP-Ti processed by warm accumulative roll bonding. *Mater. Sci. Eng. A* **2013**, *561*, 109–117. [CrossRef]
63. Lin, Z.; Wang, L. The ultrafine-grained titanium and biomedical titanium alloys processed by severe plastic deformation (SPD). *SOJ Mater. Sci. Eng.* **2013**, *1*, 1–5. [CrossRef]
64. Luo, P.; McDonald, D.T.; Palanisamy, S.; Dargusch, M.S.; Xia, K. Ultrafine-grained pure Ti recycled by equal channel angular pressing with high strength and good ductility. *J. Mater. Process. Technol.* **2013**, *213*, 469–476. [CrossRef]
65. Medvedev, A.; Ng, H.P.; Lapovok, R.; Estrin, Y.; Lowe, T.C.; Anumalasetty, V.N. Comparison of laboratory-scale and industrial-scale equal channel angular pressing of commercial purity titanium. *Mater. Lett.* **2015**, *145*, 308–311. [CrossRef]
66. Petch, N.J.; Armstrong, R.W. The tensile test. *Acta Metall. Mater.* **1990**, *38*, 2695–2700. [CrossRef]
67. Tsuji, N.; Ito, Y.; Saito, Y.; Minamino, Y. Strength and ductility of ultrafine grained aluminum and iron produced by ARB and annealing. *Scr. Mater.* **2002**, *47*, 893–899. [CrossRef]
68. Takebe, H.; Ushioda, K. Effects of Grain Size, Thickness and Tensile Direction on Yield Behavior of Pure Titanium Sheet. *J. Jpn. Inst. Met. Mater.* **2022**, *86*, 53–61. [CrossRef]
69. Figueiredo, R.B.; Langdon, T.G. Deformation mechanisms in ultrafine-grained metals with an emphasis on the Hall–Petch relationship and strain rate sensitivity. *J. Mater. Res. Technol.* **2021**, *14*, 137–159. [CrossRef]
70. Dangwal, S.; Edalati, K.; Valiev, R.Z.; Langdon, T.G. Breaks in the Hall–Petch relationship after severe plastic deformation of magnesium, aluminum, copper, and iron. *Crystals* **2023**, *13*, 413. [CrossRef]

71. Khamsuk, S.; Park, N.; Gao, S.; Terada, D.; Adachi, H.; Tsuji, N. Mechanical properties of bulk ultrafine grained aluminum fabricated by torsion deformation at various temperatures and strain rates. *Mater. Trans.* **2014**, *55*, 106–113. [CrossRef]
72. Tian, Y.Z.; Ren, Y.P.; Gao, S.; Zheng, R.X.; Wang, J.H.; Pan, H.C.; Zhang, Z.F.; Tsuji, N.; Qin, G.W. Two-stage Hall-Petch relationship in Cu with recrystallized structure. *J. Mater. Sci. Technol.* **2020**, *48*, 31–35. [CrossRef]
73. Bai, Y.; Kitamura, H.; Gao, S.; Tian, Y.; Park, N.; Park, M.-h.; Adachi, H.; Shibata, A.; Sato, M.; Murayama, M.; et al. Unique transition of yielding mechanism and unexpected activation of deformation twinning in ultrafine grained Fe-31Mn-3Al-3Si alloy. *Sci. Rep.* **2021**, *11*, 15870. [CrossRef]
74. Fu, B.; Pei, C.; Pan, H.; Guo, Y.; Fu, L.; Shan, A. Hall-Petch relationship of interstitial-free steel with a wide grain size range processed by asymmetric rolling and subsequent annealing. *Mater. Res. Express* **2020**, *7*, 116516. [CrossRef]
75. Kashyap, B.P.; Tangri, K. On the Hall-Petch relationship and substructural evolution in type 316L stainless steel. *Acta Metall. Et Mater.* **1995**, *43*, 3971–3981. [CrossRef]
76. Thompson, A.W. Effect of grain size on work hardening in nickel. *Acta Metall.* **1977**, *25*, 83–86. [CrossRef]
77. Bergström, Y.; Hallén, H. Hall–Petch relationships of iron and steel. *Met. Sci.* **1983**, *17*, 341–347. [CrossRef]
78. Lloyd, D.J. Deformation of fine-grained aluminium alloys. *Met. Sci.* **1980**, *14*, 193–198. [CrossRef]
79. Armstrong, R.W. The Yield and Flow Stress Dependence Polycrystal Grain Size. In *Yield, Flow and Fracture of Polycrystals*; Baker, T.N., Ed.; Applied Science Publishers, Ltd.: London, UK; New York, NY, USA, 1983.
80. Holt, D.L. Dislocation Cell Formation in Metals. *J. Appl. Phys.* **1970**, *41*, 3197–3201. [CrossRef]
81. Zhilyaev, A.P.; Gubicza, J.; Nurislamova, G.; Revesz, A.; Surinach, S.; Baro, M.D.; Ungar, T. Microstructural characterization of ultrafine-grained nickel. *Phys. Status Solidi A* **2003**, *198*, 263–271. [CrossRef]
82. Gubicza, J.; Fogarassy, Z.; Krállics, G.; Lábár, J.; Törköly, T. Microstructure and mechanical behavior of ultrafine-grained titanium. *Mater. Sci. Forum* **2008**, *589*, 99–104. [CrossRef]
83. Fattah-alhosseini, A.; Keshavarz, M.K.; Mazaheri, Y.; Reza Ansari, A.; Karimi, M. Strengthening mechanisms of nano-grained commercial pure titanium processed by accumulative roll bonding. *Mater. Sci. Eng. A* **2017**, *693*, 164–169. [CrossRef]
84. Meyers, M.A.; Mishra, A.; Benson, D.J. Mechanical properties of nanocrystalline materials. *Prog. Mater. Sci.* **2006**, *51*, 427–556. [CrossRef]

metals

Article

Yield Surfaces and Plastic Potentials for Metals, with Analysis of Plastic Dilatation and Strength Asymmetry in BCC Crystals

Aleksander Zubelewicz [1,*] and John D. Clayton [2]

1 Alek and Research Associates, Los Alamos, NM 87544, USA
2 Army Research Laboratory, Aberdeen, MD 21005, USA
* Correspondence: alek.zubelewicz@gmail.com

Abstract: Since the 1980s, constitutive modeling has steadily migrated from phenomenological descriptions toward approaches that are based on micromechanics considerations. Despite significant efforts, crystal plasticity remains an open field of research. Among the unresolved issues are the anomalous behavior of metals at low temperatures and the stress upturn at extreme dynamics. This work is focused on the low-temperature responses of body-centered-cubic (bcc) metals, among them, molybdenum (Mo). At these conditions, the plastic flow strength is governed by the motion of screw dislocations. The resultant non-planarity of core structures and slip causes the following: the shear stress includes non-glide components, the Schmid law is violated, there is a tension-compression asymmetry, and the yield surface and plastic potential are clearly decoupled. We find that the behavioral complexities can be explained by atomistically resolved friction coefficients in macroscopic yield and flow. The plastic flow mechanisms establish the departure point into the follow-up analysis of yield surfaces. For example, we know that while the von Mises stress is explained based on energy considerations, we will also show that the stress has a clear geometric interpretation. Moreover, the von Mises stress is just one case within a much broader class of equivalent stresses. Possible correlations among non-Schmid effects (as represented macroscopically by friction coefficients), volume change (i.e., residual elastic dilatation) from dislocation lines, and elastic anisotropy are investigated. Extensions to the shock regime are also established.

Keywords: tensor representations; friction coefficient; metal plasticity; dislocations; Schmid law

1. Introduction

The mechanisms of plastic deformation in bcc metals at low and medium temperatures are different from the observed mechanisms in face-centered cubic (fcc) and hexagonal-close-packed (hcp) metals. The plastic flow is controlled by the motion of 1/2<111> screw dislocations. The dislocations may spread out into several planes of the <111> zone [1,2]. In single crystals, the plastic flow exhibits anisotropic characteristics, and there is a tension-compression asymmetry. The non-planar structure of the dislocation cores is responsible for high friction stress (i.e., Peierls stress), there is an asymmetry of the yield stress in tension and compression, the Schmid law is violated, and the yield surface and plastic potential are clearly decoupled [3–6]. Consequently, a work-conjugate pair of the equivalent stress and the rate of plastic strain cannot be constructed. At increasing temperatures, the tension-compression asymmetry is reduced, and this non-Schmid effect nearly vanishes at room temperature [7].

In metal plasticity, most phenomenological constitutive models are formulated in the framework of von Mises (J_2) theory. Also well known, but less popular in practice, is Tresca plasticity. The obvious advantage of the Huber–von Mises yield surface is its numerical convenience. The existence of a smooth and convex yield surface makes the analysis numerically friendly. In contrast, the Tresca surface has built-in singularity points/lines, which pose issues when constructing the associated plastic flow rules. We emphasize that the two approaches were introduced over 100 years ago; the Tresca plasticity was proposed in 1864 [8], the energy-based criterion was suggested in 1904 by Huber [9,10], and the concept of plastic flow was formulated by von Mises in 1913 [11]. Tresca plasticity assumes that the plastic slip is initiated when the maximum shear stress reaches a critical magnitude. Still, we are aware that the active slip planes may deviate from the plane of maximum shear. Such misorientations are quantified by the Schmid factor. According to the Schmid law [12], plastic flow begins when the resolved shear stress on a given slip plane attains the threshold level known as the critical resolved shear stress. This law also implies that the driving force is not influenced by other components of stress. Taylor [13] and many other researchers found that the law is not applicable to bcc metals, as it has limited justification [5].

In contrast to the Tresca concept, Huber–von Mises plasticity is formulated based on energy considerations, where the plastic deformation begins when elastic energy (deviatoric part only) exceeds a certain energy barrier. In the Huber–von Mises concept, slip planes are not defined, and as explained in Section 2, the Schmid law should be used with some caution as well. It is worth mentioning Hosford's yield criterion for isotropic plasticity [14], which allows reshaping the stress envelope. In this and many other phenomenological models, emphasis is placed on the definition of the material's strength (or yield stress), while little attention is paid to the actual mechanisms of plastic flow. Also noteworthy in this context is Hershey's description of isotropic plasticity [15] and preceding seminal treatments of yielding by Taylor, Bishop, and Hill [16–18]. Although the current work focuses on macroscopically isotropic polycrystalline responses for untextured metals, prominent asymmetric yield criteria for materials of lower symmetry (e.g., orthotropy) are also noted, e.g., [19–21].

While recognizing the importance of yield stress, we focus our investigations on the mechanisms of deformation, and then, we determine whether a coupling of the flow mechanisms with the equivalent stresses exists or, just as important, whether the yield surface should be treated independently from the plastic potential. Herein, the analysis is based on the tensor representation method (TRM) developed in [22]. In Section 2, we illustrate TRM capabilities by constructing a geometric interpretation for the Huber–von Mises flow mechanism. Then, in Section 3, we focus the discussion on the flow mechanisms in bcc Mo. In what follows, we can identify two friction coefficients that capture the effects of the slip non-planarity. One of the coefficients characterizes the yield surface, and the second is used in the flow potential. We show that the coefficients properly reproduce the stress asymmetry in molybdenum at low temperatures, and the friction term can depict the change of flow mechanism at shock conditions (Appendix A). In Section 4, possible connections between the macroscopic coefficients, elastic anisotropy, and dislocation core effects are explored theoretically, with a focus on origins of local plastic dilatation.

2. Geometric Interpretation of Huber–von Mises Flow Mechanism

We begin by formulating a geometric interpretation for the Huber–von Mises plastic flow. The procedure is an important step because it explains further generalizations of the flow mechanisms and yield surfaces for bcc metals. More specifically, we want to determine the dominant slip planes, which might be associated with the J_2 stress envelope. As we know, the von Mises flow tensor $M_{ij} = \frac{\sqrt{3}\, S_{ij}}{\sqrt{J_2}}$ is defined in terms of the stress deviator $S_{ij} = \sigma_{ij} - p\delta_{ij}$, where σ_{ij} is the Cauchy stress, while the pressure $p = \sigma_{kk}/3$ (here, defined as positive in tension) and the Kronecker delta δ_{ij} complete the relation. The flow tensor M_{ij} specifies the mechanism of plastic flow such that $\dot{\varepsilon}^p_{ij} = \frac{1}{2} M_{ij}\dot{e}^p_{eq}$. When the mechanism is coupled with stress $\sigma_{ij}\dot{\varepsilon}^p_{ij} = \left(\frac{1}{2}\, M_{ij}\sigma_{ij}\right)\dot{e}^p_{eq}$, the equivalent stress becomes $\sigma_{eq} = M_{ij}\sigma_{ij}/2 = \sqrt{3J_2}$; i.e., it is the Huber–von Mises stress. We find that the flow tensor M_{ij} can be expressed in terms of three eigentensors: N^1_{ij}, N^2_{ij}, and N^3_{ij} such that the tensors are aligned with the principal stresses $\sigma_1 = N^1_{ij}\sigma_{ij}$, $\sigma_2 = N^2_{ij}\sigma_{ij}$, and $\sigma_3 = N^3_{ij}\sigma_{ij}$, where

$\sigma_1 \geq \sigma_2 \geq \sigma_3$ and $N_{ij}^1 + N_{ij}^2 + N_{ij}^3 = \delta_{ij}$. Before proceeding, it is important that we introduce the tensor representation method and show that a generic eigentensor can be uniquely expressed in terms of other second-order tensors, for example, the stress tensor.

2.1. Tensor Representations

The procedure for constructing tensor representations N_{ij}^1, N_{ij}^2, and N_{ij}^3 is described by Zubelewicz [22], where detailed derivations can be found. As stated, any symmetric second-order tensor can be represented by another second-order symmetric tensor if the original tensor and its representation produce the same invariants. Here, the generic dyadic product $N_{ij} = n_i n_j$ is constructed on a unit vector n_k. It is clear that $N_{kk} = 1$, $N_{ik}\, N_{ki} = 1$, and $N_{ik}\, N_{kl}\, N_{li} = 1$, as is true for the tensor taken to any power. We also know that any second-order symmetric tensor can be expressed in the form of three fundamental terms. Since the plastic flow is controlled by the current stress, the tensor representation of N_{ij} will be defined in terms of stress or, here, with the use of the stress deviator. In fact, the stress and stress-deviator-based representations are equivalent; hence,

$$N_{ij}^m = a_m \delta_{ij} + b_m S_{ij} + c_m S_{ik} S_{kj} \tag{1}$$

The superscript m in N_{ij}^m indicates the direction of the principal stress. As stated earlier, there are three relevant invariants: $N_{kk} = 1$, $N_{ik}N_{ki} = 1$, and $N_{ik}N_{kj}N_{ji} = 1$. The requirement is that the representation (1) retains the same invariants as the generic eigentensors. Consequently, there are three invariants and three conditions, and, upon solving the equations, we have three sets of parameters $\{a_m,\ b_m,\ c_m\}$. The first tensor N_{ij}^1 reproduces the dyadic product constructed on the unit vector pointing in the direction of the maximum tensile stress. The parameters for $m = 1$ are

$$\begin{aligned} a_1 &= \tfrac{1}{3} - \tfrac{2}{3}\cos\tfrac{\pi+\varphi}{3}\sec\varphi \\ b_1 &= \frac{\cos\left(\frac{\pi}{6} + \frac{2\varphi}{3}\right)}{\sqrt{3J_2}}\sec\varphi \\ c_1 &= \frac{\cos\frac{\pi+\varphi}{3}}{J_2}\sec\varphi \end{aligned} \tag{2}$$

The tensor N_{ij}^2 is a dyadic product associated with the second principal stress. In this case, the parameters for $m = 2$ are

$$\begin{aligned} a_2 &= \tfrac{1}{3} + \tfrac{2}{3}\cos\tfrac{\varphi}{3}\sec\varphi \\ b_2 &= \frac{\sin\frac{2\varphi}{3}}{\sqrt{3J_2}}\sec\varphi \\ c_2 &= -\frac{\cos\frac{\varphi}{3}}{J_2}\sec\varphi \end{aligned} \tag{3}$$

Lastly, the third tensor N_{ij}^3 determines the orientation of maximum compression, where for $m = 3$,

$$\begin{aligned} a_3 &= \tfrac{1}{3} - \tfrac{2}{3}\cos\tfrac{\pi-\varphi}{3}\sec\varphi \\ b_3 &= -\frac{\cos\left(\frac{\pi}{6} - \frac{2\varphi}{3}\right)}{\sqrt{3J_2}}\sec\varphi \\ c_3 &= \frac{\cos\frac{\pi-\varphi}{3}}{J_2}\sec\varphi \end{aligned} \tag{4}$$

In this construction, the second and third invariants of the stress deviator (i.e., $J_2 = S_{ij}S_{ij}/2$ and $J_3 = S_{ik}S_{kj}S_{ji}/3$) define the angle $\varphi = \sin^{-1}(A_\varphi)$, where $A_\varphi = 3\sqrt{3}J_3/\left(2J_2^{3/2}\right)$.

The angle φ varies between $\pm\pi/2$. Once again, the stress representations of the eigentensors must satisfy the condition $N^1_{ij}(S_{kl}) + N^2_{ij}(S_{kl}) + N^3_{ij}(S_{kl}) = \delta_{ij}$. In short, the TRM is a very useful tool for researchers. It has already been shown that experimentally observed flow mechanisms, at first constructed in a generic tensorial form, can be uniquely coupled with the driving tensorial stimuli [23]. In this manner, we eliminate the uncertainty about the functional form of the mechanisms.

2.2. *Atomistically Resolved Friction Coefficient*

Once again, here we attempt to construct a geometric interpretation for the flow mechanism $M_{ij} = \frac{\sqrt{3}\,S_{ij}}{\sqrt{J_2}}$. In the first step, we construct a generalized slip mechanism along three planes, where the planes are corotational with principal stresses such that

$$M_{ij} = \alpha\left(N^1_{ij} - N^3_{ij}\right) + \beta\left(N^1_{ij} - N^2_{ij}\right) - \beta\left(N^2_{ij} - N^3_{ij}\right) \tag{5}$$

Note that each plane is weighted by functions α and β. Next, we recall the definition of the flow tensor $M_{ij} = \frac{\sqrt{3}\,S_{ij}}{\sqrt{J_2}}$, where the scalar products are equal to $M_{kk} = 0$, $M_{ij}M_{ij} = 6$ and $M_{ik}M_{kj}M_{ji} = 6\ \sin\varphi$, respectively. From there, we identify the two functions α and β. The functions take the following form:

$$\begin{aligned} \alpha &= \sqrt{3}\ \cos\tfrac{\varphi}{3} \\ \beta &= \beta_0 \sin\tfrac{\varphi}{3} \end{aligned} \tag{6}$$

In the next step, the expression (5) is reorganized and presented in an equivalent form $M_{ij} = \alpha\left(N^1_{ij} - N^3_{ij}\right) + \beta\left(N^1_{ij} + N^2_{ij} + N^3_{ij} - 3N^2_{ij}\right)$, and then, the expression is presented in the final form

$$M_{ij} = \alpha\left[\left(N^1_{ij} - N^3_{ij}\right) + \mu_\varphi\left(\delta_{ij}/3 - N^2_{ij}\right)\right] \tag{7}$$

Note that $N^1_{ij} + N^2_{ij} + N^3_{ij} = \delta_{ij}$. The parameter $\mu_\varphi = 3\beta/\alpha$ is interpreted as an atomistically resolved friction coefficient. The function α varies between $3/2$ and $\sqrt{3}$; thus, the function is nearly a constant. The friction coefficient μ_φ takes values between $\mp\beta_0$. At first glance, the relation resembles the Coulomb law used in frictional materials [24]. A generic form of the flow tensor, but not a stress representation, was introduced in [25]. In the current application to bcc metals, we realize that the coefficient

$$\mu_\varphi = \sqrt{3}\,\beta_0\ \tan\varphi/3 \tag{8}$$

quantifies the slip non-planarity [1,5,26,27]. Consequently, the Huber–von Mises stress $\sigma_{eq} = \sqrt{3J_2}$ is equal to

$$\sigma_{eq} = \frac{\alpha}{2}\ \left[(\sigma_1 - \sigma_3) + \mu_\varphi(p - \sigma_2)\right] \tag{9}$$

In the flow tensor (7) and in the equivalent stress (9), the slip non-planarity is quantified in the second terms $\mu_\varphi\left(\delta_{ij}/3 - N^2_{ij}\right)$ and $\mu_\varphi(p - \sigma_2)$. In frictional materials, the second term characterizes the roughness of the slip surface, where the roughness is responsible for dilatational inelastic deformation.

In metals, too, the atomistically resolved friction coefficient μ_φ—here, directly proportional to β_0—mediates the effect of non-glide parts of the stress. However, the non-planarity does not affect the material's volume in a meaningful manner, at least at the continuum (macroscopic) scale. Therefore, plasticity is essentially volume-preserving, in an average sense, when the local volume element contains a dislocation density not exceeding $10^{15}/\mathrm{m}^2$. In Appendix A, we show that the isochoric flow assumption cannot be over-generalized to shock-loading regimes, with higher dislocation densities, rising adiabatic temperature, and local excitations; all the factors magnify core pressure.

In summary, the Huber–von Mises stress is preserved when the parameter is equal to unity, i.e., $\beta_0 = 1$, as shown by the black line in Figure 1. However, several other surfaces can be constructed. A Tresca-like criterion (blue line) is obtained for $\beta_0 = 1/2$. The true Tresca stress envelope (red line) is constructed by prescribing $\alpha = 2$ and $\beta_0 = 0$. In the Tresca-like criterion, singularity points are rounded with $A_\varphi = 3\sqrt{3}\,(1 - A_0)\; J_3/\left(2J_2^{3/2}\right)$, where $A_0 = 0.2$. Herein, the equivalent shear stress and the rate of plastic strain represent the work-conjugate pairs, that is, $\sigma_{ij}\dot{\varepsilon}_{ij}^{p} = \left(\frac{1}{2}\, M_{ij}\sigma_{ij}\right)\dot{e}_{eq}^{p}$ and $\sigma_{eq} = M_{ij}\sigma_{ij}/2$. It is worth noting that the friction parameter β_0 in (8) makes our stress envelopes (9) comparable to the Hershey–Hosford criteria for fcc polycrystals [14] with a large exponent.

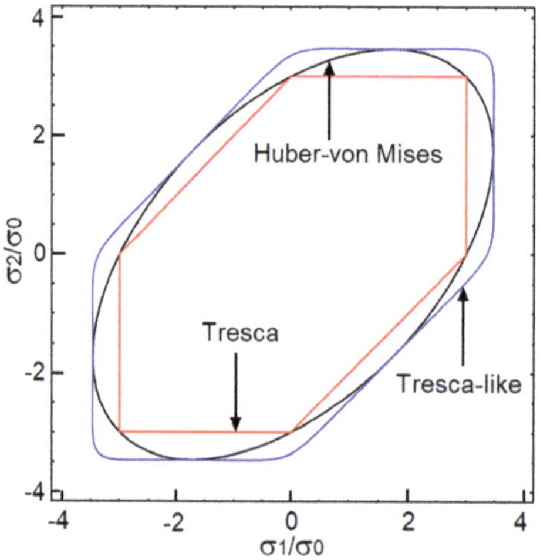

Figure 1. Yield stress plotted on the stress plane (σ_1, σ_2), where σ_3 = 0. The original Tresca criterion (red line) is obtained for α = 2 and β_0 = 0. The Huber–von Mises criterion (black line) is found from setting β_0 = 1. In the Tresca-like yield criterion β_0 = 1/2, and the line is colored blue.

3. Generalized Huber–von Mises Criterion

In fcc polycrystalline metals, slip is activated along planes somewhat misoriented with respect to the plane of maximum shear, where the Schmid factor properly quantifies the degree of misorientation. However, in bcc metals, the breakdown of the Schmid law is a known fact, as reported in many studies, e.g., [5,28,29]. Specifically, at low temperatures, plastic flow depends on the resolved shear stress and is also affected by other stress components. The non-planarity of the dislocation core structure is the main reason for the observed strength asymmetry in tension and compression. As reported in [1,26,27], the yield surface and the plastic potential are decoupled. Molecular dynamics (MD) calculations for Mo crystals support the construction of yield surfaces and plastic potentials [26,27]. A prior conclusion from [1,5,27] was that the work-conjugate pair of stress and plastic strain rate cannot be established in bcc polycrystals at low temperatures.

In bcc metals, the non-planarity of the dislocation core arises at the atomistic scale and affects the shear stress (i.e., screw dislocation core spreading onto multiple planes, with possible edge components within the core structure [5]); still, still the plastic flow remains nearly incompressible for dislocation densities far below the theoretical maximum limit (see Section 4). The friction coefficient denoted by μ_0 quantifies the core non-planarity such that

$$\mu_\varphi = \sqrt{3}\,\beta_0\;\tan\varphi/3 - \mu_0 \tag{10}$$

Again, in reference to frictional materials, the internal friction angle $\varphi/3$ characterizes the angle of asperities. This angle changes and is a function of the current stress or, more precisely, the direction of the maximum shear stress. Thus, the friction mechanism in metals is endowed with much higher configurational flexibility. We reemphasize that the internal friction parameter (8) replicates the relation used in the Coulomb law. In the case of molybdenum, studies in [27] indicate that $\beta_0 = 2/3$ properly captures the shape of the stress envelope, as seen in Figure 2. One should note that the parameter β_0 scales the non-planarity of the plastic flow. For example, in fcc metals, the parameter should be equal to $\beta_0 = 1/2$, while $\beta_0 = 1$ depicts the strongest out-of-plane contribution. In this construction, the difference between the yield stress and the plastic potential is solely controlled by the parameter μ_0. The yield stress becomes

$$\sigma_Y = \frac{\alpha}{2}\left[(\sigma_1 - \sigma_3) + \mu_\varphi^Y(p - \sigma_2)\right] \tag{11}$$

where the non-planarity coefficient entering Equation (11) is found to be $\mu_0^Y = 0.3$ for Mo. The strength differential $SD_Y = 2(\sigma_t - \sigma_c)/(\sigma_t + \sigma_c)$ introduced by Vitek et al. [1] characterizes the tension–compression asymmetry, where σ_c and σ_t are uniaxial stresses in compression and tension. Here, the differential takes the very simple form $SD_Y = \beta_0\,\mu_0^Y$. A similar differential is calculated for the plastic potential: $SD_P = \beta_0\,\mu_0^P$, where the superscript "P" is added to (10). Now, we have

$$\mu_\varphi^P = 2/\sqrt{3}\,\tan\varphi/3 - \mu_0^P \tag{12}$$

where $\mu_0^P = -0.09$ for Mo. The flow mechanism becomes

$$M_{ij}^P = \alpha\left[\left(N_{ij}^1 - N_{ij}^3\right) + \mu_\varphi^P\left(\delta_{ij}/3 - N_{ij}^2\right)\right] \tag{13}$$

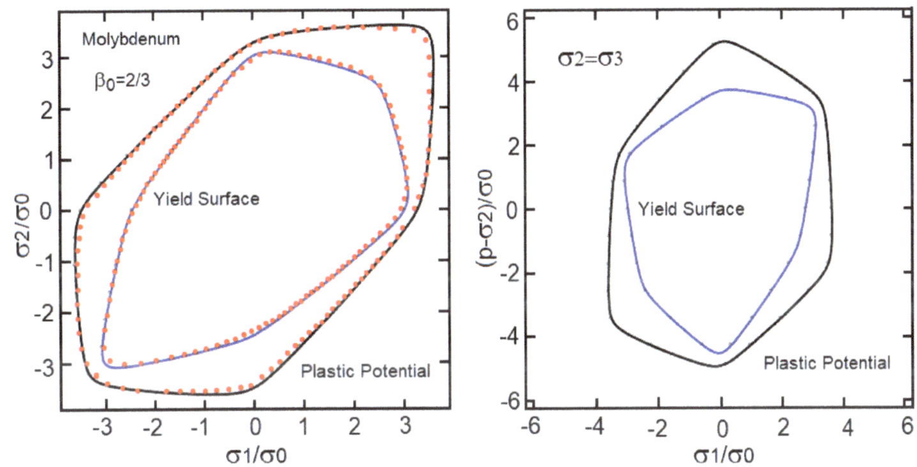

Figure 2. Yield surface (black line) and plastic potential (blue line) plotted on the plane of principal stresses σ_1 and σ_2. Red data points are based on polycrystal model predictions with slip strengths informed by MD studies of glide of screw dislocations in bcc Mo, simulated by Gröger et al. [26,27]. The original source of the red data points is ref. [27]; the blue and black lines are the new output of the model set forth in Section 3 of the current work. The right-hand-side plot depicts the surface's asymmetry on the plane of σ_1 and $(p - \sigma_2)$, where p is hydrostatic pressure.

In the absence of damage or point defects (e.g., no vacancies, interstitials, or inclusions), and at dislocation densities sufficiently low, Mo can be modeled as a plastically incompressible material: $M^P_{kk} = 0$. The plastic potential $\sigma_P = \frac{1}{2} M^P_{ij} \sigma_{ij}$ takes the following form:

$$\sigma_P = \frac{\alpha}{2} \left[(\sigma_1 - \sigma_3) + \mu^P_\varphi (p - \sigma_2) \right] \quad (14)$$

The isotropic yield surface and the plastic potential for polycrystalline Mo at low temperatures are shown in Figure 2. Red data points represent the yield and flow surfaces for Mo polycrystals, as in Figure 6 of Gröger et al. [27]. The latter are obtained from Taylor-type [16] crystal plasticity calculations on randomly orientated aggregates of Mo polycrystals [26,27]. For clarity, we used the data reported in [27] in our previous study [25] as well, but now the material model has been recalibrated and the yield surface and plastic potential are redrawn accordingly. Note that the rate of plastic work is equal to $\sigma_{ij} \dot{\varepsilon}^P_{ij} = \frac{1}{2} \left[\left(M^P_{ij} - M^Y_{ij} \right) + M^Y_{ij} \right] \sigma_{ij} \dot{e}^P_{eq}$. Consequently, the plastic power becomes

$$\sigma_{ij} \dot{\varepsilon}^P_{ij} = \left[\sigma^Y_{eq} + \left(\mu^Y_0 - \mu^P_0 \right) (p - \sigma_2) \right] \dot{e}^P_{eq} \quad (15)$$

The associated plastic flow is reestablished when $\mu^Y_0 = \mu^P_0$.

The right-hand-side plot in Figure 2 displays strong influence of the "friction" term on the plastic potential (14) and yield stress (11). The stress envelopes are clearly asymmetric. The tension–compression asymmetry is a well-established fact [1,5,26,27]. The dominant tensile loading intensifies the plastic flow, and the opposite is true for compression. The single-crystal slip system strengths of Gröger et al. [26,27] properly capture experimental trends on tension–compression asymmetries observed in experiments on Mo [30–32] at a temperature of 123 K. The current model nearly perfectly matches the yield and flow surfaces of Figure 6 in ref. [27] using just two parameters, β_0 and μ_0, where each of them has a well-defined physics interpretation; the agreement rationalizes the mechanisms-based constitutive description. Note that Figure 2 of the current work contains solid black and blue curves that are generated as the output of our new model. Figure 2 is an original figure created by the present authors; it is not a scanned reproduction of Figure 6 of ref. [27], which contains differently shaped curves from a different model.

4. Screw Dislocations in bcc Metals: Core Spreading and Volume Changes

The forthcoming analysis serves two major purposes. Firstly, the hypothesis that a correlation exists between local dilatation from dislocation lines with dislocation core spreading is examined through theoretical calculations. By the inverse argument, if all atomic motion were restricted to a single plane (i.e., no core spreading onto multiple planes), then the lattice distortion should consist only of simple shearing modes, and no volume change should occur. If a positive/dilatative volume change does occur, then external compressive pressure would work negatively against such a change at very small scales, which could induce extra glide resistance manifesting as non-Schmid effects. Possible correlations with elastic anisotropy are also newly investigated.

Secondly, the analysis theoretically predicts the maximum volume changes expected from dislocation lines for bcc metals (Mo, W, and Ta) to evaluate the proper domain of plastic incompressibility assumed in Sections 2 and 3. Calculations have been reported previously for select fcc metals and Fe [33–35] but not for Group VIB metals which show strong dislocation core spreading and non-Schmid effects and thus potentially more plastic dilatation. Backgrounds on bcc screw dislocation physics and analytical models are given in Sections 4.1 and 4.2 to set the context; new contributions follow in Section 4.3.

4.1. Background: Dislocation Core Phenomena

In bcc metals, screw dislocation mobility is generally much lower than edge dislocation mobility. Thus, the yield and flow of bcc metals are dominated by the glide resistance of screw dislocation components, which becomes the limiting factor regarding plastic strength [36,37]. Primary slip systems are <111>{110} and <111>{112}; these are the glide systems typically studied in MD investigations [5,38] and resolved in continuum crystal plasticity models of bcc metals, for example [39].

As reviewed by Duesbery and Vitek [5,38], non-Schmid effects in bcc crystals arise from two primary factors. The first is solely due to the lack of certain symmetry in the bcc crystal structure: strengths may differ when slip occurs in the twinned or anti-twinned oriented <111>{112} systems. The second is due to the unusual core structures of screw dislocations in bcc metals. The dislocation core tends to spread onto multiple planes, rather than being confined to a single {110} plane, for example. Within the core, the <111> screw dislocation (when viewed macroscopically) contains atomic-scale perturbations of both edge and screw character [5,38]. The fractional Burgers vector components of edge character in the core must sum to zero, such that the <111> dislocation remains of pure screw character macroscopically. The partial edge components contribute strongly to observed non-Schmid effects on yield and flow stresses. For shuffling and glide of the <111> screw dislocation to occur, the non-planar components must first be forced to return to a single dominant glide plane; i.e., the core spreading must be compacted. This can contribute to a rather large Peierls stress in bcc metals. In the macroscopic continuum theory of Section 3, friction coefficients depict an excessive spreading of <111> screw dislocations.

The effects of core spreading (i.e., non-planarity) are more prevalent at very low temperatures (e.g., far below room temperature, typically at or near 0 K in atomic simulations and 77–123 K in experiments [5,27,30–32]); the present discussion is focused on isothermal behavior at low temperature limits. Thermal activation, thermal expansion, and phonon drag are of no great relevance here but are applicable to shock conditions in Appendix A.

Observed non-Schmid effects vary in magnitude among bcc metals, being stronger in Group VIB crystals (e.g., Mo and W) than in Group VB crystals (e.g., Ta). Perfect crystals of the Group VIB metals have larger elastic constants than those in Group VB, which, as noted by Duesbery and Vitek [5], could exacerbate their core spreading. The above trends were deduced primarily from investigations [5,36,38] that employed empirical interatomic potentials for the behavior of bcc metals, for example, the Finnis-Sinclair potential [40]. More contemporary MD potentials [1,26,27,38] and first-principles methods (i.e., density functional theory (DFT)) such as tight binding [41] have confirmed the existence of the coupled phenomena of core spreading, non-planarity, and non-Schmid effects; details of core structures and stress differentials for asymmetric slip can differ among models [38].

4.2. Background: Volume Changes from Dislocations

The isotropic linear elastic solution for a screw dislocation [42,43] predicts that no local or global volume change manifests from its elastic fields. For edge dislocations, according to the isotropic linear elastic solution, local volume change occurs in the vicinity of the dislocation line, but no global volume change (or global shape change, for that matter) can occur for an externally unloaded (i.e., traction-free or self-equilibrated) body containing edge (or screw) dislocations in the context of pure *linear* elasticity theory, isotropic or anisotropic [44,45]. This statement is strictly true for isothermal conditions, wherein no thermal expansion or contraction from atomic vibrations in the vicinity of defects occurs.

Volume changes from glide dislocations (e.g., no vacancies from dislocation climb, and no other point defects) can arise from two notable sources in an isothermal continuum elasticity theory: nonlinear elastic effects (i.e., isothermal anharmonic effects) and dislocation core pressure. The latter can be modeled, in the context of a cylindrical annulus of elastic material enclosing a straight dislocation line, by a pressure boundary condition acting on the inner surface of the annulus, within which the core resides [46]. Atomic calculations can estimate its magnitude, for example, up to the order of 10% of the shear or bulk modulus [46,47]. However, core pressure varies inversely with the squared radial distance from the defect line [48], so its magnitude depends on the choice of core radius.

Anharmonic effects, which could induce residual lattice shape change as proven by Clayton and Bammann [43,45] in addition to volume change, can be associated with the nonlinear elastic constitutive response of the crystal, through a combination of second-

and third-order elastic constants in crystals of arbitrary symmetry. For crystals of cubic symmetry, the volume change component of average residual deformation depends on the combination of second- and third-order elastic constants and particular parts of the elastic energy density [34,43–45]. Although partitioning of the elastic energy density into the requisite components does not seem readily available for known anisotropic solutions [49,50], this partitioning may be analytically possible.

On the other hand, closed-form solutions for volume change due to dislocation lines are readily derived for the isotropic case [51] considering anharmonic terms of third order in the strain energy density, when the dislocation line energy is estimated from linear isotropic elasticity with a suitable cutoff and core radius. In the isotropic case, the requisite second- and third-order elasticity coefficients are reduced to the usual two linear (i.e., second-order) elastic constants and the ambient pressure derivatives of the shear and bulk moduli. As shown later, the normalized effect of second-order constants can be resolved solely by Poisson's ratio. Different derivations based on nonlinear continuum mechanics [43–45] or thermodynamic arguments [33,52] can arrive at similar end results.

Dilatation from edge and screw dislocations in bcc α-iron has been confirmed using MD potentials and DFT [53–55]. As noted by Clouet et al. [54,55], dilatation from <111> screw dislocation lines can also be deduced from DFT results for pure Mo [23].

Trends, e.g., expansion rather than contraction predicted for engineering metals, and of fairly small magnitude, agree with limited experimental data [33–35,43,45], as will be shown in Section 4.3. Unless the local dislocation density is extremely large, the net residual volume change from dislocations is generally considered small enough to be ignored in continuum plasticity theory for standard, as opposed to extreme, loading conditions. Yield and flow stresses can still depend on pressure even when plastic volume change is omitted in kinematics, as in Section 3 herein (but not Appendix A) or in other models [35].

4.3. Analysis: Volume Changes in bcc, fcc, and hcp Metals

A question to be investigated next is whether non-Schmid effects leading to nonzero lattice friction coefficients in the continuum plasticity theory of Section 3 correlate with dilatation from dislocation lines due to anharmonicity or whether any such correlation manifests only from an independent core pressure. Contributions of anharmonicity and core pressure to plastic volume change are expressed as follows in the isotropic elastic limit, consolidating prior nonlinear elastic and atomic-scale derivations [43] (Ch. 7), [45].

Denote the total density of dislocation lines, in dimensions of length per unit reference volume, as ρ, with b the magnitude of the full Burgers vector. Denote $f^e = \chi$ as the fraction of this density of edge character, and $f^s = 1 - \chi$ the fraction of screw character, whereby definition $f^e + f^s = 1$. Let B and G label the ambient bulk modulus and ambient shear modulus, and $B' = \mathrm{d}B/\mathrm{d}p$ and $G' = \mathrm{d}G/\mathrm{d}p$ the derivatives of bulk and shear moduli with respect to the external pressure p (here, the usual convention is positive in compression) measured in the reference state. Let p_c denote the dislocation core pressure and F^c the local volume change per unit volume induced by the core pressure. When $p_c > 0 \Rightarrow F^c > 0$, the core exerts an outward pressure on the surrounding crystal, so dilatation takes place. Negative p_c would be tensile, causing lattice contraction.

Denoting the volume change per unit reference volume of an element of crystal due to dislocations contained within as $\Delta V/V$, extending prior work [43,45], we find

$$\frac{\Delta V}{V}(\rho,\chi,p_c) = \alpha^e \chi b^2 \rho + \alpha^s (1-\chi) b^2 \rho + F^c(\rho,\chi,p_c) \tag{16}$$

$$\alpha^e = \frac{\Lambda}{(1-\nu)^2}\left[\frac{1}{2}(1-2\nu)^2(B'-1) + \frac{2}{3}\left(1-\nu+\nu^2\right)\left(G' - \frac{3(1-2\nu)}{2(1+\nu)}\right)\right],\ \alpha^s = \Lambda\left[G' - \frac{3(1-2\nu)}{2(1+\nu)}\right] \tag{17}$$

The first term on the right side of Equation (16) is due to nonlinear elastic or anharmonic effects of edge dislocation components, the second to screw components, and the third to core pressure. Dimensionless factors α^{e} and α^{s} contain the combined effects of elastic constants and their pressure derivatives on anharmonicity and dislocation line energy. Parameter Λ scales the dislocation energy per unit length [56–58], as discussed following Equation (17) below.

Equations (16) and (17) are derived by making the following substitutions in Equation (74) of ref. [45] for the energies per unit length of screw and edge dislocations, respectively, E^s and E^e, and the ratio of ambient shear to bulk modulus, G/B:

$$E^s = (1-\nu)E^e = \Lambda G b^2,\ \Lambda \approx 1,\ G/B = \frac{3(1-2\nu)}{2(1+\nu)} \tag{18}$$

The explicit effect of core pressure is also newly added in Equation (16), following its possible significance discovered in ref. [46]. As newly derived in Equation (17), the dimensionless volume change factors α^e and α^s depend only on dimensionless constants ν, B', and G'. Derivation of Equation (74) of [45] relies on the assumptions that the body is self-equilibrated with a constitutive response described by a hyperelastic energy potential, expanded to order three in the Lagrangian strain (i.e., strain energy with elastic constants of second and third orders). This body contains internal discontinuities associated with jumps in lattice displacements across slip planes, from Burgers vectors of dislocations. Traction is continuous across discontinuity surfaces. The balance of linear momentum and nonlinear elastic constitutive equations are substituted into the equation for vanishing volume-averaged stress. Further assuming isotropic elastic symmetry and algebraic manipulations produces Equation (74) in [45]. This is recast via (18) into (16) and (17) of the current work, to which F^c has been appended.

In Equation (18), the dislocation line energy is approximated as Gb^2 for screw dislocations and $Gb^2/(1-\nu)$ for edge dislocations. These are likely upper bounds among known linear elastic approximations [43,45,56,57], and they omit shielding effects from dislocation structural rearrangements [56–58]. Lower bounds would multiply the predicted dilatation of Equations (14)–(16) by a factor of Λ ranging from $1/2$ to $1/(4\pi)$ [43,58].

Listed in Table 1 are elastic coefficients for ten metals used to test the hypothesis framed for the question asked at the beginning of Section 4.3. Isotropic values are for polycrystals from Guinan and Steinberg [59]. Also shown for comparison is the Zener anisotropy factor $A = 2C_{44}/(C_{11} - C_{12})$, with $A = 1$ being isotropic. The larger the departure of A from unity, the less valid the isotropy assumption inherent in Equations (15)–(17). For hcp Mg, elastic anisotropy also depends on other combinations of elastic constants, but anisotropy is generally very low in Mg for all such combinations. Anisotropy itself is also worth considering alone for possible connections to screw core non-planarity and lattice friction.

Table 1. Anharmonic volume change factors of dislocations from Equation (15), with elastic properties. Experimental data for α_{exp} from ref. [35] on Al 1100 and from ref. [34] on Ag, Au, Cu, and Ni.

Metal	Structure	B [GPa]	G [GPa]	ν	B'	G'	A	α^e	α^s	α_{exp}	<111> Spreading
Fe	bcc	166	82	0.29	5.3	1.8	2.37	2.13	1.31	-	Yes
Mo	bcc	263	125	0.29	4.4	1.5	0.72	1.66	1.02	-	Yes
Ta	bcc	193	69	0.34	3.2	1.1	1.56	1.14	0.74	-	Yes
W	bcc	310	160	0.28	4.0	2.3	1.01	2.39	1.78	-	Yes
Al	fcc	76	26	0.35	4.4	1.8	1.22	2.14	1.46	2.04	No
Ag	fcc	103	30	0.37	6.1	1.4	3.03	1.87	1.11	1.08	No
Au	fcc	173	28	0.42	6.3	1.1	2.88	1.61	0.94	1.08	No
Cu	fcc	137	48	0.34	5.5	1.4	3.21	1.77	1.05	1.23–1.68	No
Ni	fcc	183	86	0.30	6.2	1.4	2.46	1.86	0.93	1.74–1.78	No
Mg	hcp	35	17	0.29	3.9	1.7	0.98	1.78	1.21	-	No

Most notably calculated in Table 1 are values of α^e and α^s that indicate the theoretically predicted importance of anharmonic effects on dilatation from edge and screw dislocations, respectively. For a density of perfect screw dislocations (e.g., <111> screws in a bcc metal of current interest), $f^e = \chi = 0$, and thus α^e and B' would be inconsequential. However, the core structure of a nonplanar, nominally pure screw <111> dislocation contains partial edge components that sum to zero [5]. Thus, α^e and B' could still be of relevance for correlating non-Schmid effects. As might be expected, results in Table 1 confirm $\alpha^e > \alpha^s$ for all ten metals, meaning dilatation from edge dislocations should exceed that from pure screw dislocations unless core pressure effects are larger from screw dislocations.

Experimentally obtained values of the dimensionless volume change factor, α_{exp}, are shown for comparison with model predictions of α^e and α^s in Table 1, where

$$\alpha_{exp} = G\frac{\Delta V/V}{W^c} = \frac{\Delta V/V}{\Lambda b^2 \rho} \tag{19}$$

Values of α_{exp} are calculated from Equation (19) using experimental data on volume change $\Delta V/V$ and measured stored energy per unit volume of cold work W^c [34] or measured total dislocation line density ρ [35]. Available data do not allow delineation of screw versus edge factors. Equation (18) follows from (15)–(17) with omission of the core term F^c, which also cannot be deduced from experimental data. The relatively high value of unity for Λ in (17) compensates for the neglect of a distinct core pressure. Agreement between the theory and experiment is respectable since $\alpha^s \leq \alpha_{exp} \leq \alpha^e$ for all fcc metals except Ag. Even for Ag, the lower bound of the theory, α^s, exceeds α_{exp} by only 3%. The discrepancy could be attributed to the large anisotropy of Ag (i.e., A exceeding 3), recalling that Equation (17) relies on the isotropic assumption. The theory correctly predicts that Al should have the largest dilatation among fcc metals in Table 1. Excluding anomalous results for Ag, the theory correctly predicts that Au should have the lowest dilatation, with Cu and Ni falling in between, depending on the fractions of edges and screws. Quantitative data were found in the literature only for the five fcc metals in Table 1 and not bcc or hcp. However, Spitzig and Richmond [35] stated that predictions of the analytical theory [51] agreed with experimental measurements of dilatation and dislocation densities in Fe-based metals.

Results in Table 1 show no positive correlations among observed non-Schmid effects [1,5] and any of α^s, α^e, or A. For example, for the two listed Group VIB metals, α^s is larger in W than Mo, but Mo shows stronger non-Schmid effects associated with experimentally measured strength asymmetries [5]. Some of the fcc metals and Mg (hcp) have larger values of α^s than Mo, but these do not demonstrate non-Schmid effects from <111> core spreading. The rank ordering from highest to lowest α^s is W, Al, Fe, Mg, Ag, Cu, Mo, Au, Ni, Ta. Ordering for α^e is similar but not identical: W, Al, Fe, Ag, Ni, Mg, Cu, Mo, Au, Ta.

The present analysis thus leads to the following conclusion: if dilatation from predominantly screw dislocation lines, dislocation core spreading, and non-Schmid effects are positively connected, then the dilatation must be induced from an independent core pressure, rather than anharmonic (i.e., nonlinear elastic) effects alone. A correlation with core pressure appears logical since a larger core pressure might be expected to be exerted by a more disordered non-planar core, leading, in turn, to greater dilatation. Results in Table 1 also show no correlation between elastic anisotropy factor A and non-Schmid effects. For example, the Group VB metal Ta has weaker core spreading than the Group VIB metals W and Mo [5], yet Ta is more anisotropic than W and Mo according to values of A.

The maximum magnitude of $\Delta V/V$ from either of the two anharmonic terms in Equation (14) can be estimated as follows. The closest packing of dislocation lines is limited theoretically by the lattice spacing and repulsive forces between atoms to a maximum $\rho \approx 0.01/b^2$, or approximately one dislocation line per square patch of 10×10 unit cells [60]. Thus, the absolute theoretical maximum dilatation, in percent, from anharmonicity is on the order of α^s or α^e. From Table 1 for three bcc metals, this would be a maximum expansion of 0.7% to 2.4%, depending on the particular metal and dislocation character (screw or edge). For the most extreme case of edge dislocations in W, the predicted maximum is 2.4%;

the least extreme is for screw dislocations in Ta, predicted at 0.7%. Predictions for Mo fall in between those for W and Ta. If the dislocation line energy is reduced by shielding, a theoretical maximum of around 1% expansion is expected to be more realistic for W. These theoretical predictions are of comparable magnitudes to those for five fcc metals (validated versus experimental data in Table 1) as well as hcp Mg.

Even in heavily cold-worked metals, the dislocation density is usually several orders of magnitude smaller than the above theoretical maximum, which should limit $\Delta V/V$ accordingly, except in very highly defective local regions at grain boundaries or cell walls. Experiments [35] measured the volume change in Al 1100 as 5×10^{-5} at a strain of 0.1 and dislocation density of $3 \times 10^{14}/\mathrm{m}^2$. The theoretical maximum of ρ with a typical Burgers magnitude is around $10^{17}/\mathrm{m}^2$. A dislocation density of $10^{15}/\mathrm{m}^2$ is considered quite large for a typical metal, and this in turn would give a maximum dilatation from anharmonic effects on order of 0.01%, or 10^{-4}. The pressure differential due to a dilatation from a nucleation of this density of dislocations [61] is around $10^{-4}B$, on the order of 10 MPa for the metals in Table 1. These values of residual volume change and pressure are sufficiently small to omit in the continuum plasticity theory for bcc metals under ordinary conditions (i.e., relatively low strain rates and temperatures), with Peierls yield and flow stresses on the order of 1 GPa, as modeled in Sections 2 and 3 of this work. However, for very soft metals such as Cu, 10 MPa might not be small compared to the yield stress, which can be significantly lower than 100 MPa depending on its purity [62]. The potential significance of plastic volume change under extreme loading conditions involving very high pressures, strain rates, and/or temperatures [62–64] is given further consideration in Appendix A.

5. Conclusions

At low temperatures, the yield criterion and plastic potential in bcc metals are decoupled and display a tension–compression asymmetry. The asymmetry is a consequence of screw dislocation non-planarity. Newly proposed in the current work is that the yield stress and plastic potential each include an additional term which quantifies the spreading of the dislocation core structure that leads to non-Schmid effects in plastic flow. The material parameter controlling each new term is interpreted as an atomistically resolved friction coefficient. This coefficient resembles the friction coefficient in Coulomb-type (e.g., brittle or granular) materials, but in bcc metals, the coefficient arises at the atomistic scale of the dislocation core. The theoretical analysis in Section 4 justifies the mechanisms-based considerations at the continuum scale.

The theory predicts no obvious correlation between core spreading (which affects the macroscopic friction coefficient) and dilatation from anharmonic effects under isothermal, low-temperature conditions. If a correlation between screw dislocation core spreading and residual lattice expansion exists, as logically hypothesized, such a correlation can be represented through introduction of a finite, repulsive dislocation core pressure. Theoretical predictions also justify the omission of volumetric plastic deformation arising from anharmonic defect fields in bcc metals that show strong non-Schmid effects (i.e., Mo and W) for conventional loading conditions, wherein dislocation densities are not excessive. This conclusion is consistent with theoretical predictions and experimental data on plastic dilatation and stored energy of cold work in fcc metals.

Author Contributions: Conceptualization, A.Z. and J.D.C.; formal analysis, A.Z. and J.D.C.; investigation, A.Z. and J.D.C.; writing—original draft preparation, A.Z. and J.D.C.; writing—review and editing, A.Z. and J.D.C. All authors have read and agreed to the published version of the manuscript.

Funding: This research received no external funding.

Data Availability Statement: Not applicable; manuscript contains no data.

Conflicts of Interest: The authors declare no conflict of interest.

Appendix A. Extension to Dynamic High-Pressure Regimes

The effects of residual lattice dilatation from dislocations on pressure are undeniable in shock compression experiments, as stated in other studies [62–65]. Residual dilatation has been extracted from velocity profile histories in shock compression experiments on materials of very low flow stress such as pure copper (Cu) [62,65]. Dislocation densities measured in shock-recovered samples after 35-GPa impact exceed $2 \times 10^{15}/\text{m}^2$ [66], and transient densities could be much larger [67]. For example, dislocation densities predicted by discrete dislocation dynamics and atomistic simulations of shockwaves in Cu range from $10^{16}/\text{m}^2$ to $2 \times 10^{17}/\text{m}^2$ for shock pressures from 30 GPa to 75 GPa [67,68]. The volume change from the latter could attain, from Table 1, $0.013\alpha^e \approx 2.3\%$. The pressure change associated with such significant dilatation would be around 3.2 GPa, over 5% of the impact stress [69] and much larger than the yield and flow stresses of Cu [65,68]. In cases such as this, explicit inclusion of plastic volume changes from dislocations in the finite-strain kinematics of continuum crystal plasticity theory [43,60,61,69], or later via Equation (18), is prudent.

In addition, a volume change can be a byproduct of nano-scale dynamic excitations triggered by a collective motion of dislocations [64]. Excitations distort the lattice and, we hypothesize, affect lattice stretch. Under quasi-static conditions, pressure points can be ignored; however, shocks magnify the core pressure and anharmonicity to explicitly contribute to overall dilatant behavior. Dynamic excitations triggered by a synchronized motion of dislocations generate micro-kinetic energy, which in turn may act as phono-to-phonon vibrations; that is, micro-kinetic energy expands the lattice. Dynamic behaviors magnify effects in both time and space. The magnitude of micro-kinetic energy must be large enough for these effects to be noticeable, which happens only under extreme loading conditions. A method has been set forth to calculate micro-kinetic energy [64]; thus, one can introduce additional lattice stretch, a kind of local anharmonic thermal stretch.

The tensor representation concept of Sections 2 and 3 has all the features suitable for a proper physics-based interpretation of these phenomena. The TRM analysis can be easily extended to shock conditions. The second term in Equation (8) provides a means for restraining out-of-plane spreading of dislocations at high pressures. It has been reported that single crystalline Ta, among other metals, subjected to impact loading, experiences highly localized plastic slip [70]. Dynamic loading alters the mechanisms of plastic flow. That brings us back to Equations (7) and (9), where we search for explanations of the intriguing problem. The equivalent stress σ_{eq} in (9) consists of shear stress along the glide plane and includes the non-glide contribution. One may argue that, because it tightens interatomic spaces, high shock pressure also over-constrains spreading of dislocation cores. For this reason, we modify Equation (7) by rendering the out-of-plane contribution the needed sensitivity to the changes in mass density:

$$M_{ij} = \alpha\left[\left(N_{ij}^1 - N_{ij}^3\right) + \mu_\varphi\left(\delta_{ij}/3 - \frac{\rho_0}{\rho}N_{ij}^2\right)\right] \tag{A1}$$

Here, ρ_0 and ρ are initial and current mass densities, not dislocation densities. In a linear elastic regime, the ratio ρ_0/ρ is nearly unity, so its effects can be omitted. At high pressure, equations of state quantify pressure-volume-temperature-entropy responses [62–64].

Suppose that the pressure is 50 GPa, which translates to $\rho_0/\rho \cong 0.8$. Now, temperature is high due to adiabatic shock heating and thermoelastic coupling, and, therefore, we conclude that $\mu_0 = 0$. This also means that the associated flow rules are reestablished. In Figure A1, the yield surface marked in black is unaffected by the change of mass density $(\rho_0/\rho = 1)$. The second envelope, marked in red, is plotted for $\rho_0/\rho \cong 0.8$. Shapes of the envelopes are distinctly different.

In the next step, we assume a plate impact problem where the uniaxial stress points in direction 1. We calculate the plastic strain rates along three directions, mainly $r_{12} = M_{11}/M_{22}$, $r_{13} = M_{11}/M_{33}$, and $r_{23} = M_{22}/M_{33}$. The ratios are $r_{12} = -2.172$, $r_{13} = -2.172$, and $r_{23} = 1$, respectively. We find that there is a small increase of volume $\dot{\varepsilon}_{kk}^p = 0.0367\,\dot{\varepsilon}_{eq}^p$

since the tensor in (7) is no longer traceless. Thus, a high-pressure loading generates small plastic dilatation. Experimentally observed perturbations in surface velocity profiles for shock compression of W- and Al-based metals [71,72] suggest a pressure variation, which we argue can be a result of residual dilatation from magnified core pressure and/or anharmonic effects, especially since W and Al have the largest values of dilatation factors α^e and α^s in Table 1. However, this source of perturbations cannot, in general, be uniquely separated from other phenomena related to microstructure heterogeneities.

Shock relief ($\rho_0/\rho = 1.2$) generates external tensile pressure, and the trends are reversed. Now, the ratios are smaller, $r_{12} = -1.854$, $r_{13} = -1.854$, and $r_{23} = 1$, and the rate of plastic dilatation $\dot{\varepsilon}^p_{kk} = -0.0367\, \dot{e}^p_{eq}$ opposes the elastic stretch. Thus, in tension, the plastic contraction of dislocation cores tends to absorb the already large interatomic distances. A local volume reduction can also be associated with a transient decrease in dislocation density after the shockwave has passed and the material relaxes to equilibrium.

Incompressibility is the commonly used assumption in classical constitutive models [73]. The phenomenology omits non-planarity of plastic flow and thus cannot predict phenomena associated with plastic volume changes. In the proposed continuum description, the second term in Equation (7) includes the local mass density ratio, which enables a richer description of the metal behavior at shock conditions. In conclusion, the proposed concepts can be used to study the plastic responses of bcc metals at low temperatures, as we have demonstrated for Mo in Section 3, and can be extended to extreme high-pressure conditions, as newly proposed in this appendix.

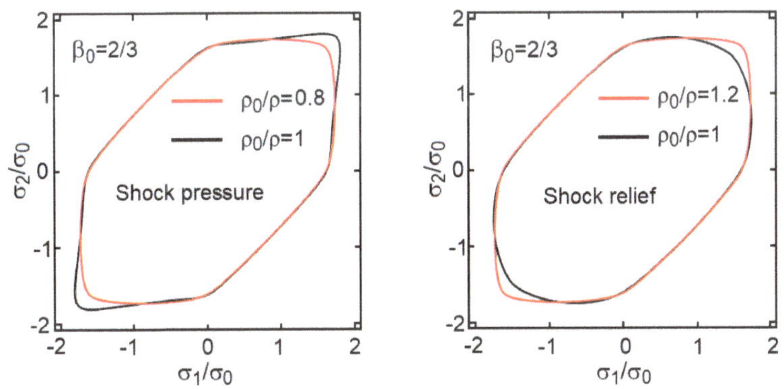

Figure A1. Yield surfaces are plotted for a non-shocked metal (black lines) and for the same metal at high shock pressure and the subsequent pressure relief (red line).

References

1. Vitek, V.; Mrovec, M.; Bassani, J.L. Influence of non-glide stresses on plastic flow: From atomistic to continuum modeling. *Mater. Sci. Eng. A* **2004**, *365*, 31–37. [CrossRef]
2. Bassani, L.J.; Racherla, V. From non-planar dislocation cores to non-associated plasticity and strain bursts. *Prog. Mater. Sci.* **2011**, *56*, 852–863. [CrossRef]
3. Qin, Q.; Bassani, J. Non-Schmid yield behavior in single crystals. *J. Mech. Phys. Solids* **1992**, *40*, 813–833. [CrossRef]
4. Qin, Q.; Bassani, J. Non-associated plastic flow in single crystals. *J. Mech. Phys. Solids* **1992**, *40*, 835–862. [CrossRef]
5. Duesbery, M.S.; Vitek, V. Plastic anisotropy in B.C.C. transition metals. *Acta Mater.* **1998**, *46*, 1481–1492. [CrossRef]
6. Vitek, V.; Paidar, V. Non-planar dislocation cores: A ubiquitous phenomenon affecting mechanical properties of crystalline materials. In *Dislocations in Solids*; Hirth, J.P., Ed.; Elsevier: Amsterdam, The Netherlands, 2008; Volume 14, Chapter 87; pp. 441–514.
7. Lim, H.; Hale, L.M.; Zimmerman, J.A.; Battaile, C.C.; Weinberger, C.R. A multi-scale model of dislocation plasticity in a-Fe: Incorporating temperature, strain rate and non-Schmid effects. *Int. J. Plasticity* **2015**, *73*, 100–118. [CrossRef]
8. Tresca, H. Mémoire sur l'écoulement des corps solides soumis à de fortes pressions. *C. R. Acad. Sci. Paris* **1864**, *59*, 754.
9. Huber, M. Specific work of strain as a measure of material effort. *Czas. Tech.* **1904**, *XXII*, 22. (In Polish)
10. Huber, M. Specific work of strain as a measure of material effort. *Arch. Mech.* **2004**, *56*, 173–190, (English Translation).
11. Von Mises, R. Mechanik der festen Körper im plastisch deformablen Zustand. *Göttin. Nachr. Math. Phys.* **1913**, *1*, 582–592.

12. Schmid, E.; Boas, W. *Kristallplastizität*; Springer: Berlin/Heidelberg, Germany, 1928.
13. Taylor, G.I.; Elam, C.F. The distortion of iron crystals. *Proc. R. Soc. A* **1926**, *112*, 337–361.
14. Hosford, W.F. A generalized isotropic yield criterion. *ASME J. Appl. Mech.* **1972**, *39*, 607–609. [CrossRef]
15. Hershey, A.V. The plasticity of an isotropic aggregate of anisotropic face-centered cubic crystals. *ASME J. Appl. Mech.* **1954**, *21*, 241–249. [CrossRef]
16. Taylor, G.I. Plastic strain in metals. *J. Inst. Metals* **1938**, *62*, 307–324.
17. Bishop, J.F.W.; Hill, R. A theory of the plastic distortion of a polycrystalline aggregate under combined stresses. *Philos. Mag.* **1951**, *42*, 414–427. [CrossRef]
18. Van Houtte, P. On the equivalence of the relaxed Taylor theory and the Bishop-Hill theory for partially constrained plastic deformation of crystals. *Mater. Sci. Eng.* **1982**, *55*, 69–77. [CrossRef]
19. Hill, R. A theory of the yielding and plastic flow of anisotropic metals. *Proc. R. Soc. A* **1948**, *193*, 281–297.
20. Cazacu, O.; Plunkett, B.; Barlat, F. Orthotropic yield criterion for hexagonal closed packed metals. *Int. J. Plasticity* **2006**, *22*, 1171–1194. [CrossRef]
21. Plunkett, B.; Cazacu, O.; Barlat, F. Orthotropic yield criteria for description of the anisotropy in tension and compression of sheet metals. *Int. J. Plasticity* **2008**, *24*, 847–866. [CrossRef]
22. Zubelewicz, A. Micromechanical study of ductile polycrystalline materials. *J. Mech. Phys. Solids* **1993**, *41*, 1711–1722. [CrossRef]
23. Zubelewicz, A. A mechanisms-based model for dynamic behavior and fracture of geomaterials. *Int. J. Rock Mech. Mining Sci.* **2014**, *72*, 277–282. [CrossRef]
24. Zubelewicz, A. Overall stress and strain rates for crystalline and frictional materials. *Int. J. Non-Linear Mechanics* **1990**, *25*, 389–392. [CrossRef]
25. Zubelewicz, A. Another perspective on elastic and plastic anisotropy of textured metals. *Proc. R. Soc. A* **2021**, *477*, 20210234. [CrossRef]
26. Gröger, R.; Racherla, V.; Vitek, V. Multiscale modeling of plastic deformation of molybdenum and tungsten: I. Atomistic studies of the core structure and glide of 1/2<111> screw dislocations at 0 K. *Acta Mater.* **2008**, *56*, 5401–5411.
27. Gröger, R.; Racherla, V.; Vitek, V. Multiscale modeling of plastic deformation of molybdenum and tungsten: II. Yield criterion for single crystals based on atomistic studies of glide 1/2<111> screw dislocations. *Acta Mater.* **2008**, *56*, 5412–5425.
28. Christian, J.W. Some surprising features of the plastic deformation of body-centered cubic metals and alloys. *Metall. Trans. A* **1983**, *14*, 1237–1256. [CrossRef]
29. Kraych, A.; Clouet, E.; Dezerald, L.; Ventelon, L.; Willaime, F.; Rodney, D. Non-glide effects and dislocation core fields in BCC metals. *NPJ Comp. Mater.* **2019**, *5*, 109. [CrossRef]
30. Hollang, L.; Hommel, M.; Seeger, A. The flow stress of ultra-high-purity molybdenum single crystals. *Phys. Stat. Solidi A* **1997**, *160*, 329–354. [CrossRef]
31. Seeger, A.; Hollang, L. The flow-stress asymmetry of ultra-pure molybdenum single crystals. *Mater. Trans.* **2000**, *41*, 141–151. [CrossRef]
32. Hollang, L.; Brunner, D.; Seeger, A. Work hardening and flow stress of ultrapure molybdenum single crystals. *Mater. Sci. Eng. A* **2001**, *319*, 233–236. [CrossRef]
33. Holder, J.; Granato, A.V. Thermodynamic properties of solids containing defects. *Phys. Rev.* **1969**, *182*, 729–741. [CrossRef]
34. Wright, T.W. Stored energy and plastic volume change. *Mech. Mater.* **1982**, *1*, 185–187. [CrossRef]
35. Spitzig, W.A.; Richmond, O. The effect of pressure on the flow stress of metals. *Acta Metall.* **1984**, *32*, 457–463. [CrossRef]
36. Vitek, V. Computer simulation of the screw dislocation motion in bcc metals under the effect of the external shear and uniaxial stresses. *Proc. R. Soc. A* **1976**, *352*, 109–124.
37. Ismail-Beigi, S.; Arias, T.A. Ab initio study of screw dislocations in Mo and Ta: A new picture of plasticity in bcc transition metals. *Phys. Rev. Lett.* **2000**, *84*, 1499–1503. [CrossRef] [PubMed]
38. Vitek, V. Core structure of screw dislocations in body-centered cubic metals: Relation to symmetry and interatomic bonding. *Philos. Mag.* **2004**, *84*, 415–428. [CrossRef]
39. Clayton, J.D. Dynamic plasticity and fracture in high density polycrystals: Constitutive modeling and numerical simulation. *J. Mech. Phys. Solids* **2005**, *53*, 261–301. [CrossRef]
40. Finnis, M.W.; Sinclair, J.E. A simple empirical N-body potential for transition metals. *Philos. Mag. A* **1984**, *50*, 45–55. [CrossRef]
41. Li, J.; Wang, C.Z.; Chang, J.P.; Cai, W.; Bulatov, V.V.; Ho, K.M.; Yip, S. Core energy and Peierls stress of a screw dislocation in bcc molybdenum: A periodic-cell tight-binding study. *Phys. Rev. B* **2004**, *70*, 104113. [CrossRef]
42. Teodosiu, C. *Elastic Models of Crystal Defects*; Springer: Berlin/Heidelberg, Germany, 1982.
43. Clayton, J.D. *Nonlinear Mechanics of Crystals*; Springer: Dordrecht, The Netherlands, 2011.
44. Toupin, R.; Rivlin, R. Dimensional changes in crystals caused by dislocations. *J. Math. Phys.* **1960**, *1*, 8–15. [CrossRef]
45. Clayton, J.D.; Bammann, D.J. Finite deformations and internal forces in elastic-plastic crystals: Interpretations from nonlinear elasticity and anharmonic lattice statics. ASME J. *Eng. Mater. Tech.* **2009**, *131*, 041201. [CrossRef]
46. Clayton, J.D. Defects in nonlinear elastic crystals: Differential geometry, finite kinematics, and second-order analytical solutions. *ZAMM-J. Appl. Math. Mech.* **2015**, *95*, 476–510. [CrossRef]
47. Webb, E.B.; Zimmerman, J.A.; Seel, S.C. Reconsideration of continuum thermomechanical quantities in atomic scale simulations. *Math. Mech. Solids* **2008**, *13*, 221–266. [CrossRef]

48. Henager, C.H., Jr.; Hoagland, R.G. Dislocation core fields and forces in FCC metals. *Scripta Mater.* **2004**, *50*, 1091–1095. [CrossRef]
49. Foreman, A.J.E. Dislocation energies in anisotropic crystals. *Acta Metall.* **1955**, *3*, 322–330. [CrossRef]
50. Steeds, J.W. *Introduction to Anisotropic Elasticity Theory of Dislocations*; Clarendon Press: Oxford, UK, 1973.
51. Seeger, A.; Haasen, P. Density changes of crystals containing dislocations. *Philos. Mag.* **1958**, *3*, 470–475. [CrossRef]
52. Zener, C. Theory of lattice expansion introduced by cold work. *Trans. Am. Inst. Mining Metall. Engrs.* **1942**, *147*, 361–368.
53. Sinclair, J.; Gehlen, P.; Hoagland, R.; Hirth, J. Flexible boundary conditions and nonlinear geometric effects in atomic dislocation modeling. *J. Appl. Phys.* **1978**, *49*, 3890–3897. [CrossRef]
54. Clouet, E.; Ventelon, L.; Willaime, F. Dislocation core energies and core fields from first principles. *Phys. Rev. Lett.* **2009**, *102*, 055502. [CrossRef]
55. Clouet, E.; Ventelon, L.; Willaime, F. Dislocation core field II. *Screw Dislocation in Iron. Phys. Rev. B* **2011**, *84*, 224107. [CrossRef]
56. Kocks, U.F.; Argon, A.S.; Ashby, M.F. Thermodynamics and kinetics of slip. *Prog. Mater. Sci.* **1975**, *19*, 1–288.
57. Hull, D.; Bacon, D.J. *Introduction to Dislocations*, 3rd ed.; Butterworth-Heinemann: Oxford, UK, 1984.
58. Ninomiya, T. Theory of melting, dislocation model. I. *J. Phys. Soc. Japan* **1978**, *44*, 263–268. [CrossRef]
59. Guinan, M.W.; Steinberg, D.J. Pressure and temperature derivatives of the isotropic polycrystalline shear modulus for 65 elements. *J. Phys. Chem. Solids* **1974**, *35*, 1501–1512. [CrossRef]
60. Clayton, J.D. An alternative three-term decomposition for single crystal deformation motivated by non-linear elastic dislocation solutions. *Q. J. Mech. Appl. Math.* **2014**, *67*, 127–158. [CrossRef]
61. Clayton, J.D. A continuum description of nonlinear elasticity, slip and twinning, with application to sapphire. *Proc. R. Soc. A* **2009**, *465*, 307–334. [CrossRef]
62. Clayton, J.D. *Nonlinear Elastic and Inelastic Models for Shock Compression of Crystalline Solids*; Springer: Berlin/Heidelberg, Germany, 2019.
63. Horie, Y. Thermodynamics of dislocations and shock compression of solids. *Phys. Rev. B* **1980**, *21*, 5549–5557. [CrossRef]
64. Zubelewicz, A. Thermodynamics description of dynamic plasticity in metals. *Forces Mech.* **2022**, *9*, 100121. [CrossRef]
65. Clayton, J.D. Nonlinear thermomechanics for analysis of weak shock profile data in ductile polycrystals. *J. Mech. Phys. Solids* **2019**, *124*, 714–757. [CrossRef]
66. Rohatgi, A.; Vecchio, K.S.; Gray, G.T., III. A metallographic and quantitative analysis of the influence of stacking fault energy on shock-hardening in Cu and Cu–Al alloys. *Acta Mater.* **2001**, *49*, 427–438. [CrossRef]
67. Bringa, E.M.; Caro, A.; Wang, Y.; Victoria, M.; McNaney, J.M.; Remington, B.A.; Smith, R.F.; Torralva, B.R.; Van Swygenhoven, H. Ultrahigh strength in nanocrystalline materials under shock loading. *Science* **2005**, *309*, 1838–1841. [CrossRef] [PubMed]
68. Shehadeh, M.A.; Bringa, E.M.; Zbib, H.M.; McNaney, J.M.; Remington, B.A. Simulation of shock-induced plasticity including homogeneous and heterogeneous dislocation nucleations. *Appl. Phys. Lett.* **2006**, *89*, 171918. [CrossRef]
69. Elkhodary, E.I.; Zikry, M.A. A fracture criterion for finitely deforming crystalline solids-the dynamic fracture of single crystals. *J. Mech. Phys. Solids* **2011**, *59*, 2007–2022. [CrossRef]
70. Lim, H.; Carroll, J.D.; Battaile, C.C.; Chen, S.R.; APMoore, A.P.; Lane, J.D. Anisotropy and strain localization in dynamic impact experiments of tantalum single crystals. *Sci. Rep.* **2018**, *8*, 5540. [CrossRef] [PubMed]
71. Vogler, T.J.; Clayton, J.D. Heterogeneous deformation and spall of an extruded tungsten alloy: Plate impact experiments and crystal plasticity modeling. *J. Mech. Phys. Solids* **2008**, *56*, 297–335. [CrossRef]
72. Lloyd, J.T.; Clayton, J.D.; Becker, R.; McDowell, D.L. Simulation of shock wave propagation in single crystal and polycrystalline aluminum. *Int. J. Plasticity* **2014**, *60*, 118–144. [CrossRef]
73. Hill, R. *The Mathematical Theory of Plasticity*; Clarendon Press: Oxford, UK, 1950.

metals

MDPI

Article

Study of the Thermomechanical Behavior of Single-Crystal and Polycrystal Copper

Sudip Kunda [1], Noah J. Schmelzer [1], Akhilesh Pedgaonkar [1], Jack E. Rees [1], Samuel D. Dunham [1], Charles K. C. Lieou [2], Justin C. M. Langbaum [1] and Curt A. Bronkhorst [1,*]

1 Department of Mechanical Engineering, University of Wisconsin-Madison, Madison, WI 53706, USA; skunda@wisc.edu (S.K.); langbaum@wisc.edu (J.C.M.L.)
2 Department of Nuclear Engineering, University of Tennessee-Knoxville, Knoxville, TN 37996, USA; clieou@utk.edu
* Correspondence: cbronkhorst@wisc.edu

Abstract: This research paper presents an experimental, theoretical, and numerical study of the thermomechanical behavior of single-crystal and polycrystal copper under uniaxial stress compression loading at varying rates of deformation. The thermomechanical theory is based on a thermodynamically consistent framework for single-crystal face-centered cubic metals, and assumes that all plastic power is partitioned between stored energy due to dislocation structure evolution (configurational) and thermal (kinetic vibrational) energy. An expression for the Taylor–Quinney factor is proposed, which is a simple function of effective temperature and is allowed by second-law restrictions. This single-crystal model is used for the study of single- and polycrystal copper. New polycrystal thermomechanical experimental results are presented at varying strain rates. The temperature evolution on the surface of the polycrystal samples is measured using mounted thermocouples. Thermomechanical numerical single- and polycrystal simulations were performed for all experimental conditions ranging between 10^{-3} and 5×10^3 s^{-1}. A Taylor homogenization model is used to represent polycrystal behavior. The numerical simulations of all conditions compare reasonable well with experimental results for both stress and temperature evolution. Given our lack of understanding of the mechanisms responsible for the coupling of dislocation glide and atomic vibration, this implies that the proposed theory is a reasonably accurate approximation of the single-crystal thermomechanics.

Keywords: dislocations; crystals; polycrystals; stress–strain; copper; Taylor–Quinney factor

1. Introduction

Some of the earliest studies on the conversion of mechanical work to heat in the plastic deformation of metals were by [1,2]. Such energetic considerations are important in the study of the dynamic response of metals in applications such as armor systems and crash testing of vehicles. Study of the thermomechanics of metals has important implications for studying phenomena such as phase transformations [3] and twinning [4], as well as for metal forming and machining. The fraction of the plastic work on a metal that is converted to heat is known as the Taylor–Quinney factor. The complement of this is called the stored energy of cold work.

A common assumption for the fraction of plastic work converted to thermal energy is that it is a constant equal to approximately 0.9, in accordance with the results in [1]. A constant value of 1.0 has also been chosen in the literature [5]. However, experimental results for various pure metals and alloys indicate dependence of the Taylor–Quinney factor on the strain and strain rate. For example, dynamic experiments using the Kolsky bar [6,7] and the dislocation dynamics simulation in [8] indicate that the dislocation accumulation pattern, along with the dislocation density, is important in determining the Taylor–Quinney factor. Study of the thermomechanics of commercially pure titanium reveals that the dissipation of mechanical work as heat may strongly depend on the loading conditions [9]. At high strain rates ($\approx$3000 s^{-1}), the authors found that commercially pure titanium shows a profusion of twinning in compression and dominant shear, but not in tension. This corresponds to markedly higher values of the Taylor–Quinney factor for the compression and shear load paths as compared to tension. Quasi-static and dynamic experiments on pure iron [10], polycrystal tantalum [11], single-crystal tantalum [12], and both single-crystal and polycrystal copper [13] provide further experimental evidence of the strain and strain rate dependence of the Taylor–Quinney factor.

The study of crystal mechanics started with work that explicitly embedded the geometry of the slip systems of the material into the kinematical part of the model [14] and by the multiplicative decomposition of the deformation gradient into elastic and plastic parts [15–17]. The elastic part of the deformation gradient captures the distortion of the lattice, while the plastic part of deformation gradient captures the transfer of mass along the slip planes of the material. The rotational part of the elastic part of the deformation gradient also contributes to the transfer of mass. The use of such kinematic assumptions has allowed for study of phenomena such as strain localization [18], latent hardening and secondary slip [19,20], and texture evolution [21–23], with the combination of these aspects of plastic deformation studied in [24]. Certainly, dislocations behave very differently with crystal atomic structure, which is reflected in the development of continuum crystal mechanics theories [25–44]. Face-centered cubic materials are generally the best understood, with dislocation interaction representing the dominant resistance to motion for most materials. Body-centered cubic materials are very interesting in that the behavior of screw dislocations dominates motion and due to questions about the role of kink–pair nucleation mechanisms in the physics of dislocation motion [45]. Hexagonal close-packed materials differ substantially with material type, as the c/a ratio and corresponding Burgers vector for different slip systems change dramatically [46]. There are many physical factors which impact the thermomechanical response of crystalline materials.

There have also been many important contributions to the development of thermodynamically consistent theories for the description of thermomechanics in crystalline solids, among which are [47–72]. As we develop theories for the description of inelastic processes such as dislocation slip, deformation twinning, structural phase transformation, and damage, accurate partitioning of the energy will provide a stronger physical basis. This forms natural restrictions to our crystal mechanics formulations and provides greater quantitative authority for physical interpretation of experimental results. This naturally affords the need to draw in thermodynamics information from experiments and focused physics calculations. More sophisticated boundary conditions are also necessary in order to document experiments and for application to numerical simulations. Experimental design must then account for the need to characterize thermal boundary conditions and transport properties of materials. Certainly, experiments with higher deformation rate can be assumed to be adiabatic; however, experimental diagnostics are also more limited at higher deformation rates.

This article provides a thermodynamically consistent crystal mechanics framework for solving combined thermomechanical problems in the single and polycrystal mechanics of metallic materials. The fundamental concepts that form the basis of this framework are laid out in [73,74]. On the basis of this theory, a computational framework was developed for polycrystalline isotropic materials and used in [75–77]. Further development led to a material model for metal single crystals [78], which is the basis of the computation in this work. The above provides a single-crystal model which captures the strain rate dependence of the thermomechanical behavior. In this model, both the amount of plastic work and the fraction of it converted to heat is dependent on the strain rate. The differential form of the Taylor–Quinney factor is used in this work. The computational model for the polycrystal is a Taylor model, in which the contribution of each individual grain is calculated using the aforementioned single-crystal model. Welded thermocouple beads are used to study the temperature evolution of the sample.

Details of the nomenclature used in this work are provided in Table 1. A list of symbols used in the presented theory is provided in Table 2.

Table 1. Nomenclature table.

Type of Quantity	Description of Symbol	Direct Notation	Indicial Notation
Scalars	Italicized small/cap letters without subscripts or superscripts	a, b, c, A, B, C	a, b, c, A, B, C
Matrices or Vectors	Bold upright letters, using capital letters for vectors in the reference configuration and small letters for vectors in the current configuration.	$\mathbf{u}, \mathbf{U}$	u_i, U_i
Second order tensors	Bold upright letters with underlines, with capital letters for objects in the reference configuration and and small letters for objects in the current configuration.	$\underline{\mathbf{s}}, \underline{\mathbf{S}}$	s_{ij}, S_{ij}
Fourth order tensors	Blackboard bold capital letters.	$\mathbb{C}$	C_{ijkl}

Table 2. List of theory variable symbols.

Variable Symbol	Definition or Meaning
$\underline{\mathbf{F}}, \underline{\mathbf{F}}^e, \underline{\mathbf{F}}^p$	Total, elastic, and plastic deformation gradients
$\underline{\mathbf{L}}, \underline{\mathbf{L}}^e, \underline{\mathbf{L}}^p$	Total, elastic, and plastic velocity gradients
$\dot{\gamma}^\alpha$	Resolved plastic strain rate on slip system α
$\mathbf{s}^\alpha, \mathbf{m}^\alpha$	Unit slip direction vector and normal to slip system α
$\underline{\mathbf{T}}$	Cauchy stress tensor
$\underline{\mathbf{S}}$	Second Piola–Kirchhoff stress tensor
$\mathbb{C}$	Anisotropic fourth-order tensor of elastic constants
C_{11}, C_{12}, C_{44}	Independent crystallographic moduli for fcc lattice
μ	Shear modulus
$\mathcal{W}$	Jacobian matrix of stress versus strain
U_K, S_K	Kinetic–vibrational (thermal) energy and entropy density
U_C, S_C	Configurational energy and entropy density
U_D, S_D	Dislocation energy and entropy
U_1, S_1	Residual configurational energy and entropy density
F_C	Configurational free energy density
e_D	Dislocation line energy
$\mathbf{q}_C, \mathbf{q}_K$	Configurational and thermal fluxes
χ, χ_0	Effective temperature and initial effective temperature
θ, T	Thermal temperature (in units of energy and Kelvin)
ρ^α	Dislocation density on slip system α
ρ^α_{ss}	Steady-state dislocation density on slip system α
v^α	Mean dislocation velocity on slip system α
χ_{ss}	Steady-state effective temperature (in units of e_D)
β	Taylor–Quinney factor
c_p	Specific heat capacity
κ^α_ρ	Dislocation storage rate
κ_χ	Effective temperature increase rate
a	Minimum separation between dislocations
b	Burgers vector
t_0	Atomic time scale
α_T	Stress scale parameter
τ^α	Resolved shear stress on slip system α
s^α	Slip resistance due to dislocation interaction on slip system α
s_l	Intrinsic lattice resistance to dislocation motion
ρ^α	Dislocation density corresponding to slip system α
l^α	Dislocation mean free path on slip system α
t^α	Dislocation depinning time on slip system α
T_P	Dislocation depinning barrier (in units of Kelvin)

Table 2. *Cont.*

Variable Symbol	Definition or Meaning
$a^{\alpha\beta}$	Dislocation interaction tensor
$d^{\alpha\beta}$	Slip interaction tensor
k_c, k_{nc}	Mean free path parameters

2. Materials and Methods

2.1. Theory

This section presents a coupled thermomechanical theory describing the physical behavior of single crystals, which is applied here to face-centered cubic metals and copper in particular.

2.1.1. Plasticity

Based on the theory developed in [15–17], it is assumed that the total deformation gradient may be decomposed into elastic and plastic components.

$$\underline{\mathbf{F}} = \underline{\mathbf{F}}^e\underline{\mathbf{F}}^p. \tag{1}$$

It is assumed that the body starts motion in the reference configuration. It is mapped to the current configuration by the total deformation gradient. The plastic part of the deformation gradient maps the reference configuration to the lattice configuration. The lattice configuration is then mapped to the current configuration by the elastic deformation gradient. The rate of change of plastic deformation in the metal is captured using the plastic velocity gradient, which is a function of the orientations of the all the slip systems and the slip rates on the corresponding slip systems:

$$\underline{\mathbf{L}}^p = \dot{\underline{\mathbf{F}}}^p(\underline{\mathbf{F}}^p)^{-1} = \sum_{\alpha} \dot{\gamma}^\alpha \mathbf{s}^\alpha \otimes \mathbf{m}^\alpha, \tag{2}$$

where $\mathbf{s}^\alpha$ is a unit vector parallel to the direction of slip on slip system α and $\mathbf{m}^\alpha$ is a unit vector normal to the slip plane for slip system α. The symbol $\dot{\gamma}^\alpha$ is the rate of slip. The quantities $\mathbf{s}^\alpha$, $\mathbf{m}^\alpha$ and $\underline{\mathbf{L}}^p$ all inhabit the lattice configuration.

The second Piola–Kirchhoff stress is used as a stress measure in this model, denoted by the symbol $\underline{\mathbf{S}}$ and defined by

$$\underline{\mathbf{S}} = J(\underline{\mathbf{F}}^e)^{-1}\underline{\mathbf{T}}(\underline{\mathbf{F}}^e)^{-T}, \tag{3}$$

where $\underline{\mathbf{T}}$ is the Cauchy stress and $J = \det \underline{\mathbf{F}}$ is the Jacobian of the total deformation gradient, which is also equal to $\det \underline{\mathbf{F}}^e$ by the assumption of isochoric plastic deformation. The Cauchy stress inhabits the current configuration, whereas the second Piola–Kirchhoff stress is defined in the lattice configuration.

The stress response is provided by

$$\underline{\mathbf{S}} = \frac{1}{2}\mathbb{C} : ((\underline{\mathbf{F}}^e)^T\underline{\mathbf{F}}^e - \underline{\mathbf{1}}) = \mathbb{C} : \underline{\mathbf{E}}^e, \tag{4}$$

where $\mathbb{C}$ and $\underline{\mathbf{E}}^e$ are quantities in the lattice configuration, $\mathbb{C}$ is the fourth-order elastic stiffness tensor, and $\underline{\mathbf{1}}$ is the second-order unit tensor. Because copper is a FCC metal, $\mathbb{C}$ has only three independent components. These three elastic moduli describe the behavior of a single crystal that has its <100> directions aligned with the x, y, and z directions of the coordinate axis. The elastic stiffness tensor of a single crystal of any given orientation is calculated by first computing the elasticity tensor of the reference orientation from knowledge of the elastic moduli and then using appropriate tensor transformations to calculate the elasticity tensor for the given orientation. In this work, the temperature dependence of the elastic moduli is accounted for by assuming a linear dependence on temperature as

$$C_{c,ijkl}(T) = C_{c0,ijkl} + m_{ijkl}T, \tag{5}$$

where $C_{0,ijkl}$ are the components of the stiffness tensor at 0K and T is the absolute temperature. Of importance to the theory is the resolved shear stress on slip system α, denoted by τ^{α} (in the lattice configuration) and provided by

$$\tau^{\alpha} = (\underline{\mathbf{F}}^{e})^{T}\underline{\mathbf{F}}^{e}\underline{\mathbf{S}} : (\mathbf{s}^{\alpha} \otimes \mathbf{m}^{\alpha}) \approx \underline{\mathbf{S}} : (\mathbf{s}^{\alpha} \otimes \mathbf{m}^{\alpha}), \tag{6}$$

where it is assumed that the elastic deformation in the material is small relative to the plastic deformation. With this assumption, the volume density of plastic power, defined as $\underline{\mathbf{T}} : \underline{\mathbf{F}}^{e}\underline{\mathbf{L}}^{p}(\underline{\mathbf{F}}^{e})^{-1}$, resolves to

$$\underline{\mathbf{T}} : \underline{\mathbf{F}}^{e}\underline{\mathbf{L}}^{p}(\underline{\mathbf{F}}^{e})^{-1} = \sum_{\alpha}\tau^{\alpha}\dot{\gamma}^{\alpha}. \tag{7}$$

2.1.2. Thermomechanics

Here, we provide an outline of the crystal mechanics theory. The thermodynamic basis of this work is found in [59,73,74]. After the development of these ideas into isotropic plasticity theories in [75–77], a single-crystal version of this theory was presented in [78]. The basic assumption of the theory is that the energy of thermal vibrations is too small to create new dislocations without an applied stress. As a result, it is justifiable to separate the metal into two subsystems: a kinetic–vibrational subsystem, and a configurational subsystem. The latter is characterized by the mean positions of the atoms (and defects) in the metal crystal, while the former is characterized by the motion (vibration) of the atoms about those mean positions. The total volume densities of the system's energy and entropy are denoted by the symbols U_{tot} and S_{tot}, respectively. The assumption described above allows these symbols to be respectively decomposed as $U_{tot} = U_K + U_C$ and $S_{tot} = S_K + S_C$, where the letter K denotes quantities belonging to the kinetic–vibrational subsystem and the letter C denotes quantities belonging to the configurational subsystem. As the configurational subsystem is concerned with the positions of atoms and defects in the lattice, U_C and S_C must depend on the current dislocation density. Denoting the dislocation density on slip system α by ρ^{α}, we write

$$\begin{aligned} U_C(S_C, \rho^{\alpha}) &= U_D(\rho^{\alpha}) + U_1(S_1), \\ S_C(U_C, \rho^{\alpha}) &= S_D(\rho^{\alpha}) + S_1(U_1), \end{aligned} \tag{8}$$

where the right-hand sides represent a division of the configurational subsystem into a part dependent on the dislocation population and a part dependent on other sources of imperfection, such as point defects. The following quantity is important in the theory

$$\chi = \frac{\partial U_C}{\partial S_C}, \tag{9}$$

where χ is known as the effective temperature. It characterizes the configurational subsystem by quantifying the atomic disorder in the material relative to a perfect crystal.

In this setup, the local form of the first law of thermodynamics reads $\underline{\mathbf{T}} : \underline{\mathbf{L}} = \dot{U}_C + \dot{U}_K$, where heating contributes to the evolution of U_K. Explicitly accounting for heat production and heat fluxes results in

$$\begin{aligned} \underline{\mathbf{T}} : \underline{\mathbf{L}} &= \dot{U}_C + \dot{U}_K \\ &= \chi\dot{S}_C + \nabla \cdot \mathbf{q}_C + \left(\frac{\partial U_C}{\partial t}\right)_{S_C, \rho^{\alpha}} + \sum_{\alpha}\left(\frac{\partial U_C}{\partial \rho^{\alpha}}\right)_{S_C}\dot{\rho}^{\alpha} + \theta\dot{S}_K + \nabla \cdot \mathbf{q}_K. \end{aligned} \tag{10}$$

Here,

$$\mathbf{q}_C = -\kappa_C \nabla \chi, \quad \mathbf{q}_K = -\kappa \nabla \theta, \tag{11}$$

are the effective and ordinary heat fluxes, respectively, with κ_C and κ as the corresponding conductivities. The symbol $\theta = k_B T$ denotes the ordinary temperature in energy units.

The symbol $\underline{\mathbf{L}}$ denotes the velocity gradient, which may be separated into its elastic and plastic components:

$$\underline{\mathbf{L}} = \underline{\mathbf{L}}^e + \underline{\mathbf{F}}^e \underline{\mathbf{L}}^p \underline{\mathbf{F}}^{e-1}. \tag{12}$$

With this notation, we write

$$\left(\frac{\partial U_C}{\partial t}\right)_{S_C, \rho^\alpha} = \underline{\mathbf{T}} : \underline{\mathbf{L}}^e. \tag{13}$$

The left-hand side of Equation (13) represents the time rate of change of the internal energy of the system when the configurational state is held constant. By definition, this is the rate of change of the elastic part of the energy. Therefore, the remaining terms on the right-hand side of Equation (10) must add to the plastic power. This is written as

$$\underline{\mathbf{T}} : \underline{\mathbf{F}}^e \underline{\mathbf{L}}^p \underline{\mathbf{F}}^{e-1} = \chi \dot{S}_C - \kappa_C \nabla^2 \chi + \sum_\alpha \left(\frac{\partial U_C}{\partial \rho^\alpha}\right)_{S_C} \dot{\rho}^\alpha + \theta \dot{S}_K - \kappa \nabla^2 \theta. \tag{14}$$

The second law of thermodynamics in the current context takes the following form

$$\dot{S}_{\text{tot}} = \dot{S}_C + \dot{S}_K + \nabla \cdot \frac{\mathbf{q}_C}{\chi} + \nabla \cdot \frac{\mathbf{q}_K}{\theta} \geq 0. \tag{15}$$

To investigate the consequences of the second law, it is multiplied by χ; quantity $\chi \dot{S}_C$ in Equation (15) is eliminated using Equation (14) to obtain the following inequality

$$\begin{aligned} &\underline{\mathbf{T}} : \underline{\mathbf{F}}^e \underline{\mathbf{L}}^p \underline{\mathbf{F}}^{e-1} - \sum_\alpha \left(\frac{\partial U_C}{\partial \rho^\alpha}\right)_{S_C} \dot{\rho}^\alpha \\ &+ (\chi - \theta)\left(\dot{S}_K - \frac{\kappa}{\theta} \nabla^2 \theta\right) + \frac{\kappa}{\theta} (\nabla \theta)^2 + \frac{\kappa_c}{\chi} (\nabla \chi)^2 \geq 0. \end{aligned} \tag{16}$$

The consequences of this inequality can be studied by applying the Coleman–Noll procedure [79]. This procedure allows us to conclude that if a sum of independent terms satisfies an inequality (such as Inequality (16)), each of the terms in the sum must individually satisfy the inequality. Therefore, we write

$$\begin{aligned} \underline{\mathbf{T}} : \underline{\mathbf{F}}^e \underline{\mathbf{L}}^p \underline{\mathbf{F}}^{e-1} &\geq 0, \\ -\sum_\alpha \left(\frac{\partial U_C}{\partial \rho^\alpha}\right)_{S_C} \dot{\rho}^\alpha &\geq 0, \\ (\chi - \theta)\left(\dot{S}_K - \frac{\kappa}{\theta} \nabla^2 \theta\right) &\geq 0, \\ \frac{\kappa}{\theta} (\nabla \theta)^2 &\geq 0, \\ \frac{\kappa_c}{\chi} (\nabla \chi)^2 &\geq 0. \end{aligned} \tag{17}$$

The first of these inequalities is guaranteed by Equations (7) and (26). The last two inequalities hold because of the nature of the quantities. The consequences of the third inequality, discussed in detail in [76], are that the Taylor–Quinney factor must be a function of the effective temperature; in addition, as the material approaches a state of maximal disorder, all of the input work turns to thermal energy and the Taylor–Quinney factor must tend to 1.0. Further development of the second inequality shows a derivative of the configurational free energy of the system. Application of the chain rule of calculus to Equations (8) results in

$$\left(\frac{\partial U_C}{\partial \rho^\alpha}\right)_{S_C} = \frac{\partial U_D}{\partial \rho^\alpha} - \frac{\partial U_1}{\partial S_1}\frac{\partial S_D}{\partial \rho^\alpha}. \tag{18}$$

The derivative $\frac{\partial U_1}{\partial S_1}$ can be identified as the effective temperature of a part of the configurational subsystem. Because the configurational subsystem has an effective temperature of χ, we have $\frac{\partial U_1}{\partial S_1} = \chi$. Therefore, we may write

$$\left(\frac{\partial U_C}{\partial \rho^\alpha}\right)_{S_C} = \frac{\partial U_D}{\partial \rho^\alpha} - \chi\frac{\partial S_D}{\partial \rho^\alpha} = \frac{\partial (U_D - \chi S_D)}{\partial \rho^\alpha}. \tag{19}$$

Here, $F_C \equiv U_D - \chi S_D$ is the configurational free energy. This means that as the dislocation density increases, the free energy tends to minimize and reaches a minimum when $\dot{\rho}^\alpha = 0$. Thus, the steady-state dislocation density minimizes the configurational free energy.

2.1.3. Constitutive Model

Here, we provide details of the computational framework, which was developed in [78] from the theory in [73] described in the previous section. We start from the Orowan relation as

$$\dot{\gamma}^\alpha = \rho^\alpha b v^\alpha, \tag{20}$$

where ρ^α is the mean mobile dislocation density on slip system α, b is the magnitude of the Burger's vector, and v^α is the mean dislocation velocity on slip system α.

In the physical picture of dislocation motion assumed in this work, the obstacles faced by the dislocation on slip system α have a mean spacing of l^α. The mean time required by the dislocation to cross this distance is t^α, meaning that the mean dislocation velocity is $v^\alpha = l^\alpha / t^\alpha$. This expression for the velocity changes the Orowan relation as

$$\dot{\gamma}^\alpha = \rho^\alpha b l^\alpha / t^\alpha. \tag{21}$$

This "depinning" of dislocations from obstacles is assumed to be stress-driven and thermally activated. This is accounted for in the expression for the time scale t^α as

$$t^\alpha = t_0 \exp\left[\frac{T_p}{T}\left\{1 - \left(\frac{\tau^\alpha - s_l}{s^\alpha}\right)^p\right\}^q\right]. \tag{22}$$

Here, t_0 is approximately the inverse of the Debye frequency, T is the absolute temperature, and T_p is a measure of the energy barrier to dislocation depinning. The actual energy barrier is provided by $k_B T_p$, where k_B is the Boltzmann constant. The symbol τ^α denotes the resolved shear stress on slip system α, whereas s_l is the intrinsic lattice resistance. The symbol s^α denotes the slip resistance due to dislocation interaction on slip system α, and is expressed as

$$s^\alpha = \alpha_T \mu b \sqrt{\sum_\beta a^{\alpha\beta} \rho^\beta}, \tag{23}$$

where α_T is the Taylor factor, accounting for the uncertainty of microstructure and chemical composition of the material. The symbol μ denotes the shear modulus, which may be calculated from the elastic moduli C_{11}, C_{12}, and C_{44} using the following expression:

$$\mu = \sqrt{C_{44}\left(\frac{C_{11} - C_{12}}{2}\right)}. \tag{24}$$

The symbol $a^{\alpha\beta}$ is a tensor accounting for dislocation interaction. A similar dislocation interaction tensor $d^{\alpha\beta}$ is used to determine the mean free path of the dislocation l^α:

$$l^\alpha = \frac{1}{\sqrt{\sum_\beta d^{\alpha\beta}\rho^\beta}}. \tag{25}$$

The parameters $d^{\alpha\beta}$ are interaction parameters. According to [80], they may be calculated using $a^{\alpha\beta}$, k_{inter}, and k_{copl}, using the expressions $d^{\alpha\beta} = \frac{a^{\alpha\beta}}{k^2_{\text{inter}}}$ for intersecting slip systems and $d^{\alpha\beta} = \frac{a^{\alpha\beta}}{k^2_{\text{copl}}}$ for self-interaction and coplanar slip systems. The constants k_{inter} and k_{copl} are described in [80], and are related to the inverse proportionality between the mean free path length l^α and the resolved shear stress on slip system α. Using these expressions, we obtain the following expressions for the slip rate on slip system α as

$$\dot{\gamma}^\alpha_{|\tau^\alpha|>0} = \frac{\rho^\alpha b}{t_0\sqrt{\sum_\beta d^{\alpha\beta}\rho^\beta}} \exp\left[-\frac{T_p}{T}\left\{1 - \left(\frac{\tau^\alpha - s_l}{s^\alpha}\right)^p\right\}^q\right] \text{sign}(\tau^\alpha). \tag{26}$$

Equation (26) is an Arrhenius-type expression. Physically, it corresponds to the idea that dislocations in which atoms have higher kinetic energy (measured by temperature T) have a higher probability of crossing the barrier to dislocation motion (measured by T_p). Such flow rules have been used to study localization in polycrystals [81], texture evolution [82], and interface stability in metallic multilayered composites [83] and bicrystals [34]. The effective temperature χ (with the initial value denoted by χ_0) evolves according to

$$\frac{\dot{\chi}}{e_D} = \frac{\kappa_\chi}{\mu}\left(1 - \frac{\chi}{\chi_{ss}}\right)\sum_\beta \tau^\beta\dot{\gamma}^\beta, \tag{27}$$

where κ_χ is a dimensionless parameter. The fraction $\beta = \frac{\chi}{\chi_{ss}}$ is the Taylor–Quinney factor, and quantifies the fraction of the plastic work that is converted to kinetic vibrational energy. The quantity χ_{ss} is the steady-state effective temperature, which accounts for temperature and strain rate dependence through the following equation:

$$\chi_{ss} = \chi_{ss0}\left[1 + \frac{1}{N}\sum_\alpha\left(\frac{|\dot{\gamma}^\alpha|}{\dot{\gamma}_0}\right)^{k_B T/A}\right] \tag{28}$$

where χ_{ss0} is a material parameter, $\dot{\gamma}_0$ is a reference strain rate, A is an activation energy, and N is the number of slip systems ($N = 12$ for FCC). The effect of the strain rate is introduced in Equation (28) by raising a strain rate ratio to an exponent of the form $k_B T/A$, following similar choices made in [34,81–84]. The dislocation density (with the initial value denoted by ρ_0) evolves according to the equation

$$\dot{\rho}^\alpha = \frac{\kappa^\alpha_\rho}{a^2}\frac{\tau^\alpha\dot{\gamma}^\alpha}{\mu}\left(1 - \frac{\rho^\alpha}{\rho^\alpha_{ss}}\right), \tag{29}$$

where a is a length scale characterizing the critical distance of dislocation interaction and κ_ρ^α denotes a parameter for hardening the slip system, calculated according to

$$\kappa_\rho^\alpha = \frac{\kappa_\rho^0}{\nu_\alpha^2} \frac{\sqrt{\rho_{ss}} - \sqrt{\rho^\alpha}}{\sqrt{\rho_{ss}} - \sqrt{\rho_0}}, \tag{30}$$

where κ_ρ^0 is a material hardening parameter and ρ_{ss} is the steady-state dislocation density, which is calculated from the effective temperature through the equation

$$\rho_{ss} = \frac{1}{a^2} \exp\left(-\frac{1}{\chi}\right). \tag{31}$$

Equation (30) introduces a dependence of the hardening behavior of a slip system on the dislocation density of that slip system. The steady-state dislocation density in Equation (31) is also dependent on the deformation rate and temperature through the variable χ. The nominal material hardening is captured by κ_ρ^0. At the beginning of deformation, the post-multiplying factor $\frac{\sqrt{\rho_{ss}}-\sqrt{\rho^\alpha}}{\sqrt{\rho_{ss}}-\sqrt{\rho_0}}$ (varying between 0 and 1) equals 1 for all slip systems, as ρ^α starts with the value ρ_0 for all slip systems. It decreases as the deformation progresses and dislocation density increases. Therefore, Equation (30) captures the idea that the nature of the material associated with a slip system changes as the dislocation density on that slip system increases. In accordance with [59,60], the parameter ν_α is used to calculate κ_ρ^α as

$$\nu_\alpha = \ln\left(\frac{T_p}{T}\right) - \ln\left[\ln\left(\frac{b\rho^\alpha}{t_0|\dot{\gamma}_0^\alpha|\sqrt{\sum_\beta d^{\alpha\beta}\rho^\beta}}\right)\right], \tag{32}$$

where $\dot{\gamma}_0^\alpha$ is the total shear rate on slip system α.

The temperature then evolves according to the equation

$$\dot{T} = \frac{1}{\rho C_p}\left(\beta \sum_\alpha \tau^\alpha \dot{\gamma}^\alpha + \kappa \nabla^2 \theta\right), \tag{33}$$

where ρ is the mass density and C_p is the specific heat.

2.2. Computation

All of the simulations were performed using the FEM software ABAQUS/Standard 2023. The material model was specified by writing a UMAT subroutine. The problem domain for all the simulations (single-crystal and polycrystal) was a right circular cylinder. The axis of the cylinder was aligned with the z-axis for the single crystal simulations and aligned with the y-axis for the polycrystal simulations. In both cases, the cylinder was compressed at a constant true strain rate using the ABAQUS user subroutine DISP.

For the single-crystal simulations, the cylinder had a diameter of 6 mm and a height of 6 mm. The crystal was oriented so that the <123> direction of the crystal was initially parallel to the z-axis (the axis along which compression was applied). This was to concentrate most of the slip activity on a single slip plane. Because this asymmetry of sample orientation with respect to the slip systems for copper will not produce an axisymmetric deformation field, full three-dimensional simulations were performed. For single-crystal simulations, the cylinders were compressed at true strain rates of 0.1 s^{-1}, 2.0 s^{-1}, 3000 s^{-1}, and 4800 s^{-1}. The element C3D20T from the ABAQUS/Standard element library was chosen to mesh the cylinder domain for the single crystal simulations. These are 20-noded hexahedral thermomechanical elements that use quadratic shape functions. The mesh was comprised of 5120 such elements, and is shown in Figure 1a. This mesh was the smallest element size which would complete a simulation within a reasonable time. The temperature response from the single-crystal simulations was taken as the average of all free surface node temperatures.

For polycrystal simulations, the cylinder had a height of 19.05 mm and diameter of 12.7 mm. The cylinders were compressed at true strain rates of 10^{-3} s^{-1}, 10^{-2} s^{-1}, 10^{-1} s^{-1}, and 10^{0} s^{-1}. A Taylor model was adopted for the polycrystal simulations. Each element integration point of the Taylor model was assumed to be a composite of 500 randomly oriented grains. All of the grains had distinct randomly distributed orientations, simulating the aggregate response of a random collection of grains at each element integration point. The deformed polycrystal cylindrical samples remained cylindrical during deformation, and there is nothing about the Taylor model representation of the polycrystalline material that would break the axisymmetry. Therefore, an axisymmetric model was used to represent the experiments using two-fold symmetry on both the cylinder axis and length. The behavior of each grain was governed by the same UMAT as that used in the single-crystal simulations.

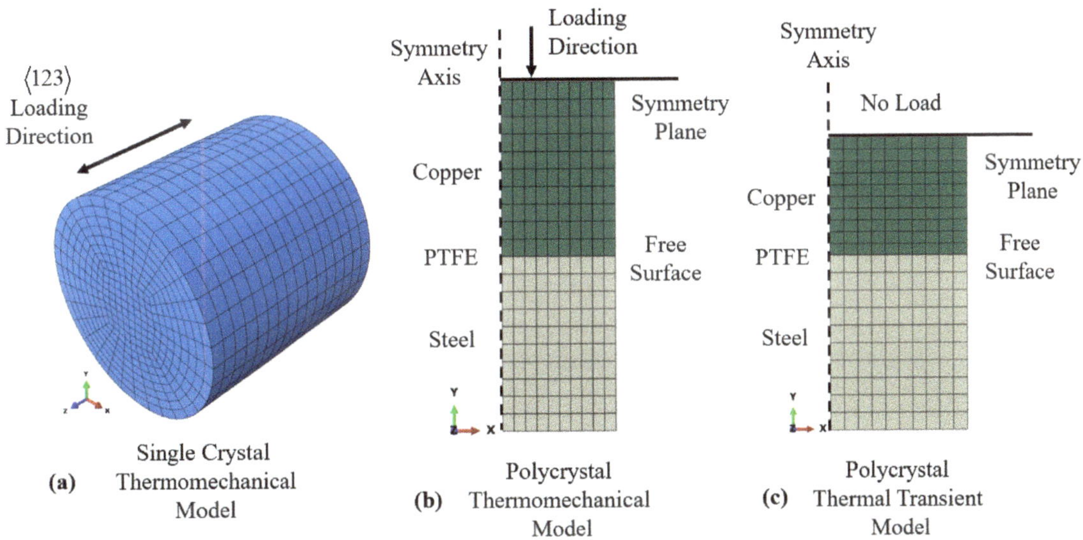

Figure 1. Meshes representing initial configurations: (**a**) single-crystal, sample size of 6mm diameter and 6 mm length; (**b**) axisymmetric and half-length thermomechanical polycrystal mesh (sample size of 12.7 mm diameter and 19.05 mm length), where the left edge is the axis of symmetry, the top edge is the sample's center, the bottom (tan) region corresponds to steel platens (9.34 mm height, 6.35 mm width), the top (green) region corresponds to copper (9.525 mm height), and a thin (0.0127 mm) PTFE layer is between the steel and copper; (**c**) axisymmetric and half-length thermal transient model.

A half-height axisymmetric model was used for the polycrystal simulations. The left edge of the rectangular mesh shown in Figure 1b is the axis of symmetry. The top edge of the mesh is the axial center of the sample. The mesh was divided into three regions for each of the represented materials (the copper sample, PTFE film lubricant, and tool steel compression platen). The green mesh region of the elements in Figure 1b represents copper. The height of this region was 9.525 mm, with its bottom on the x axis. The width of this region (as well as the other two) was 6.35 mm. The length and diameter of the physical copper samples was 19.05 mm and 12.7 mm, respectively. Just below the copper is a region of elements that is one element in height, representing the lubricating PTFE disk, which is too thin to appear in the figure. The elements in this layer were all 0.0127 mm tall. The remainder of the mesh below the PTFE layer represents the steel compression platens with a height of 9.34 mm (the thickness of the physical steel compression platens). The layer of nodes at the bottom of the copper (top of the PTFE) region was restrained from motion in the y direction, and the user-defined boundary condition was applied to the top of the copper region. Note that the PTFE layer and steel compression platens were included in the polycrystal simulation model for only the thermal transport part of the problem. Neither material played a role in the mechanical part of the problem. Only the copper region of the numerical model contributed to the calculation of the copper stress–strain curves.

A Taylor polycrystal model was used to represent the polycrystal copper material. The element CAX8T was used to mesh the rectangular area shown in Figure 1b, with 100 elements representing both the copper and steel and 10 elements representing the PTFE film. Doubling the number of elements did not lead to any noticeable difference in results. These are axisymmetric and thermomechanical eight-node elements with quadratic shape functions. The mesh consisted of 200 such elements. Every integration point in the model was assumed to be a composite of 500 randomly-oriented grains. At every integration point, the function encoding the material model for the single crystal was called 500 times with the same deformation gradient but different grain orientations. The output of these 500 function calls was averaged and returned to ABAQUS as the output of the computation at the integration point. Because the presence of both PTFE and steel in the polycrystal simulations was for heat transport only, both were modeled as isotropic elastic materials with extremely low stiffness, ensuring that they could expand freely with the copper cylinder. Realistic thermal properties were used to represent both materials. Both the copper/PTFE and PTFE/steel interface surfaces defined by connecting nodes were forced to remain fixed in the y-direction.

Following mechanical deformation with the thermomechanical mesh shown in Figure 1b, the cooling transient was simulated using a heat transfer model of the deformed sample shown in Figure 1c. The DCAX8 element was used for this problem, which was initialized for temperature at each node from the results of the thermomechanical problem for the appropriate strain rate condition. This enabled evaluation of the thermal transport film coefficient for the outer surface of the model to represent the thermal transport transient measured experimentally for each condition. Note that the simulated temperature for both the thermomechanical and heat transfer analyses was taken as the position of the thermocouple, which is the node in the top right-hand corner of each of the two axisymmetric meshes.

2.2.1. Mechanical Tangent Operators

To calculate nodal forces and displacements with a user-defined constitutive model at the end of time step $n+1$, ABAQUS/Standard requires the the user to estimate the Cauchy stress ($\underline{\mathbf{T}}$) and Jacobian matrix of the constitutive model $\mathbb{W}_{EP} = \frac{\partial \underline{\mathbf{T}}}{\partial \underline{\mathbf{e}}}$. These quantities are calculated using information from time step n, which is supplied to the user by ABAQUS. In the following, a quantity with a subscript n refers to a quantity calculated in the previous resolved time step, while subscript $n+1$ refers to a quantity involved in the current unresolved time step calculation. The relative deformation gradient $\underline{\mathbf{F}}_{\text{rel}}$ is defined as $\underline{\mathbf{F}}_{n+1} = \underline{\mathbf{F}}_{\text{rel}}\underline{\mathbf{F}}_n$. The symbol $\underline{\mathbf{e}} = \ln \underline{\mathbf{U}}$, where $\underline{\mathbf{U}}$ is the relative stretch portion of the polar decomposition of the relative deformation gradient. By ABAQUS convention, $\underline{\mathbf{T}}$ and $\underline{\mathbf{e}}$ are both defined as 6×1 column vectors with the following forms:

$$\underline{\mathbf{T}} = \begin{bmatrix} T_{11,n+1} \\ T_{22,n+1} \\ T_{33,n+1} \\ T_{12,n+1} \\ T_{13,n+1} \\ T_{23,n+1} \end{bmatrix}, \underline{\mathbf{e}} = \begin{bmatrix} e_{11,n+1} \\ e_{22,n+1} \\ e_{33,n+1} \\ e_{12,n+1} \\ e_{13,n+1} \\ e_{23,n+1} \end{bmatrix}.$$

Here, we show the method of deriving the fourth-rank tensor $\mathbb{W}_{EP} = \frac{\partial \underline{\mathbf{T}}}{\partial \underline{\mathbf{e}}}$. The 6×6 matrix required by ABAQUS can then be calculated using the following formula:

$$W_{EP} = \begin{bmatrix} W_{1111} & W_{1122} & W_{1133} & (W_{1112}+W_{1121})/2 & (W_{1113}+W_{1131})/2 & (W_{1123}+W_{1132})/2 \\ W_{2211} & W_{2222} & W_{2233} & (W_{2212}+W_{2221})/2 & (W_{2213}+W_{2231})/2 & (W_{2223}+W_{2232})/2 \\ W_{3311} & W_{3322} & W_{3333} & (W_{3312}+W_{3321})/2 & (W_{3313}+W_{3331})/2 & (W_{3323}+W_{3332})/2 \\ W_{1211} & W_{1222} & W_{1233} & (W_{1212}+W_{1221})/2 & (W_{1213}+W_{1231})/2 & (W_{1223}+W_{1232})/2 \\ W_{1311} & W_{1322} & W_{1333} & (W_{1312}+W_{1321})/2 & (W_{1313}+W_{1331})/2 & (W_{1323}+W_{1332})/2 \\ W_{2311} & W_{2322} & W_{2333} & (W_{2312}+W_{2321})/2 & (W_{2313}+W_{2331})/2 & (W_{2323}+W_{2332})/2 \end{bmatrix}. \quad (34)$$

We follow the Jacobian derivation method provided in [85]. We make the assumption that the incremental stretch is small, which is justified because the time steps taken by ABAQUS are reasonably small. This allows us to make the following mathematical simplification: $\underline{\mathbf{e}} = \ln \underline{\mathbf{U}} \approx \underline{\mathbf{U}} - 1 \implies \delta\underline{\mathbf{e}} = \delta\underline{\mathbf{U}}$. To derive the material Jacobian, we note that the Cauchy stress can be written in terms of the second Piola–Kirchhoff stress $\underline{\mathbf{S}}$:

$$\underline{\mathbf{T}}_{n+1} = \frac{1}{\det \underline{\mathbf{F}}^e_{n+1}} \left[\underline{\mathbf{F}}^e_{n+1} \underline{\mathbf{S}}^e_{n+1} \underline{\mathbf{F}}^{eT}_{n+1} \right].$$

Taking a derivative of this expression with respect to $\underline{\mathbf{e}}$ and introducing the tensors $\mathbb{X} = \frac{\partial \underline{\mathbf{F}}^e}{\partial \underline{\mathbf{e}}}$ and $\mathbb{Q} = \frac{\partial \underline{\mathbf{S}}}{\partial \underline{\mathbf{e}}}$, we obtain the following equation:

$$W_{ijkl} = \frac{1}{\det \underline{\mathbf{F}}^e} \left[X_{imkl} S_{mn} F^{eT}_{nj} + F^e_{im} Q_{mnkl} F^{eT}_{nj} + F^e_{im} S_{mn} X_{jnkl} - F^e_{im} S_{mn} F^{eT}_{nj} (X_{pqkl} F^{e-1}_{qp}) \right]. \quad (35)$$

Using the small stretch assumption made above, we write $\mathbb{X} \approx \frac{\partial \underline{\mathbf{F}}^e}{\partial \underline{\mathbf{U}}}$ and $\mathbb{Q} \approx \frac{\partial \underline{\mathbf{S}}}{\partial \underline{\mathbf{U}}}$. The following steps, taken from [85], are used to compute the tangent modulus $\mathbb{W}_{EP}$. This symbol is written in blackboard bold font to emphasize that it is a fourth-order tensor and that its components are to be used to construct the actual 6×6 matrix (using Equation (34)) that is passed back to ABAQUS.

1. Compute the fourth-rank tensor $\mathbb{L}$:

$$L_{ijkl} = F^{eT}_{ik,n} U_{lm,n+1} F^e_{mj,n} + F^{eT}_{im,n} U_{mk,n+1} F^e_{lj,n}. \quad (36)$$

2. Compute the elastic stiffness tensor ($\mathbb{C}_c$ denotes the elastic stiffness tensor in the crystal frame). To calculate the elastic stiffness tensor for the current grain orientation, the rotation tensor from the reference orientation to the current orientation is denoted by $\underline{\mathbf{Q}}$, and is used to calculate the elastic stiffness tensor using the expression $C_{ijkl} = Q_{ip} Q_{jq} Q_{kr} Q_{ls} C_{c,pqrs}$.
3. Use the quantities from steps 1 and 2 to calculate the fourth-rank tensor $\mathbb{D}$:

$$D_{ijkl} = \frac{1}{2} C_{ijmn} L_{mnkl}. \quad (37)$$

4. For each slip system α, use the Schmid tensor $\mathbf{S}^\alpha = \mathbf{s}^\alpha \otimes \mathbf{m}^\alpha$ to compute the fourth-rank tensors $\mathbb{G}^\alpha$ and $\mathbb{J}^\alpha$ and the second-rank tensor $\underline{\mathbf{B}}^\alpha$:

$$G^\alpha_{mnkl} = L_{mpkl} S^\alpha_{pn} + L_{pnkl} S^\alpha_{pm}, \quad (38)$$

$$J^\alpha_{ijkl} = \frac{1}{2} C_{ijmn} G^\alpha_{mnkl}, \quad (39)$$

$$B^\alpha_{ij} = \frac{1}{2} \Delta t \frac{\partial \dot{\gamma}^\alpha_n}{\partial \tau^\alpha} \left(S^\alpha_{ij} + S^\alpha_{ji} \right). \quad (40)$$

5. Compute the fourth-rank tensors $\mathbb{K}$ and $\mathbb{Q}$, which requires summing over all slip systems:

$$K_{ijkl} = I_{ijkl} + \sum_{\alpha} C^{\alpha}_{ij} B^{\alpha}_{kl}, \tag{41}$$

$$Q_{ijkl} = K^{-1}_{ijmn}\left(D_{mnkl} - \sum_{\alpha} \dot{\gamma} \Delta t J^{\alpha}_{mnkl}\right). \tag{42}$$

6. Compute the following quantities:

$$\begin{aligned} R^{\alpha}_{ij} &= B^{\alpha}_{kl} Q_{klij}, \\ X_{ijkl} &= R_{ik}\left(F^{e}_{lj}(t_n) - F^{e}_{lp}(t_n) \sum_{\alpha} \dot{\gamma}^{\alpha} \Delta t S^{\alpha}_{pj}\right) - R_{im} U_{mn} F^{e}_{np}(t_n) \sum_{\alpha} R^{\alpha}_{kl} S^{\alpha}_{pj}. \end{aligned} \tag{43}$$

The R_{ik} appearing in the computation of $\mathbb{X}$ represents the rotation components of the relative deformation gradient.

7. As the final step, compute the Jacobian for the mechanical behavior of the material:

$$W_{ijkl} = \frac{1}{\det \underline{\mathbf{F}}^{e}}\left[X_{imkl} S_{mn} F^{e}_{jn} + F^{e}_{im} Q_{mnkl} F^{e}_{jn} + F^{e}_{im} S_{mn} X_{jnkl} - F^{e}_{im} S_{mn} F^{e}_{jn}(X_{pqkl} F^{e-1}_{qp})\right]. \tag{44}$$

2.2.2. Thermal Tangent Operators

ABAQUS/Standard forms the tangent matrix for the thermal transport problem using certain quantities provided by the user. The calculation methodology for these quantities is shown in this section. Several derivatives are important in these calculations. The first of these is the variation of $\dot{\gamma}^{\alpha}$ with respect to τ^{α}:

$$\begin{aligned} \frac{\partial \dot{\gamma}^{\alpha}}{\partial \tau^{\alpha}} = &\frac{p q \rho^{\alpha} l^{\alpha} b T_p}{s^{\alpha} t_0 T} \exp\left[-\frac{T_p}{T}\left(1 - \left(\frac{\tau^{\alpha} - s_l}{s^{\alpha}}\right)^{p}\right)^{q}\right] \times \\ &\left[\left(1 - \left(\frac{\tau^{\alpha} - s_l}{s^{\alpha}}\right)^{p}\right)^{q-1}\right]\left[\left(\frac{\tau^{\alpha} - s_l}{s^{\alpha}}\right)^{p-1}\right]. \end{aligned} \tag{45}$$

The above must be multiplied by the sign of τ^{α}. The derivative of $\dot{\gamma}^{\alpha}$ with respect to the temperature is provided by

$$\begin{aligned} \frac{\partial \dot{\gamma}^{\alpha}}{\partial T} = &\frac{\rho^{\alpha} l^{\alpha} b}{t_0} \frac{T_p}{T^2} \exp\left[-\frac{T_p}{T}\left(1 - \left(\frac{\tau^{\alpha} - s_l}{s^{\alpha}}\right)^{p}\right)^{q}\right] \times \\ &\left[\left(1 - \left(\frac{\tau^{\alpha} - s_l}{s^{\alpha}}\right)^{p}\right)^{q}\right]. \end{aligned} \tag{46}$$

Using these derivatives we can calculate the following four quantities, which are required by ABAQUS for calculating the tangent operators connected to the thermal part of the thermomechanical problem:

- r_{pl}: The portion of the mechanical work done on the material that is transformed to heat. It is calculated here as a portion of the plastic work done on the material. It is calculated as $\beta \sum_{\alpha} \tau^{\alpha} \dot{\gamma}^{\alpha}$, where β is the Taylor–Quinney factor, α denotes the index for one of 12 slip systems in the metal, τ^{α} is the resolved shear stress on slip system α, and $\dot{\gamma}^{\alpha}$ is the slip rate on the same slip system.
- The variation of the Cauchy stress with respect to temperature: In this work, we assume that the thermal variation of the Cauchy stress arises from the thermal vari-

ation of the elastic moduli of the single crystal. This expression is derived from the definition of the second Piola–Kirchhoff stress:

$$\frac{\partial \underline{\mathbf{T}}}{\partial T} = \frac{1}{\det \underline{\mathbf{F}}^e} \underline{\mathbf{F}}^e \frac{\partial \underline{\mathbf{S}}}{\partial T} \underline{\mathbf{F}}^{eT}, \tag{47}$$

$$\frac{\partial S_{ij}}{\partial T} = Q_{ip} Q_{jq} Q_{ks} Q_{lr} m_{pqrs} E^e_{kl}. \tag{48}$$

The temperature dependence of the elastic moduli, and consequently of the Cauchy stress, is obtained from Equation (5).

- The variation of r_{pl} with respect to temperature:

$$\begin{aligned} r_{pl} &= \beta \sum_{\alpha} \tau^{\alpha} \dot{\gamma}^{\alpha} \\ \frac{\partial r_{pl}}{\partial T} &= \beta \sum_{\alpha} \left[\frac{\partial \tau^{\alpha}}{\partial T} \dot{\gamma}^{\alpha} + \tau^{\alpha} \frac{\partial \dot{\gamma}^{\alpha}}{\partial T} \right], \end{aligned} \tag{49}$$

$$\frac{\partial \tau^{\alpha}}{\partial T} = \frac{\partial \underline{\mathbf{S}}}{\partial T} : (\mathbf{s}^{\alpha} \otimes \mathbf{m}^{\alpha}). \tag{50}$$

- The variation of r_{pl} with respect to the strain:

$$\begin{aligned} \frac{\partial r_{pl}}{\partial E_{ij}} &\approx \frac{\partial r_{pl}}{\partial U_{ij}} \\ &= \beta \sum_{\alpha} \left[\frac{\partial \tau^{\alpha}}{\partial U_{ij}} \dot{\gamma}^{\alpha} + \tau^{\alpha} \frac{\partial \dot{\gamma}^{\alpha}}{\partial U_{ij}} \right], \end{aligned} \tag{51}$$

$$\frac{\partial \tau^{\alpha}}{\partial U_{ij}} = \frac{\partial \underline{\mathbf{S}}}{\partial U_{ij}} : (\mathbf{s}^{\alpha} \otimes \mathbf{m}^{\alpha}). \tag{52}$$

The derivatives of $\underline{\mathbf{S}}$ and $\dot{\gamma}^{\alpha}$ with respect to $\underline{\mathbf{U}}$ are denoted by the symbols $\mathbb{Q}$ and $\underline{\mathbf{J}}$, respectively, and appear in the derivation of the mechanical Jacobian.

2.2.3. Mechanical and Thermal Quantities for the Taylor Polycrystal Model

The quantities described in Sections 2.2.1 and 2.2.2 are those calculated for describing the behavior of single crystals. For the Taylor polycrystal, these quantities are calculated for each of 500 grain orientations assigned to each integration point of the Taylor model aggregate. Each of the 500 calculations assumes the deformation gradient assigned to the integration point by ABAQUS. From the UMAT of the Taylor model, the mean of these 500 quantities is returned to the ABAQUS global equation solver:

- Cauchy stress.
- Tangent stiffness moduli in Equation (34).
- Rate of thermal energy generated.
- Derivative of the Cauchy stress with respect to the temperature.
- Derivative of the rate of thermal energy generation with respect to the temperature.
- Derivative of the rate of thermal energy generation with respect to the strain.

2.3. *Experiment*

2.3.1. Single-Crystal Experiments

Stress–strain and temperature evolution data were respectively taken from Figures 2b and 5a of [13] for comparison to the single-crystal simulations.

The samples in these experiments were copper right-circular cylinders with diameter 6 mm and height 4–6 mm. They were machined from stock cylinders which had the <123> direction of the single crystal oriented parallel to the axis of the cylinder. The reason for this alignment was to concentrate the plastic slip, at least at the onset of plastic flow, onto a single slip system.

The single-crystal samples were then compressed in the quasi-static regime using an MTS system (Eden Prairie, MN, USA) under displacement control and in the dynamic regime using Kolsky bars of diameter 12.7 mm made of C300 maraging steel. For the quasi-static regime, the temperature of the sample was measured by recording thermal images of the sample as it was being compressed using a thermal imaging camera. For the dynamic regime, thermal detection was carried out using a liquid nitrogen-cooled MCT (mercury–cadmium–telluride) detector. This detector measures temperatures over an area of 100 μm × 100 μm.

To minimize issues related to surface roughness and surface oxidation, the samples were thermally conditioned by preheating to 200 °C for 1 min to ensure that all samples had similar surface conditions.

2.3.2. Polycrystal Experiments

The polycrystal experiments consisted of uniaxial stress compression tests in which the stress, strain, and temperature evolution of the sample were measured simultaneously. Temperature measurements were taken using thermocouples (J type) welded to the outside center of the sample. The samples were right-circular cylinders machined from half-inch round bar stock of OFHC polycrystalline copper. The slenderness ratio of all the samples was equal to or less than 1.5. As such, the lengths of the samples were less than or equal to 0.75 inches (19.05 mm) and their diameters were 0.5 inches (12.7 mm). After machining, the samples were annealed in an Argon-flushed furnace by holding the samples at 800 °C for 1 hour and allowing the samples to cool naturally in the same furnace. Thermocouple beads were welded onto the annealed samples using a thermocouple spot welder. Displacement of the sample height reduction was measured using an MTS extensometer mounted directly to the radial surface of cylindrical high-strength steel compression platens 6 inches in diameter and 2 inches thick. These experiments were conducted using a high-capacity Instron servo-hydraulic test system with displacement control on the extensometer to impose a constant true strain rate on the sample for the duration of the experiment. The flat ends of the sample were lubricated for the duration of the compression test by inserting circular pieces of PTFE film of thickness 0.0127 mm and diameter 0.5 inches (the same as the samples).

2.4. Material Parameter Evaluation

The starting point for material parameter evaluation was the list provided in Table 2 of [78]. However, the provided material parameters cannot be used as-is, since the current model differs from [78] in Equations (22) and (28). The list of material parameters used here is provided in Table 3. All parameters are used for both single- and polycrystal calculation except where different polycrystal values are provided in parentheses. The elastic constants $C_{11,0}$, $C_{12,0}$, and $C_{44,0}$ characterize FCC Copper at 0 K. As mentioned in describing Equation (5), it is assumed that the elastic moduli decrease linearly with the temperature. These variation parameters are m_{11}, m_{12}, and m_{44}. These elastic moduli and their temperature variation parameters were estimated from [86]. For a given temperature T, the three elastic moduli are provided by

$$\begin{aligned} C_{11}(T) &= C_{11,0} + m_{11}T, \\ C_{12}(T) &= C_{12,0} + m_{12}T, \\ C_{44}(T) &= C_{44,0} + m_{44}T. \end{aligned} \tag{53}$$

Using these elastic constants, the elastic stiffness tensor at a temperature T of the single-crystal orientation used as reference is

$$\begin{aligned} C_{c,ijkl} &= C_{11}(T) \text{ if } i = j = k = l, \\ \text{otherwise, } C_{c,ijkl} &= C_{12}(T)\delta_{ij}\delta_{kl} + C_{44}(T)(\delta_{ik}\delta_{jl} + \delta_{il}\delta_{kj}). \end{aligned} \tag{54}$$

The values of the matrix entries $a^{\alpha\beta}$ used in Equation (23) depend on relative orientations of the slip normals and slip directions of the slip system corresponding to α and β:

- $a^{\alpha\alpha} = a_{\text{self}}$: Captures the interaction of a dislocation with other dislocations in the same slip system.
- $a^{\alpha\beta} = a_{\text{copl}}$, if $\mathbf{m}^{\alpha} = \mathbf{m}^{\beta}$ and $\mathbf{s}^{\alpha} \neq \mathbf{s}^{\beta}$: This corresponds to the dislocation interactions that lead to the formation of dipoles.
- $a^{\alpha\beta} = a_{\text{hirth}}$, if $\mathbf{m}^{\alpha} \neq \mathbf{m}^{\beta}$ and $\mathbf{s}^{\alpha} \perp \mathbf{s}^{\beta}$: This corresponds to the formation of a Hirth lock.
- $a^{\alpha\beta} = a_{\text{colli}}$, if $\mathbf{m}^{\alpha} \neq \mathbf{m}^{\beta}$ and $\mathbf{s}^{\alpha} = \mathbf{s}^{\beta}$: This corresponds to collinear interactions between dislocations.
- $a^{\alpha\beta} = a_{\text{gliss}}$, if $\mathbf{m}^{\alpha} \neq \mathbf{m}^{\beta}$ and $\mathbf{s}^{\alpha} \neq \mathbf{s}^{\beta}$: This corresponds to the formation of a glissile junction.
- $a^{\alpha\beta} = a_{\text{lomer}}$ for other configurations of slip normals and slip directions: Corresponds to the formation of a Lomer lock.

Table 3. List of material parameters for single-crystal simulations. For the polycrystal simulations, all values are the same except for those quantities in parentheses. The symbol ρ_0 denotes the initial dislocation density on each slip system.

Material Parameter Symbol	Value
ρ	8960 Kg/m^3
C_p	380 J/Kg-K
K_T	394 W/m-K
$C_{11,0}$	179,500 MPa
$C_{12,0}$	126,400 MPa
$C_{44,0}$	82,500 MPa
m_{11}	−36.3 MPa/K
m_{12}	−16.4 MPa/K
m_{44}	−25.7 MPa/K
α_T	2.75 (2.0)
b	2.57×10^{-7} mm
ρ_0	1×10^5 mm^{-2} (2×10^5 mm^{-2})
χ_0	0.20 (0.19)
$\chi_{ss,0}$	0.25 (0.115)
a_{self}	0.122
a_{copl}	0.122
a_{hirth}	0.070
a_{colli}	0.625
a_{gliss}	0.137
a_{lomer}	0.122
e_D	1.0
t_0	1×10^{-12} s
T_p	40,800 K
k_B	1.38×10^{-23} J/K
a	$10b$
κ_ρ^0	60 (68)
κ_χ	3.3 (4.3)
$\dot{\gamma}_0$	10^5 s^{-1}
k_{copl}	15
k_{inter}	200
A	17.1×10^{-19} J
s_l	0 MPa
p	0.33
q	1.66

The values a_{self}, a_{copl}, a_{hirth}, a_{colli}, a_{gliss}, and a_{lomer} were calculated from the DD (Dislocation Dynamics) simulations in [37,87,88]. According to [80], the parameters $d^{\alpha\beta}$ needed in Equation (25) can be calculated using $a^{\alpha\beta}$, k_{inter} and k_{copl}, using the expressions

$d^{\alpha\beta} = \frac{a^{\alpha\beta}}{k^2_{\text{inter}}}$ for intersecting slip systems and $d^{\alpha\beta} = \frac{a^{\alpha\beta}}{k^2_{\text{copl}}}$ for self-interaction and coplanar slip systems. The same approach is followed in the present work.

The material parameter A introduced in this work is an energy barrier term, and controls the temperature and rate sensitivity of χ_{ss}. Because the slip rate on any slip system is usually less than the reference slip rate $\dot{\gamma}_0$, a lower value of A corresponds to a higher value of χ_{ss}. Because χ is a measure of the disorder of the system, a higher value of χ_{ss} corresponds to a higher value of the saturation stress of the material. Therefore, a lower value of A corresponds to a strong material. The symbol χ_0 denotes the initial value of χ.

Saturation flow stress increases with strain rate [81,84]. The state variable χ is a measure of the configurational disorder in the system. Therefore, its saturation value χ_{ss} should decrease with temperature. From [84], it is apparent that saturation flow stress decreases as temperature increases. For a FCC material, the saturation flow stress is strongly controlled by the saturation dislocation density. The dislocation content is a measure of the configurational disorder of a system. Therefore, based on experimental data on the saturation value of flow stress, we can make inferences about the saturation value of the configurational disorder of the system. This notion is captured by Equation (28) for variation of χ_{ss}.

Several of the material parameters in this work are different from the material parameters in [78] such as κ_χ, κ^0_ρ, k_{copl}, k_{inter}, and α_T. Differences in the first two quantities are related to differences in the effective temperature and dislocation evolution equations. The remaining three may be related to manufacturing differences in the materials being simulated.

Both PTFE and tool steel were used in the polycrystal simulations and represented for thermal transport characteristics only. For PTFE, $\rho = 2200$ Kg/m^3, $K_T = 0.25$ W/m-K, and $C_p = 1500$ J/Kg-K; For steel, $\rho = 7850$ Kg/m^3, $K_T = 45$ W/m-K, and $C_p = 420$ J/Kg-K.

2.5. Free Surface Heat Transfer Coefficient Parameter Evaluation

The film coefficient for air governing the transport of heat across the curved surface of the cylinder is assumed to be equal to 10.0 W/m^2-K for the single crystal simulations. This is consistent with prior computational results for room-temperature experiments conducted in air. There is uncertainty in the value of the film coefficient, as seen in [77], which uses a value of 3.0 W/m^2-K, and [89], which shows a wide range of values for the film coefficient (up to 13.0 W/m^2-K). For the polycrystal experiments presented here, each sample was deformed to the target strain at a constant true strain rate and held under load while measuring the thermal transient cooling. This allowed for determination of the film coefficient for the conditions of these experiments. A value of 0.18 W/m^2-K was determined to be suitable using results from the 0.1 s^{-1} experiments, as demonstrated in Figure 2.

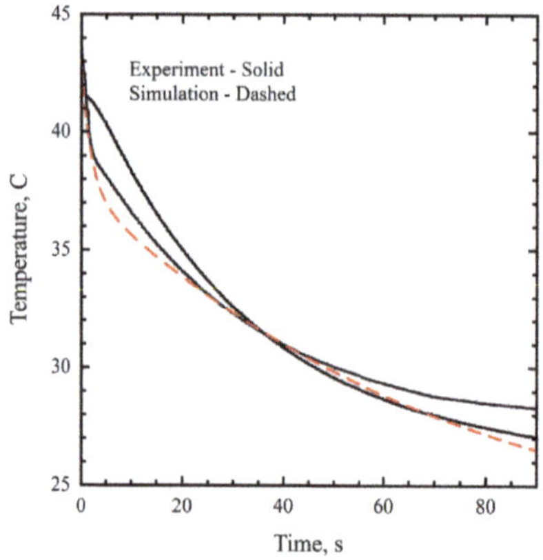

Figure 2. Temperature transient results used to determine a numerical film coefficient for the air–solid interface of 0.18 W/m^2-K. The temperature measured by the thermocouples and simulated temperature are taken from the center surface node.

3. Results

A comparison of the experimental and simulation stress–strain curves for single-crystal compression is shown in Figure 3a, demonstrating reasonable agreement between simulation and experiment. There is divergence between the simulation and experimental results for the two lower deformation rates at large strain values. Because the deformed sample shapes were not reported in [13], it is difficult to diagnose the observed softening response. As the simulations suggest strong sample distortion with deformation, it is possible that strong shear localization occurred in the experiments but was not captured by the mesh used in the simulations. Similarly, for the polycrystal case, Figure 4a shows a reasonable representation of the experimental stress–strain curves by the simulations. However, the deformation rate sensitivity of the material in the simulations is too strong in comparison with the experiments.

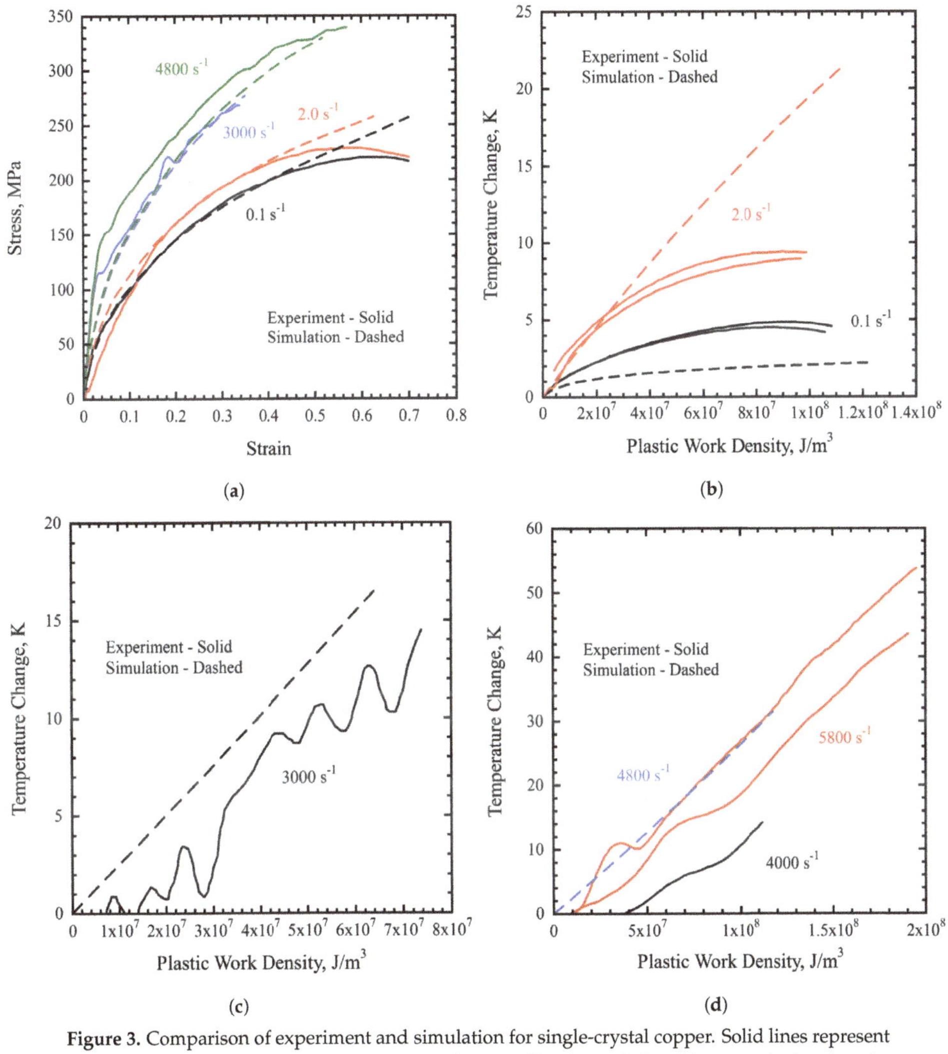

Figure 3. Comparison of experiment and simulation for single-crystal copper. Solid lines represent experiments and dashed lines represent simulations. Plastic work densities are calculated as the area under the stress–strain curve. The simulation temperature change was taken from the center surface node. (**a**) Single crystal stress–strain; (**b**) 0.1 s^{-1} and 2.0 s^{-1} temperature change; (**c**) 3000 s^{-1} temperature change; (**d**) 4000 s^{-1} and 5800 s^{-1} temperature changes.

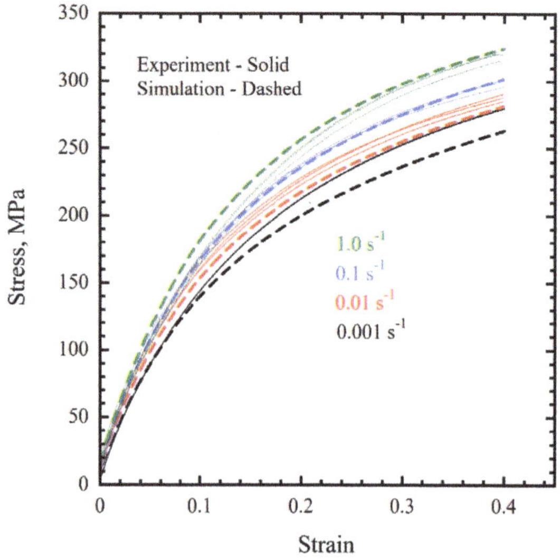

(a)

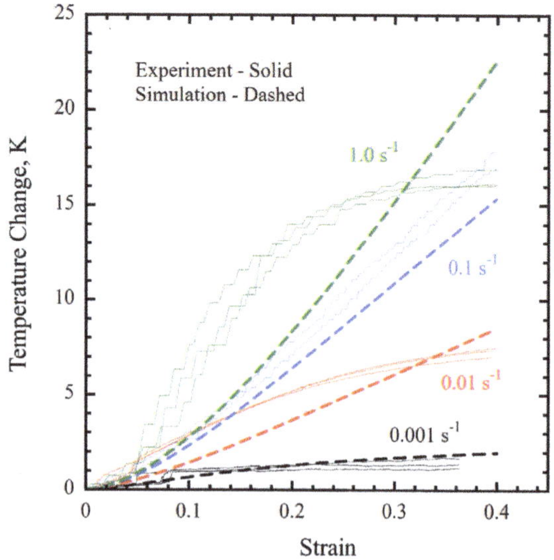

(b)

Figure 4. Comparison of polycrystal experiment and simulation results: (**a**) stress-strain curve for polycrystal copper and (**b**) temperature evolution for polycrystal copper. The experimental temperature was measured by thermocouples, while the simulated temperature was taken from the center surface node.

As expected, the single-crystal temperature curves in Figure 3b–d show high dependence of the temperature evolution behavior on the strain rate. The images corresponding to the two higher strain rate simulations (3000 s^{-1} and 4800 s^{-1}) are shown in Figure 3c,d respectively. The polycrystal temperature curves in Figure 4b for the simulations show a reasonably accurate prediction of the thermocouple-measured temperature evolution in the experiments. However, a variation in the curvature of each result can be seen as a function of evolved strain. There is qualitative agreement in the strain rate sensitivity of the thermal behavior of the model, and it is again in reasonably good agreement with the experimental results. The difference in the character of the simulation and experimental temperature evolution curves suggests that some uncertainty remains in the temperature measurement as well as in our physical understanding of this thermodynamic process. Nonetheless, order-of-magnitude agreement can be seen with the simple theory used here.

The deformed shapes of the domains of the single crystal simulations for the four different strain rates are shown in Figure 5 (showing von Mises stress) and Figure 6 (showing absolute temperature). The deformed shape of the domain and contour plots for the von Mises stress and temperature show the anisotropy and heterogeneity of the deformation. Anisotropy is shown by the cross-section of the cylindrical domain, which has transformed from a circular to an elliptical shape. Heterogeneity is shown by the shearing visible in the deformed shape domains, made clearer by the patterns of the contour plots.

The predicted evolution of the Taylor–Quinney factor for both single-crystal models (Figure 7a) and polycrystal models (Figure 7b) demonstrates mild influence of the strain rate. There is clearly a strain effect on the Taylor–Quinney factor, with the Taylor–Quinney factor increasing with strain as expected. A trend appears with respect to the strain rate, however, where the Taylor–Quinney factor is lower for higher strain rates at the beginning of the simulation. With deformation, the evolution of the Taylor–Quinney factor for the higher strain rates catches up to and then exceeds the values for lower strain rates. The evolution of χ and χ_{ss} can be seen in Figure 7c,d, respectively.

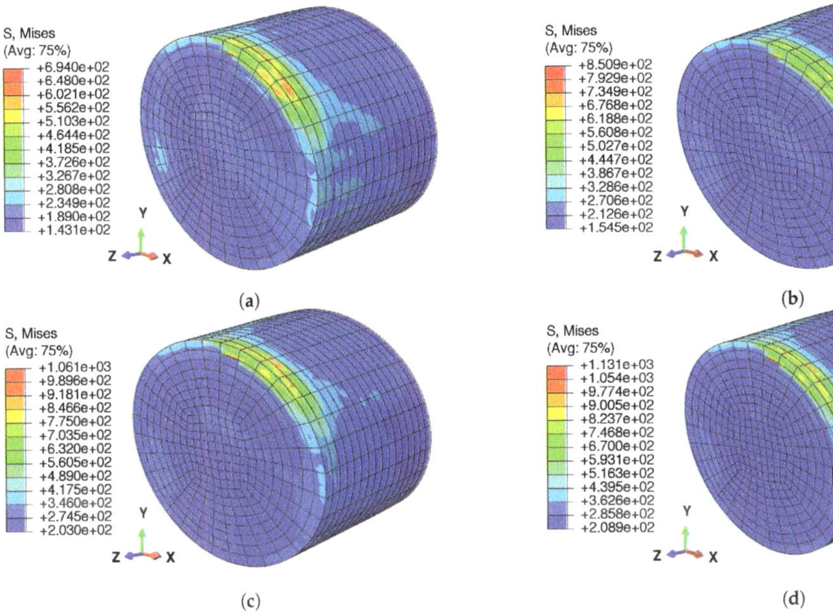

Figure 5. Deformed shapes of single-crystal simulations, showing the von Mises stress for different strain rates: (**a**), $\dot{\epsilon} = 0.1\ s^{-1}$; (**b**) $\dot{\epsilon} = 2\ s^{-1}$; (**c**) $\dot{\epsilon} = 3000\ s^{-1}$; (**d**) $\dot{\epsilon} = 4800\ s^{-1}$. All images were taken at an axial compressive strain of 0.35. The von Mises stress, also known as equivalent stress, is defined as $\sigma_{VM} = \sqrt{3/2\underline{\mathbf{T}}' : \underline{\mathbf{T}}'}$.

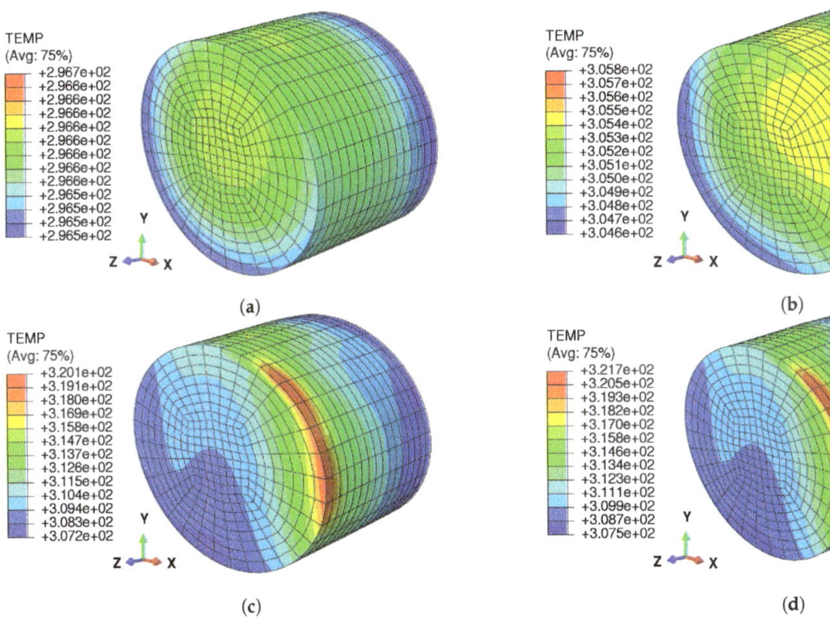

Figure 6. Deformed shapes of single-crystal simulations showing the temperature for different strain rates: (**a**) $\dot{\epsilon} = 0.1\ s^{-1}$; (**b**) $\dot{\epsilon} = 2\ s^{-1}$; (**c**) $\dot{\epsilon} = 3000\ s^{-1}$; (**d**) $\dot{\epsilon} = 4800\ s^{-1}$. The initial temperature for all simulations was 295 K. All images were taken at an axial compressive strain of 0.35.

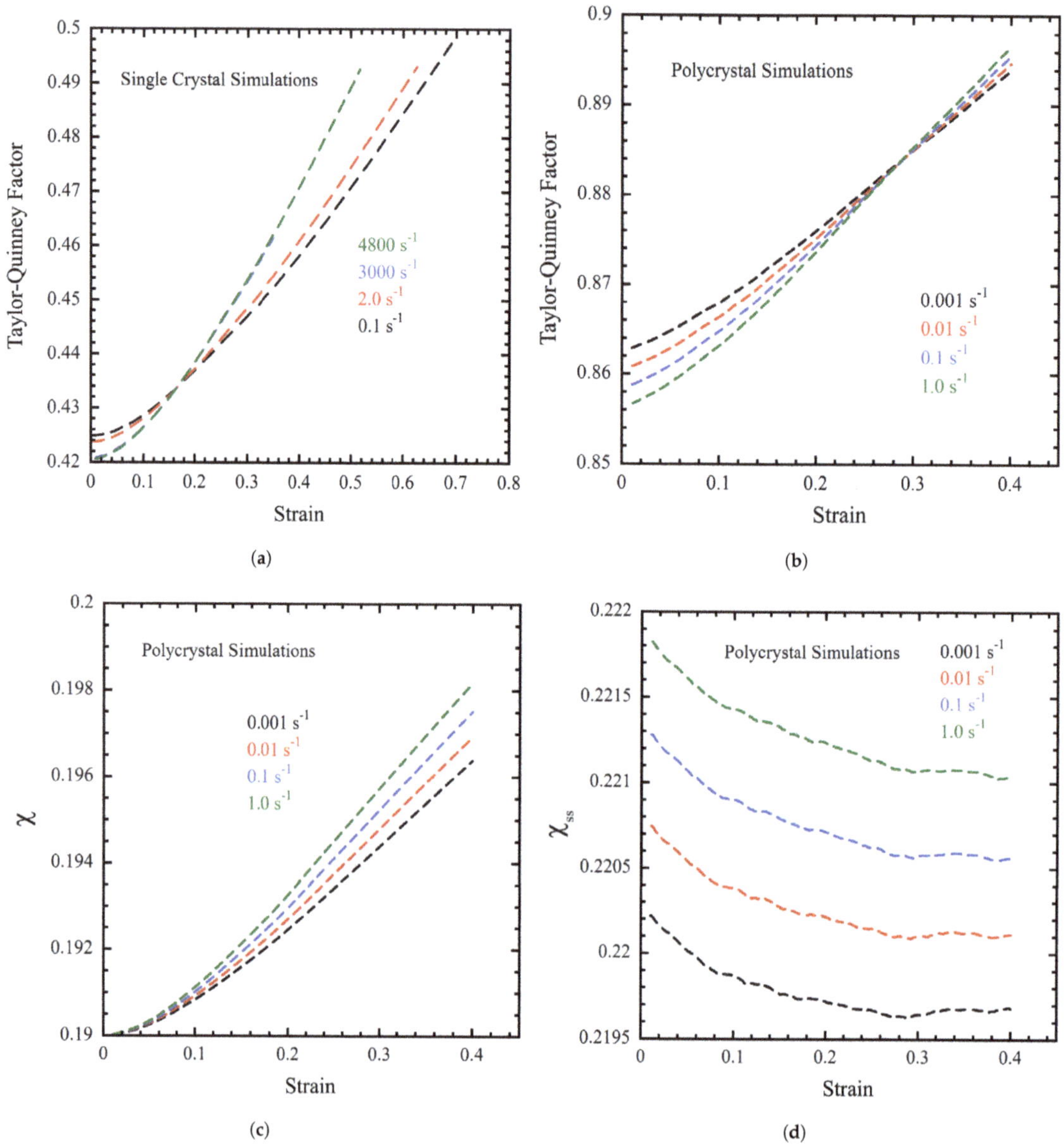

Figure 7. Evolution of Taylor-Quinney state variables for single and polycrystal simulations: (**a**) single-crystal Taylor-Quinney factor; (**b**) polycrystal Taylor-Quinney factor; (**c**) polycrystal effective temperature; (**d**) polycrystal steady-state effective temperature.

4. Discussion

The finite deformation theory presented here has demonstrated some success in representing the experimental results. It represents dislocation interactions which are specific to the interacting slip system dislocations for face-centered cubic materials, an approach motivated by the study of these interactions with discrete dislocation dynamics and molecular dynamics calculations [36–38,43,80,87,88]. This is represented by a tensorial interaction within the classical Taylor expression for interaction resistance to glide. The evolution of material state in this thermodynmaic theory is also driven by plastic power, as opposed to simply the kinematic quantity of plastic slip or plastic strain rate [59–62,73,75,76]. The partitioning of plastic power into configurational and kinetic–vibrational categories affords the opportunity to suggest the coupling between plastic power and the proportion going to increasing the temperature of the material. Especially for high deformation rate loading conditions, the change in material temperature during deformation cannot be ignored; our physical mechanistic understanding of this process is quite weak. It follows that the Taylor–Quinney factor expression used here is only suggested by thermodynamics restrictions as one of the most simple admissible approaches. While such an expression is not void of physical insight, it represents a simple expression given the complexity of dislocation mechanics and the partitioning of energy in such material systems. Certainly, atomistic tools are an important avenue for gaining more physical insight into thermodynamics and physics of dislocation motion. There is also a possibility of comparing the results from the current theory with molecular dynamics simulation results, considering the advanced potential of recent developments [90]. However, such a direct comparison is difficult, as there is a large difference between the high strain rates typically necessary for MD calculations and the significantly lower strain rates of achievable polycrystal experiments and simulations. The metadynamics method developed by [91,92] may also be able to sample the potential energy landscape much more efficiently and extend the feasible strain rate regime down by many orders of magnitude. The continuum model in this work uses an Arrhenius-type flow rule based on thermal activation, and is not currently formulated for phonon drag regimes of dislocation velocity, though this could certainly be added as required [40]. As a result, the mechanical response of the continuum model is expected to match that of molecular dynamics simulation for face-centered cubic materials. The possibility of agreement of the thermal response is less certain; this flow rule physically describes the mechanical behavior of mobile dislocations but not their thermal coupling, as the exact physical basis of this coupling is not understood. However, there is much physical insight to be gained by controlled atomistic studies of this process, given the wealth of results which have already been produced.

From the polycrystal temperature evolution curves in Figure 4b, we see qualitative agreement in the strain rate behavior between experiment and simulation. The differences in curvature between the curves raise questions about the estimation of the thermal power generated in the model and the thermal energy lost to the surroundings through the boundary. It is also reasonable to expect that deformation may bring about changes in the thermocouple contact with the sample, as the bead weld area on the sample is deforming as well. Again, this highlights the challenges associated with measuring temperature for large deformation conditions, especially for high deformation rates. For the Taylor polycrystal model, the thermal power at an element integration point was calculated by averaging over the thermal power generation of all 500 single-crystal orientations active at the element integration point. What is missed in this simple homogenization approach is intergranular interactions with deformation and corresponding inhomogeneous deformation field at the single-crystal and lower length scales. This is thought to produce a greater degree of local plastic deformation, which may be missed by homogenization. On the other hand, the Taylor homogenization technique is known to be an upper bound for stress response. However, as the single-crystal and polycrystal stress–strain response is well represented, the homogenized energy should be consistent with that imposed experimentally. Then, the question is to what extent the local inhomogeneous deformation alters the energy partitioning process. Although the present results reasonably represent the experimental data, they do not address questions of the role this may play in the thermodynamic process and observations of temperature change. This has practical implication for how to approach these questions, as combined thermomechanics calculations are costly and performing sample-sized simulations (even very small samples) with realistic microstructures would be quite resource challenging even for advanced computational architectures.

The calculation of the single-crystal Taylor–Quinney factor for Figure 7a was performed by taking the average of the Taylor–Quinney factor of all the integration points in the 3D cylindrical model. This was necessary due to the highly anisotropic nature of the deformation field, as shown in Figures 5 and 6. This is also consistent with the experiments due to using the entire field of thermal imaging [13]. For the polycrystal model, the deformation field is uniform, and as such the Taylor–Quinney factor was determined by averaging over all 500 crystals at the integration point nearest to the top right-hand corner node, which corresponded to the thermocouple location on the experimental samples. The polycrystal Taylor–Quinney factor evolution is provided in Figure 7b. In general, the Taylor–Quinney factor increases with deformation; however, the actual values in Figure 7a,b show that there is not much change in the predicted value over the considered range of deformations. It is, however, interesting to note that the difference in magnitude of the Taylor–Quinney factor between the single-crystal and polycrystal predictions is larger for the polycrystal by a factor of two. As mentioned earlier, the orientation of the single crystal was chosen to ensure that something close to single slip could be achieved. This produced a flow stress of 200 MPa at a strain of -0.4, in comparison with 300 MPa for the polycrystal samples for a strain rate of 0.1 s^{-1} for both datasets. The polycrystal temperature prediction for that strain rate represents the experimental curves rather well. The single-crystal prediction of temperature for that strain rate is approximately half of the experimental magnitude, indicating that perhaps the predicted single-crystal Taylor–Quinney factor is too low. The single-crystal temperature significantly overpredicts that of the experiment for a strain rate of 2.0 s^{-1}. It should be noted that the boundary conditions for the single-crystal experiments were not reported by [13], and consequently those employed here for modeling these results may not be quite correct.

Although mechanical size effects are becoming more prominent features in advanced crystal mechanics theories, this was not considered in the present work. As already discussed, coupled thermomechanics problems have an implicit size effect which appears in the thermal transport component of the problem. Mechanical size effects may be introduced in various ways. Size effects related to plasticity may be introduced by calculating the dislocation density at an element integration point from the plastic part of the deformation gradient. This would necessarily require calculation of the gradient of the plastic part of the deformation gradient, and would introduce a size effect into the continuum model. Size effects related to the elastic part of the response can be introduced by assuming the material to be a micromorphic continuum. These are additional valid elements of physics which may contribute to this thermomechanical problem, and should be pursued; however, as alluded to already, adding more length scales to this study would also add substantial computational expense, and must be done with specific physics questions in mind. Given our significant remaining questions about the thermomechanical response of materials and partitioning of mechanical power, we consider it premature to include mechanical size effects at the present time, as there are higher-priority outstanding questions.

The results presented here suggest that there remains some uncertainty in the accurate measurement of temperature for the loading conditions considered here. As demonstrated in Figure 4b, the temperature evolution curves demonstrate differences in character between the different strain rate conditions. While there is good repeatability between the duplicate experiments at the same loading conditions, the curvature of the curves (with the exception of the 0.1 s^{-1} conditions, which compare well with the simulations) are not understood at present. It is also clearly important to quantify the thermal boundary conditions, which remains an opportunity for further development. This will again require close examination in the future. This is clearly an area for opportunity, as measured temperature is a critical state variable in this area of study.

5. Conclusions

We have presented an integrated study of the thermomechanical behavior of copper which includes new theoretical, computational, and experimental results. The primary variable of interest in this work is the strain rate (or time, in the context of thermal transport), which spanned the range from 10^{-3} to 5×10^{3} s^{-1}. All conditions were initially at room temperature. The theory and corresponding thermomechanical simulations demonstrated reasonably good agreement with the experimental results for both stress and temperature evolution, given the simplicity of the model and that there is much that we do not yet understand about the coupled thermomechanics of plastic deformation, mechanistic energy partitioning, and thermal energy conversion. The theory is formulated with large deformation kinematics, is thermodynamically consistent, and accounts for plastic power partitioning to stored energy of cold work within the dislocation structure and thermal energy of the material system. The theory accounts for advanced dislocation interaction via the continuum theory and plastic power driving structural evolution, rather than kinematical quantities such as slip rate. The dislocation interaction energy is not yet directly represented, instead being implicitly represented through material hardening as a result of dislocation interactions. Configurations of dislocation structure development such as dislocation subcells are not yet directly represented in this theory.

The single-crystal results presented here using experiments found in the literature [13] are a very important component in developing an understanding of polycrystalline thermomechanical behavior. The deformation fields at the single-crystal length scale within polycrystalline aggregates are more complex, and there does not yet exist a good way to experimentally quantify the influence of multiple grain boundary types on the deformation behavior of individual single crystals. More coupled diagnostic thermomechanical experiments performed on single crystals are needed.

The theory presented here has been used against the presented polycrystalline dataset by employing the Taylor model to homogenize the response of a representative set of 500 crystals of annealed copper with random initial crystallographic orientations. In such a treatment, each grain experiences the same imposed deformation gradient for each time step. However, depending upon its crystallographic orientation with respect the problem boundary conditions, each grain responds in a different way kinetically. The homogenized result for the simulations is then the numerical average kinetic response of each crystal, as done here, although weighted averages can also be taken. This is a reasonable first step with this new theory, and the numerical results provide a reasonable representation. Of course, intergranular interactions are an important part of the deformation response of polycrystalline metallic materials, and this remains an important question to explore.

Author Contributions: Methodology, software, investigation, data curation: N.J.S., A.P., S.D.D., J.E.R., J.C.M.L. and C.K.C.L.; software, simulation, validation, data curation, writing—original draft preparation: S.K.; conceptualization, supervision, project administration, writing—review/editing/revision, software, funding acquisition: C.A.B. All authors have read and agreed to the published version of the manuscript.

Funding: C.A. Bronkhorst acknowledges funding from NSF project number 2051355 and Army Research Laboratory contract W911NF2320073 for the Center of Extreme Events in Structurally Evolving Materials.

Data Availability Statement: The raw data supporting the conclusions of this article will be made available by the authors on request.

Conflicts of Interest: S. Kunda and C. A. Bronkhorst declare that this manuscript has not been submitted elsewhere and that there exist no financial or practical conflicts of interest which have bearing upon this work and its publication.

References

1. Taylor, G.I.; Quinney, H. The latent energy remaining in a metal after cold working. *Proc. R. Soc. Lond. Ser. A Contain. Pap. A Math. Phys. Character* **1934**, *143*, 307–326.
2. Farren, W.S.; Taylor, G.I. The heat developed during plastic extension of metals. *Proc. R. Soc. Lond. Ser. A Contain. Pap. A Math. Phys. Character* **1925**, *107*, 422–451.
3. Barton, N.; Benson, D.; Becker, R. Crystal level continuum modelling of phase transformations: The $\alpha \leftrightarrow \epsilon$ transformation in iron. *Model. Simul. Mater. Sci. Eng.* **2005**, *13*, 707. [CrossRef]
4. Feng, B.; Bronkhorst, C.A.; Addessio, F.L.; Morrow, B.M.; Li, W.; Lookman, T.; Cerreta, E.K. Coupled nonlinear elasticity, plastic slip, twinning, and phase transformation in single crystal titanium for plate impact loading. *J. Mech. Phys. Solids* **2019**, *127*, 358–385. [CrossRef]

5. Bronkhorst, C.A.; Gray, G.; Addessio, F.L.; Livescu, V.; Bourne, N.; McDonald, S.; Withers, P. Response and representation of ductile damage under varying shock loading conditions in tantalum. *J. Appl. Phys.* **2016**, *119*, 085103. [CrossRef]
6. Mason, J.; Rosakis, A.; Ravichandran, G. On the strain and strain rate dependence of the fraction of plastic work converted to heat: An experimental study using high speed infrared detectors and the Kolsky bar. *Mech. Mater.* **1994**, *17*, 135–145. [CrossRef]
7. Rittel, D.; Wang, Z. Thermo-mechanical aspects of adiabatic shear failure of AM50 and Ti6Al4V alloys. *Mech. Mater.* **2008**, *40*, 629–635. [CrossRef]
8. Benzerga, A.; Bréchet, Y.; Needleman, A.; Van der Giessen, E. The stored energy of cold work: Predictions from discrete dislocation plasticity. *Acta Mater.* **2005**, *53*, 4765–4779. [CrossRef]
9. Rittel, D.; Zhang, L.; Osovski, S. The dependence of the Taylor–Quinney coefficient on the dynamic loading mode. *J. Mech. Phys. Solids* **2017**, *107*, 96–114. [CrossRef]
10. Rittel, D.; Ravichandran, G.; Venkert, A. The mechanical response of pure iron at high strain rates under dominant shear. *Mater. Sci. Eng. A* **2006**, *432*, 191–201. [CrossRef]
11. Rittel, D.; Bhattacharyya, A.; Poon, B.; Zhao, J.; Ravichandran, G. Thermomechanical characterization of pure polycrystalline tantalum. *Mater. Sci. Eng. A* **2007**, *447*, 65–70. [CrossRef]
12. Rittel, D.; Silva, M.; Poon, B.; Ravichandran, G. Thermomechanical behavior of single crystalline tantalum in the static and dynamic regime. *Mech. Mater.* **2009**, *41*, 1323–1329. [CrossRef]
13. Rittel, D.; Kidane, A.; Alkhader, M.; Venkert, A.; Landau, P.; Ravichandran, G. On the dynamically stored energy of cold work in pure single crystal and polycrystalline copper. *Acta Mater.* **2012**, *60*, 3719–3728. [CrossRef]
14. Asaro, R.J.; Rice, J. Strain localization in ductile single crystals. *J. Mech. Phys. Solids* **1977**, *25*, 309–338. [CrossRef]
15. Asaro, R.J. Crystal plasticity. *J. Appl. Mech.* **1983**, *50*, 921–934. [CrossRef]
16. Asaro, R.J. Micromechanics of crystals and polycrystals. *Adv. Appl. Mech.* **1983**, *23*, 1–115.
17. Asaro, R.J.; Needleman, A. Overview no. 42 texture development and strain hardening in rate dependent polycrystals. *Acta Metall.* **1985**, *33*, 923–953. [CrossRef]
18. Bassani, J.L. Plastic flow of crystals. *Adv. Appl. Mech.* **1993**, *30*, 191–258.
19. Bassani, J.L.; Wu, T.Y. Latent hardening in single crystals. II. Analytical characterization and predictions. *Proc. R. Soc. Lond. Ser. A Math. Phys. Sci.* **1991**, *435*, 21–41.
20. Wu, T.Y.; Bassani, J.L.; Laird, C. Latent hardening in single crystals-I. Theory and experiments. *Proc. R. Soc. Lond. Ser. A Math. Phys. Sci.* **1991**, *435*, 1–19.
21. Kalidindi, S.R.; Bronkhorst, C.A.; Anand, L. Crystallographic texture evolution in bulk deformation processing of FCC metals. *J. Mech. Phys. Solids* **1992**, *40*, 537–569. [CrossRef]
22. Bronkhorst, C.A.; Kalidindi, S.; Anand, L. Polycrystalline plasticity and the evolution of crystallographic texture in FCC metals. *Philos. Trans. R. Soc. Lond. Ser. A Phys. Eng. Sci.* **1992**, *341*, 443–477.
23. Harren, S.; Asaro, R. Nonuniform deformations in polycrystals and aspects of the validity of the Taylor model. *J. Mech. Phys. Solids* **1989**, *37*, 191–232. [CrossRef]
24. Harren, S.; Lowe, T.C.; Asaro, R.J.; Needleman, A. Analysis of large-strain shear in rate-dependent face-centred cubic polycrystals: Correlation of micro-and macromechanics. *Philos. Trans. R. Soc. Lond. Ser. A Math. Phys. Sci.* **1989**, *328*, 443–500.
25. Acharya, A.; Tang, H.; Saigal, S.; L. Bassani, J. On boundary conditions and plastic strain-gradient discontinuity in lower-order gradient plasticity. *J. Mech. Phys. Solids* **2004**, *52*, 1793–1826. [CrossRef]
26. Aifantis, E.C. On the role of gradients in the localization of deformation and fracture. *Int. J. Eng. Sci.* **1992**, *30*, 1279–1299. [CrossRef]
27. Anand, L.; Gurtin, M.; Lele, S.; Gething, C. A one-dimensional theory of strain-gradient plasticity: Formulation, analysis, numerical results. *J. Mech. Phys. Solids* **2005**, *53*, 1789–1826. [CrossRef]
28. Arsenlis, A.; Parks, D. Crystallographic aspects of geometrically-necessary and statistically-stored dislocation density. *Acta Mater.* **1999**, *47*, 1597–1611. [CrossRef]
29. Busso, E.P.; McClintock, F.A. A dislocation mechanics-based crystallographic model of a B2-type intermetallic alloy. *Int. J. Plast.* **1996**, *12*, 1–28. [CrossRef]
30. Busso, E.; Meissonnier, F.; O'Dowd, N. Gradient-dependent deformation of two-phase single crystals. *J. Mech. Phys. Solids* **2000**, *48*, 2333–2361. [CrossRef]
31. Gerken, J.M.; Dawson, P.R. A crystal plasticity model that incorporates stresses and strains due to slip gradients. *J. Mech. Phys. Solids* **2008**, *56*, 1651–1672. [CrossRef]
32. Gurtin, M.E. On the plasticity of single crystals: Free energy, microforces, plastic-strain gradients. *J. Mech. Phys. Solids* **2000**, *48*, 989–1036. [CrossRef]
33. Gurtin, M.E.; Needleman, A. Boundary conditions in small-deformation, single-crystal plasticity that account for the Burgers vector. *J. Mech. Phys. Solids* **2005**, *53*, 1–31. [CrossRef]
34. Mayeur, J.; Beyerlein, I.; Bronkhorst, C.; Mourad, H. Incorporating interface affected zones into crystal plasticity. *Int. J. Plast.* **2015**, *65*, 206–225. [CrossRef]
35. Zhu, H.; Zbib, H. On the role of strain gradients in adiabatic shear banding. *Acta Mech.* **1995**, *111*, 111–124. [CrossRef]
36. Dequiedt, J.; Denoual, C.; Madec, R. Heterogeneous deformation in ductile FCC single crystals in biaxial stretching: The influence of slip system interactions. *J. Mech. Phys. Solids* **2015**, *83*, 301–318. [CrossRef]

37. Devincre, B.; Kubin, L.; Hoc, T. Physical analyses of crystal plasticity by DD simulations. *Scr. Mater.* **2006**, *54*, 741–746 . [CrossRef]
38. Devincre, B.; Hoc, T.; Kubin, L. Dislocation Mean Free Paths and Strain Hardening of Crystals. *Science* **2008**, *320*, 1745–1748 . [CrossRef]
39. Grilli, N.; Janssens, K.; Nellessen, J.; Sandlöbes, S.; Raabe, D. Multiple slip dislocation patterning in a dislocation-based crystal plasticity finite element method. *Int. J. Plast.* **2018**, *100*, 104–121. [CrossRef]
40. Hansen, B.; Beyerlein, I.; Bronkhorst, C.; Cerreta, E.; Dennis-Koller, D. A dislocation-based multi-rate single crystal plasticity model. *Int. J. Plast.* **2013**, *44*, 129–146. [CrossRef]
41. Hansen, B.; Bronkhorst, C.; Ortiz, M. Dislocation subgrain structures and modeling the plastic hardening of metallic single crystals. *Model. Simul. Mater. Sci. Eng.* **2010**, *18*, 055001. [CrossRef]
42. Lee, S.; Cho, H.; Bronkhorst, C.A.; Pokharel, R.; Brown, D.W.; Clausen, B.; Vogel, S.C.; Anghel, V.; Gray, G.T.; Mayeur, J.R. Deformation, dislocation evolution and the non-Schmid effect in body-centered-cubic single- and polycrystal tantalum. *Int. J. Plast.* **2023**, *163*, 103529. [CrossRef]
43. Madec, R.; Kubin, L.P. Dislocation strengthening in FCC metals and in BCC metals at high temperatures. *Acta Mater.* **2017**, *126*, 166–173. [CrossRef]
44. Nguyen, T.; Fensin, S.J.; Luscher, D.J. Dynamic crystal plasticity modeling of single crystal tantalum and validation using Taylor cylinder impact tests. *Int. J. Plast.* **2021**, *139*, 102940. [CrossRef]
45. Butler, B.G.; Paramore, J.D.; Ligda, J.P.; Ren, C.; Fang, Z.Z.; Middlemas, S.C.; Hemker, K.J. Mechanisms of deformation and ductility in tungsten—A review. *Int. J. Refract. Met. Hard Mater.* **2018**, *75*, 248–261. [CrossRef]
46. Feng, B.; Bronkhorst, C.; Addessio, F.; Morrow, B.; Cerreta, E.; Lookman, T.; Lebensohn, R.; Low, T. Coupled elasticity, plastic slip, and twinning in single crystal titanium loaded by split-Hopkinson pressure bar. *J. Mech. Phys. Solids* **2018**, *119*, 274–297. [CrossRef]
47. Acharya, A. New inroads in an old subject: Plasticity, from around the atomic to the macroscopic scale. *J. Mech. Phys. Solids* **2010**, *58*, 766–778. [CrossRef]
48. Anand, L.; Gurtin, M.E.; Reddy, B.D. The stored energy of cold work, thermal annealing, and other thermodynamic issues in single crystal plasticity at small length scales. *Int. J. Plast.* **2015**, *64*, 1–25. [CrossRef]
49. Arora, R.; Acharya, A. Dislocation pattern formation in finite deformation crystal plasticity. *Int. J. Solids Struct.* **2020**, *184*, 114–135 . [CrossRef]
50. Berdichevsky, V. Why is classical thermodynamics insufficient for solids? In Proceedings of the 2018 AIAA/ASCE/AHS/ASC Structures, Structural Dynamics, and Materials Conference, Kissimmee, FL, USA, 8–12 January 2018. [CrossRef]
51. Berdichevsky, V. A continuum theory of edge dislocations. *J. Mech. Phys. Solids* **2017**, *106*, 95–132. [CrossRef]
52. Berdichevsky, V. Entropy and temperature of microstructure in crystal plasticity. *Int. J. Eng. Sci.* **2018**, *128*, 24–30. [CrossRef]
53. Berdichevsky, V. Beyond classical thermodynamics: Dislocation-mediated plasticity. *J. Mech. Phys. Solids* **2019**, *129*, 83–118. [CrossRef]
54. del Castillo, P.R.D.; Huang, M. Dislocation annihilation in plastic deformation: I. Multiscale irreversible thermodynamics. *Acta Mater.* **2012**, *60*, 2606–2614. [CrossRef]
55. Roy Chowdhury, S.; Roy, D. A non-equilibrium thermodynamic model for viscoplasticity and damage: Two temperatures and a generalized fluctuation relation. *Int. J. Plast.* **2019**, *113*, 158–184. [CrossRef]
56. Hochrainer, T. Thermodynamically consistent continuum dislocation dynamics. *J. Mech. Phys. Solids* **2016**, *88*, 12–22. [CrossRef]
57. Jafari, M.; Jamshidian, M.; Ziaei-Rad, S. A finite-deformation dislocation density-based crystal viscoplasticity constitutive model for calculating the stored deformation energy. *Int. J. Mech. Sci.* **2017**, *128–129*, 486–498. [CrossRef]
58. Jiang, M.; Devincre, B.; Monnet, G. Effects of the grain size and shape on the flow stress: A dislocation dynamics study. *Int. J. Plast.* **2019**, *113*, 111–124. [CrossRef]
59. Langer, J.; Bouchbinder, E.; Lookman, T. Thermodynamic theory of dislocation-mediated plasticity. *Acta Mater.* **2010**, *58*, 3718–3732. [CrossRef]
60. Langer, J.S. Statistical thermodynamics of strain hardening in polycrystalline solids. *Phys. Rev. E* **2015**, *92*, 032125. [CrossRef]
61. Le, K.; Tran, T.; Langer, J. Thermodynamic dislocation theory of adiabatic shear banding in steel. *Scr. Mater.* **2018**, *149*, 62–65. [CrossRef]
62. Le, K. Thermodynamic dislocation theory for non-uniform plastic deformations. *J. Mech. Phys. Solids* **2018**, *111*, 157–169. [CrossRef]
63. Le, K. Thermodynamic dislocation theory: Finite deformations. *Int. J. Eng. Sci.* **2019**, *139*, 1–10. [CrossRef]
64. Le, K.C.; Tran, T.M.; Langer, J.S. Thermodynamic dislocation theory of high-temperature deformation in aluminum and steel. *Phys. Rev. E* **2017**, *96*, 013004. [CrossRef] [PubMed]
65. Levitas, V.I.; Javanbakht, M. Thermodynamically consistent phase field approach to dislocation evolution at small and large strains. *J. Mech. Phys. Solids* **2015**, *82*, 345–366. [CrossRef]
66. Nieto-Fuentes, J.; Rittel, D.; Osovski, S. On a dislocation-based constitutive model and dynamic thermomechanical considerations. *Int. J. Plast.* **2018**, *108*, 55–69. [CrossRef]
67. Po, G.; Huang, Y.; Ghoniem, N. A continuum dislocation-based model of wedge microindentation of single crystals. *Int. J. Plast.* **2019**, *114*, 72–86. [CrossRef]

68. Roy, A.; Acharya, A. Finite element approximation of field dislocation mechanics. *J. Mech. Phys. Solids* **2005**, *53*, 143–170. [CrossRef]
69. Roy, A.; Acharya, A. Size effects and idealized dislocation microstructure at small scales: Predictions of a Phenomenological model of Mesoscopic Field Dislocation Mechanics: Part II. *J. Mech. Phys. Solids* **2006**, *54*, 1711–1743. [CrossRef]
70. Shizawa, K.; Kikuchi, K.; Zbib, H.M. A strain-gradient thermodynamic theory of plasticity based on dislocation density and incompatibility tensors. *Mater. Sci. Eng. A* **2001**, *309-310*, 416–419 . [CrossRef]
71. Clayton, J. Nonlinear thermomechanics for analysis of weak shock profile data in ductile polycrystals. *J. Mech. Phys. Solids* **2019**, *124*, 714–757. [CrossRef]
72. Clayton, J.D. *Nonlinear Mechanics of Crystals*; Springer Science & Business Media: Berlin/Heidelberg, Germany, 2010; Volume 177.
73. Langer, J.S. Thermal effects in dislocation theory. *Phys. Rev. E* **2016**, *94*, 063004. [CrossRef] [PubMed]
74. Langer, J. Statistical thermodynamics of dislocations in solids. *Adv. Phys.* **2021**, *70*, 445–467. [CrossRef]
75. Lieou, C.K.; Bronkhorst, C.A. Dynamic recrystallization in adiabatic shear banding: Effective-temperature model and comparison to experiments in ultrafine-grained titanium. *Int. J. Plast.* **2018**, *111*, 107–121. [CrossRef]
76. Lieou, C.K.; Mourad, H.M.; Bronkhorst, C.A. Strain localization and dynamic recrystallization in polycrystalline metals: Thermodynamic theory and simulation framework. *Int. J. Plast.* **2019**, *119*, 171–187. [CrossRef]
77. Lieou, C.K.; Bronkhorst, C.A. Thermomechanical conversion in metals: Dislocation plasticity model evaluation of the Taylor-Quinney coefficient. *Acta Mater.* **2021**, *202*, 170–180. [CrossRef]
78. Lieou, C.K.; Bronkhorst, C.A. Thermodynamic theory of crystal plasticity: Formulation and application to polycrystal fcc copper. *J. Mech. Phys. Solids* **2020**, *138*, 103905. [CrossRef]
79. Noll, W.; Coleman, B.D.; Noll, W. The thermodynamics of elastic materials with heat conduction and viscosity. In *The Foundations of Mechanics and Thermodynamics: Selected Papers*; Springer: Berlin/Heidelberg, Germany, 1974; pp. 145–156.
80. Kubin, L.; Devincre, B.; Hoc, T. Modeling dislocation storage rates and mean free paths in face-centered cubic crystals. *Acta Mater.* **2008**, *56*, 6040–6049. [CrossRef]
81. Bronkhorst, C.; Hansen, B.; Cerreta, E.; Bingert, J. Modeling the microstructural evolution of metallic polycrystalline materials under localization conditions. *J. Mech. Phys. Solids* **2007**, *55*, 2351–2383. [CrossRef]
82. Hansen, B.; Carpenter, J.; Sintay, S.; Bronkhorst, C.; McCabe, R.; Mayeur, J.; Mourad, H.; Beyerlein, I.; Mara, N.; Chen, S.; et al. Modeling the texture evolution of Cu/Nb layered composites during rolling. *Int. J. Plast.* **2013**, *49*, 71–84. [CrossRef]
83. Mayeur, J.; Beyerlein, I.; Bronkhorst, C.; Mourad, H.; Hansen, B. A crystal plasticity study of heterophase interface character stability of Cu/Nb bicrystals. *Int. J. Plast.* **2013**, *48*, 72–91. [CrossRef]
84. Kocks, U. Laws for work-hardening and low-temperature creep. *J. Eng. Mater. Technol.* **1976**, *98*, 76–85. [CrossRef]
85. Balasubramanian, S. Polycrystalline Plasticity: Application to Deformation Processing of Lightweight Metals. Ph.D. Thesis, Massachusetts Institute of Technology, Cambridge, MA, USA, 1998.
86. Simmons, G.; Wang, H. *Single Crystal Elastic Constants and Calculated Aggregate Properties: A Handbook*; The M.I.T. Press: Cambridge, MA, USA, **1971**.
87. Madec, R.; Devincre, B.; Kubin, L.P. From dislocation junctions to forest hardening. *Phys. Rev. Lett.* **2002**, *89*, 255508. [CrossRef] [PubMed]
88. Madec, R.; Devincre, B.; Kubin, L.; Hoc, T.; Rodney, D. The role of collinear interaction in dislocation-induced hardening. *Science* **2003**, *301*, 1879–1882. [CrossRef] [PubMed]
89. Awbi, H.B. Calculation of convective heat transfer coefficients of room surfaces for natural convection. *Energy Build.* **1998**, *28*, 219–227. [CrossRef]
90. Owen, C.J.; Naghdi, A.D.; Johansson, A.; Massa, D.; Papanikolaou, S.; Kozinsky, B. Unbiased Atomistic Predictions of Crystal Dislocation Dynamics using Bayesian Force Fields. *arXiv* **2024**, arXiv:2401.04359.
91. Fan, Y.; Kushima, A.; Yip, S.; Yildiz, B. Mechanism of Void Nucleation and Growth in bcc Fe: Atomistic Simulations at Experimental Time Scales. *Phys. Rev. Lett.* **2011**, *106*, 125501. [CrossRef]
92. Fan, Y.; Yip, S.; Yildiz, B. Autonomous basin climbing method with sampling of multiple transition pathways: Application to anisotropic diffusion of point defects in hcp Zr. *J. Phys. Condens. Matter* **2014**, *26*, 365402. [CrossRef]

metals

MDPI

Article

Modeling and Simulation of Fatigue Crack Initiation Process Based on Field Theory of Multiscale Plasticity (FTMP): Part I: PSB Ladder Formation and Verification

Xinping You [1] and Tadashi Hasebe [2,*]

1 Department of Mechanical Engineering, Graduate School, Kobe University, 1-1 Rokkodai-cho, Kobe 657-8501, Japan; you.xinping@mail.mm4.scitec.kobe-u.ac.jp
2 Department of Mechanical Engineering, Kobe University, 1-1 Rokkodai-cho, Kobe 657-8501, Japan
* Correspondence: hasebe@mech.kobe-u.ac.jp; Tel.: +81-0788036113

Abstract: In this study, we successfully reproduced the persistent slip band (PSB) with laddered patterning, showcasing the predictive capability of the framework of Field Theory of Multiscale Plasticity (FTMP) without relying on ad hoc models, intricate mathematical models, or elaborate finite element discretization. The FTMP-incorporated CP-FEM simulation not only reasonably replicates the experimentally observed laddered morphology and PSB but also effectively simulates surface roughening and grooving, independent of vacancy formation and diffusion. These results highlight the significance of laddered morphology and set the stage for further investigations into the effects of vacancy formation, as extended in the subsequent paper. Leveraging incompatibility tensor-based degrees of freedom, the FTMP framework offers exceptional capabilities for natural modeling dislocation substructures typically overlooked in conventional approaches, positioning it as a transformative tool for advancing our understanding of the mechanisms that dictate slip band-fatigue crack transitions.

Keywords: fatigue; persistent slip band; crack initiation; crystal plasticity; field theory; non-Riemannian plasticity; finite element method; vacancy diffusion

1. Introduction

Research into metal fatigue has a long-standing history, with significant contributions dating back over half a century. Early investigations by Thompson et al. [1] revealed that fatigue cracks initiate within slip bands of a single grain, highlighting the critical role of microscopic dislocation structures in fatigue crack formation. Building on this foundation, Woods et al. [2] emphasized the presence of persistent slip bands (PSBs), characterized by uniformly arranged dislocation walls, often referred to as ladder walls. Polák et al. [3–6] further explored the relationship between these dislocation substructures and the resultant surface morphology. Essmann et al. [7,8] conducted extensive studies on irreversible deformation associated with the ladder structure of PSBs and its influence on crack initiation at the PSB–matrix interface. They examined the crucial role of vacancies commonly found in fatigued metals [9,10], including their agglomeration into crack nucleation sites.

Subsequent studies have continued to enhance our understanding of the relationship between persistent slip band (PSB) ladder structures and fatigue crack initiation, an inquiry that remains ongoing. Historically, crack initiation modeling continues to be an unsatisfactory endeavor due to its inherent complexities as a whole. A significant portion of the research has focused on interactions with singular interfaces, such as grain and twin boundaries in various practical alloy materials [11–13]. Meanwhile, simulations of crack initiation and subsequent propagation processes have often been limited to scenarios involving artificial notches and stress concentration sites [14], as observed in compact tension

(CT) specimens [15–17], or even from a pre-existing cracks [15,18–20], relying heavily on phenomenological elasto-plasticity and established fracture mechanics.

In recent years, there has been a notable increase in the application of crystal plasticity finite element analysis (CP-FEM) [21–24], which integrates detailed geometric information about microstructures and crystallography. This has been complemented by rapidly advancing experimental observation techniques, including micro-beam X-ray Laue diffraction [25], in situ and ex situ SEM-EBSD (electron backscattering diffraction) [23,26], ECCI (electron channeling contrast imaging) [22], and HR-DIC (high-resolution digital image correlation) [14] analyses. Furthermore, significant progress has been made in coupling simulations with metallurgical microstructure modeling, particularly through the application of discrete dislocation dynamics (DDDs) [24,27–29] and phase-field (PF) [30,31] techniques as powerful numerical tools. An extensive review by Sangid [32] highlights the potential of high-precision experimental measurements and computational simulations; however, it does not focus on crack nucleation or the role of dislocation substructures in crack growth and propagation. The interrelationships between dislocation substructures, particularly cell formation, have been suggested in various studies [33–35].

Despite the advances with DDD and PF methods, progress in accurately reproducing the fatigue crack initiation process purely from deformation fields remains limited, without relying on hard spots or similar features. CP-FEM analyses, in particular, are often employed as supporting tools rather than predictive ones, due to inherent limitations in their predictive capabilities, which indicates a lack of standalone competency. This inadequacy in predicting deformation-induced dislocation substructures highlights a significant drawback of conventional CP-FEM, as the slip bands responsible for fatigue crack initiation are closely linked to dislocation substructures that are spontaneously induced during deformation. This limitation is further underscored by the inability to reproduce even simple intragranular banded substructuring with misorientation [21] or to capture fundamental clues to dislocation substructure formation under cyclic straining without resorting to overly complex models [22].

The significance of deformation-induced patterning in plasticity is paramount for understanding the multiscale nature of material mechanics. This importance arises from the dynamic changes in dislocation substructures influenced by long-range stress fields and misorientation, particularly in materials that yield dislocation structures with cellular morphology [36,37]. These dislocation patterns play a crucial role in the storage and release of elastic strain energy, influencing not only the deformation capacity but also the strength and toughness of materials. A notable exception is the ladder structure observed in PSBs, which predominantly facilitates deformation during cyclic loading and typically does not produce a long-range stress field due to its composition of edge dislocation dipoles [38]. This PSB ladder structure significantly impacts the mechanisms of fatigue crack initiation, as previously described [39–43]. However, the underlying reasons for this patterning and its critical roles remain undefined, partly due to limitations in conventional approaches to plasticity.

While analogy-based methods, such as reaction–diffusion equations grounded in the concept of dissipative structures, can address patterning-related issues, they often provide limited insights beyond mere analogy [44,45]. These methods do not elucidate the mechanisms involved in crack initiation. Recent research, fueled by advances in computational capabilities, has increasingly relied on DDDs [27,46,47] and molecular dynamics (MD) simulations [48], employing multiscale methods to model these phenomena. However, these studies often focus narrowly on specific aspects, such as passing stress [38,49], rather than capturing the complete picture that includes the roles of PSB ladders and the processes of crack nucleation, e.g., [48,50].

Modeling fatigue presents significant challenges, particularly concerning crack initiation, as it must consider the effects of dislocation substructures, that is, the laddered morphology in PSBs. The foundational work by Repetto and Ortiz [51] on finite element simulations of fatigue crack initiation emphasized the role of vacancy-induced elongation in the PSB region, which leads to surface protrusion as a key factor; however, their study did not account for the laddered morphology explicitly. In contrast, Nakai et al. [52–54] reported that cracks primarily evolve from intrusions rather than extrusions. Their research identified a potential critical condition for the transition from intrusions to cracks, based on detailed observations made using atomic force microscopy (AFM) combined with interrupted fatigue tests. According to their findings, extrusions and/or surface reliefs do not develop into cracks. Through their systematic series of studies on several materials, they successfully identified a potential critical condition for the transition from intrusions to cracks, with further details presented in Section 2.

Focusing on the transition from slip bands to cracks, particularly once this critical condition is reached, may provide a more effective framework and minimize ambiguities in the modeling process. Integrating this critical condition with our Field Theory of Multiscale Plasticity (FTMP)-based approach [55–62] can significantly reduce computational costs, as FTMP does not always require intricate fine mesh divisions to reproduce deformation-induced substructural patterns accurately. This versatile theoretical framework facilitates the study of deformation-induced patterning, which is crucial for controlling mechanical properties across multiscale contexts [55,56]. By incorporating a new underlying degree of freedom related to the incompatibility tensor into the hardening law, we enable the autonomous emergence and spontaneous evolution of dislocation patterns that closely resemble experimental observations, as demonstrated in Ref. [57], through straightforward deformation analyses using CP-FEM. This mechanism effectively redistributes excessive elastic strain energy generated during deformation, promoting system-wide equilibrium. In this context, the incompatibility tensor serves as a critical facilitator of energy release, driving the formation of complex dislocation substructures.

This study aims to elucidate the roles of the laddered structure in PSBs on the fatigue crack initiation process through FTMP-incorporated crystal plasticity (CP)-based finite element method (FEM) simulations. Part I of the paper presents a preliminary series of simulations utilizing a simplified vacancy model to examine surface roughening behavior and the subsequent grooving processes.

2. Background

Two pivotal research findings have catalyzed the current series of studies. First, Yokoi et al. [63,64] demonstrated that the addition of copper (Cu) to steels can dramatically alter dislocation substructures, transforming them from cellular to planar configurations. This transformation results in finer surface intrusions and extrusions, significantly delaying the initiation of fatigue cracks, as illustrated in Figure 1A. This finding underscores the crucial role of dislocation substructures in influencing the fatigue crack initiation process.

Second, Nakai et al. [52–54] investigated the transition process from slip bands to cracks through a systematically designed series of interrupted AFM observations. They revealed that surface grooving—an evident manifestation of slip bands—evolves into fatigue cracks when the slip distance (or groove depth) exceeds a critical threshold, which varies among different materials, as shown in Figure 1B. This finding highlights a key mechanism that dictates the transition from slip banding to fatigue cracking. Notably, surface reliefs (bottom left) do not contribute to crack formation; only grooves serve as precursors to crack nucleation, at least according to their studies. In this context, the effect of Cu addition on the core structure of screw dislocations in α-Fe has been studied separately through ab initio calculations [65]. Cross slip by screw dislocations is considered critical for dynamic recovery, which drives cell formation. We reported a transition of the core structure from isotropic to fully extended, a change that tends to inhibit cross slip, thereby hindering dynamic recovery and the resulting cell formation. This provides one possible explanation for why the addition of Cu leads to vein-like or planar dislocation substructures, rather than cellular morphology.

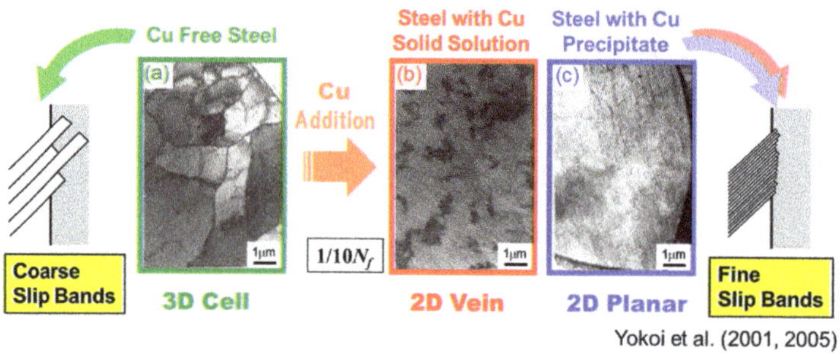

(A)

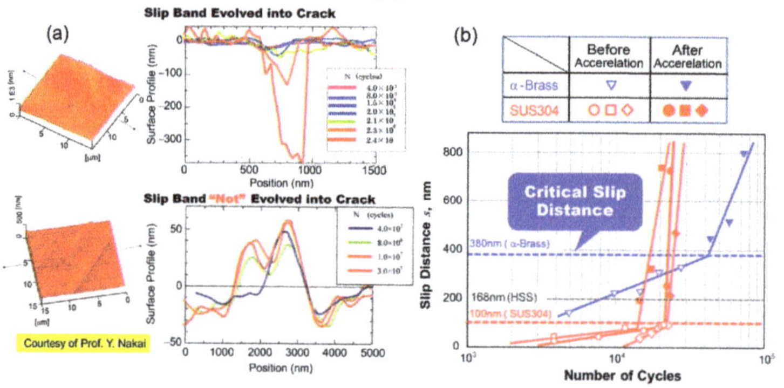

(B)

Figure 1. Informative excremental results for modeling fatigue crack initiation process. (**A**) Effect of dislocation substructures on slip banding and fatigue cracking, adapted from Yokoi et al. [63,64] (2001, 2005). (**B**) Transition of slip bands into cracks, measured by critical slip distance (a) compares surface profile variations with stress cycling and AFM micrographs; (b) shows slip distance variation with cycle number for SUS304 (red) and α-Brass (blue), with open and solid plots representing data before and after slip distance acceleration, adapted from Nakai et al. [52–54] (Courtesy of Y. Nakai).

By explicitly simulating these critical aspects through deformation analyses, we aim to provide groundbreaking insights that enhance our understanding of the fatigue crack initiation process in practical applications. For the former, the case of copper solid solution yielding a 2D vein structure represents the most fundamental scenario. Here, we confine our focus to the widely observed PSB ladder structure, with other matrix morphologies left for future investigation.

3. Theory and Kinematics

3.1. Brief Description About FTMP

The Field Theory of Multiscale Plasticity (FTMP) [55–62] is a comprehensive framework encompassing three critical aspects: (i) evolution, (j) description, and (k) cooperation, as illustrated in Figure 2a. It primarily addresses "inhomogeneously evolving" deformation fields, which typically manifest as dislocation substructures (Figure 2b: simulated examples). Notably, the theory's ability to describe these aspects of evolution distinguishes it from many others. Furthermore, the interaction formalism facilitates explicit treatments of (k) cooperation across multiple scales (Figure 2f).

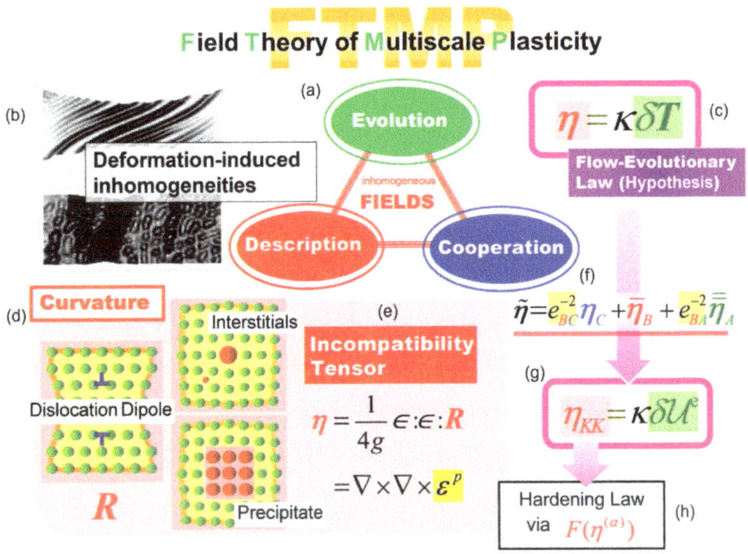

Figure 2. Overview of FTMP as a new theory of multiscale plasticity. (**a**) Key concepts of the FTMP-based framework. (**b**) Representative simulation results obtained via CP-FEM. (**c**) The flow-evolutionary law (FEL) as the central concept. (**d**) Schematic representation of the curvature tensor in differential geometry for describing defect fields in general. (**e**) Continuum mechanics-based interpretation of the curvature tensor. (**f**) The interaction formalism represented by the incompatibility tensor field. (**g**) A specific form of the FEL. (**h**) The corresponding incompatibility term to be incorporated additively into the hardening law of the CP constitutive equation for CP-FEM simulations.

For the mathematical description of these inhomogeneous fields, differential geometric concepts are predominantly employed, particularly the curvature tensor $R_{klm}^{\cdot\cdot\cdot n}$ (Figure 2d) and the torsion tensor $S_{kl}^{\cdot\cdot j}$, defined respectively as

$$\begin{cases} S_{kl}^{\cdot\cdot j} = \Gamma_{[kl]}^{j} \\ R_{klm}^{\cdot\cdot\cdot n} = 2\left[\partial_{[k}\Gamma_{l]m}^{n} + \Gamma_{[k|p|}^{n}\Gamma_{l]m}^{p}\right] \end{cases} \tag{1}$$

where Γ_{ij}^{k} represents the coefficient of connection. The significance of utilizing these differential geometric quantities lies in the fact that all imperfections in crystalline space can be completely expressed through them. These tensors correspond to the incompatibility tensor η_{ij} (Figure 2e) and the dislocation density tensor, respectively, drawing on the non-Riemannian plasticity framework advocated by K. Kondo [66,67]. Both can be contracted into second-rank tensors without loss of information by applying the alternating tensor $\in_{ijk}$ as follows:

$$\begin{cases} \alpha_{ij} = \frac{1}{2} \in_{ikl} S_{kl}^{\cdot\cdot j} \\ \eta_{ij} = \frac{1}{4g} \in_{ikl}\in_{jmn} R_{klm}^{\cdot\cdot\cdot n}\left(g = \det(g_{ij})\right) \end{cases} \tag{2}$$

with g_{ij} being the metric tensor. The incompatibility tensor $\eta_{ij}= \boldsymbol{\eta}$ is further defined as the double curl of the plastic strain tensor $\varepsilon_{ij}^{p}= \boldsymbol{\varepsilon}^{p}$ (Figure 2e), while the dislocation density tensor $\alpha_{ij}= \boldsymbol{\alpha}$ is derived from the curl of the plastic distortion tensor $\beta_{ij}^{p} = \boldsymbol{\beta}^{p}$.

$$\begin{cases} \alpha_{ij} = - \in_{ikl} \partial_k\beta_{lj}^{p} \\ \eta_{ij} = \in_{ikl}\in_{jmn} \partial_k\partial_m\varepsilon_{ln}^{p} \end{cases} \text{or} \begin{cases} \boldsymbol{\alpha} = -\nabla \times \boldsymbol{\beta}^{p} \\ \boldsymbol{\eta} = \nabla \times \nabla \times \boldsymbol{\varepsilon}^{p} \end{cases} \tag{3}$$

Consequently, strain gradients are intrinsically integrated into the theory. It is important to note that the dislocation density tensor, which represents the first gradient of strain, encompasses the concept of geometrically necessary dislocations (GNDs) [58,60,68]. The component-wise expression of Equation (3) is presented in Appendix A.

At the core of FTMP is the concept known as the "flow-evolutionary law (FEL)", which serves as a working hypothesis (Figure 2c) [57,59].

$$\eta_{ij} = \kappa \delta T_{ij} \tag{4}$$

where κ refers to the duality coefficient. This relationship connects the incompatibility tensor η_{ij} with the fluctuation of the energy-momentum tensor $\delta T_{ij}= \delta \boldsymbol{T}$, where the fluctuation is defined as the deviation from the spatial average, i.e., $\delta T_{ij} \equiv T_{ij} - \langle T_{ij} \rangle$, with $\langle \cdot \rangle \equiv (1/V)\int(\cdot)dV$. The incompatibility tensor η_{ij} in this context has been extended to four-dimentional (4D) spacetime (see also Appendix A). The FEL was initially derived in [57] by equating the incompatible displacement with the inhomogeneous force, which are the corresponding conserved quantities of η_{ij} and δT_{ij}, respectively. Details on the energy-momentum tensor and a brief historical background leading to the formulation of the FEL are provided in Appendix B.

The FTMP framework extensively utilizes the incompatibility tensor, η_{ij}, among other elements. Its enhanced applications, presented in Figure 3, are categorized into three aspects: an extended definition to four-dimensional (4D) spacetime, the flow-evolutionary hypothesis (Equation (4)), and the interaction formalism applicable across multiple scales. The 4D-extended definition of η_{ij} is expressed as

$$\eta_{ij} = \in_{iklp} \in_{jmnp} \partial_k \partial_m \varepsilon^p_{ln} \tag{5}$$

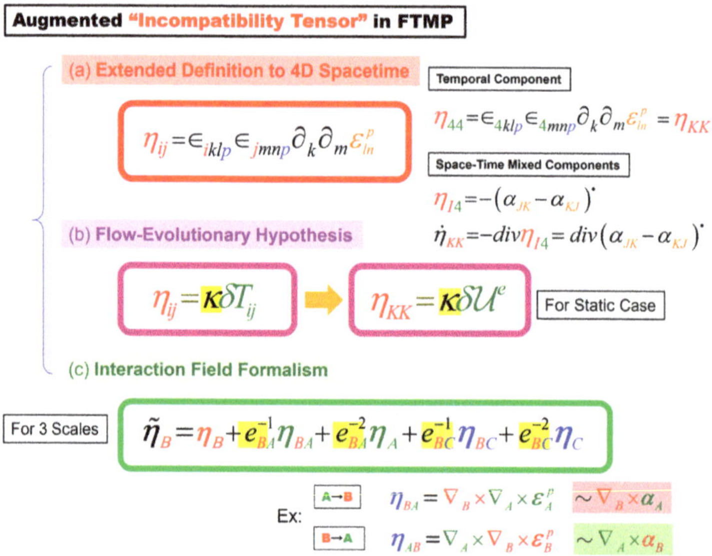

Figure 3. Augmented "incompatibility tensor" in FTMP: (**a**) extended definition of the incompatibility tensor into 4D spacetime, with two specific components of pure temporal and space-time mixed, (**b**) flow-evolutionary law as working hypothesis, with the specific form for duality diagram presentation scheme (pointed by right arrow), and (**c**) interaction formalism, presented for three scale problem with Scales A, B and C, with explicit examples for Scale A to B or Scale B to A interactions.

In this context, the indices indicated by lowercase letters denote four-dimensional spacetime (1, 2, 3 for spatial dimensions and 4 for time), while capital letters are used for spatial components when necessary for clarity. The pure temporal component of the 4D-extended incompatibility tensor η_{44} results in the spatial trace, represented as

$$\eta_{44} = \in_{4klp} \in_{4mnp} \partial_k \partial_m \varepsilon^p_{ln} = \eta_{KK} = tr^{sp}\boldsymbol{\eta} \tag{6}$$

A detailed process for achieving the above relationship is provided in Appendix A. Thus, the pure temporal component of Equation (4) becomes

$$\eta_{KK} = \kappa\delta(U^e + \mathcal{K}) \tag{7}$$

where U^e denotes the elastic strain energy, and $\mathcal{K}$ represents the kinetic energy. In static conditions, as is often the case, this simplifies to

$$\eta_{KK} = \kappa\delta U^e \tag{8}$$

as presented in Figure 2g. Equation (8) visualizes the dynamic interrelationships between excessive strain energy being converted or redistributed into the incompatibility-related degrees of freedom, which drive the field evolutions during the course of elasto-plastic deformation.

In practice, there is no need to directly solve Equation (8). Instead, we can incorporate the incompatibility-based underlying degrees of freedom into the hardening law of the constitutive equation used in CP-FEM simulations, as detailed separately below (also see Figure 2h). This integration allows the system to autonomously manage excessive storage of elastic strain energy, leading to significant field evolutions, such as substructure formation. In this context, the evolution of dislocation substructures arises naturally from system-wide accommodations facilitated by incompatibility. If this approach is correct, the necessary dislocation patterning can spontaneously emerge, provided that we properly integrate the incompatibility-related degrees of freedom into the simulation; otherwise, it may not occur. Thus, we anticipate that the laddered patterning in PSBs can be effectively reproduced without the need for any ad hoc models, underscoring the robustness of the FTMP framework.

The divergence-free condition for the incompatibility tensor in the 4D spacetime reads

$$div^{4D}\boldsymbol{\eta} = 0 \Leftrightarrow \dot{\eta}_{44} + div\eta_{A4} = 0 \tag{9}$$

Here, $\dot{\eta}_{44} = \dot{\eta}_{KK}$ according to Equation (3), while the space-temporal mixed component η_{4A} is given as

$$\eta_{A4} = -(\alpha_{BC} - \alpha_{CB})^{\bullet} \tag{10}$$

Ultimately, we establish the following relationship between the incompatibility rate and the edge dislocation density flux by substituting Equation (10) into Equation (9).

$$\dot{\eta}_{KK} = div(\alpha_{BC} - \alpha_{CB})^{\bullet} \tag{11}$$

This relationship allows multiple interpretations regarding pair-wise dislocation creation and annihilation processes, making it useful for modeling various detailed processes related to dislocation fields. For a detailed derivation of Equation (11) from Equation (9), refer to Appendix C. The present study utilizes this model to investigate vacancy sources arising from the pair annihilation of edge dislocations in Part II [69].

3.2. Kinematics of Crystal Plasticity

The FTMP concepts can be integrated with the conventional kinematics framework of crystal plasticity [70,71], serving as a foundational vehicle for their application. We essentially adhere to this framework, beginning with Lee's elastic-plastic decomposition of the deformation gradient tensor to accurately capture the finite deformation conditions.

$$\boldsymbol{F} = \boldsymbol{F}^{*} \cdot \boldsymbol{F}^{p} \tag{12}$$

where F^* accounts for elastic deformation, including lattice rotation, and F^p serves as its plastic counterpart.

The constitutive equation for elasticity in a rate form, suitable for finite crystal plasticity, is expressed as

$$(\tau^*_{(J)})^{\bigcirc} = C^e : d^* \text{with } (\tau^*_{(J)})^{\bigcirc} = \sigma^{\bigcirc} + \sigma \cdot trd \tag{13}$$

where $(\tau^*_{(J)})^{\bigcirc}$ is the Jaumann rate of the Kirchhoff Stress tensor, viewed from an observer on the rotating lattice. σ expresses the Cauchy stress tensor, C^e is the elasticity tensor, and d^* represents the elastic part of the strain rate tensor d. The relation $\dot{J} = trd = trd^*$ has been used in the second equation, with $\dot{J}$ denoting the Jacobian rate.

Ultimately, we reach

$$\tau^{\bigcirc}_{(J)} = C^e : d - \sum_{\alpha=1}^{N} R^{(\alpha)} \dot{\gamma}^{(\alpha)} \text{with} \begin{cases} R^{(\alpha)} = C^e : P^{(\alpha)} + \beta^{(\alpha)} \\ \beta^{(\alpha)} = W^{(\alpha)} \cdot \sigma - \sigma \cdot W^{(\alpha)} \end{cases} \tag{14}$$

Here, $P^{(\alpha)} = \left(s^{(\alpha)} \otimes m^{(\alpha)}\right)_{sym}$ and $W^{(\alpha)} = \left(s^{(\alpha)} \otimes m^{(\alpha)}\right)_{skew}$ are evaluated using the unit vectors for the slip direction $s^{(\alpha)}$ and slip plane normal $m^{(\alpha)}$, with (α) referring to the slip systems. The plastic distortion tensor β^p and plastic strain tensor ε^p are evaluated, respectively, as

$$\begin{cases} \beta^p = \sum\limits_{\alpha=1}^{N} \left(s^{(\alpha)} \otimes m^{(\alpha)}\right) \gamma^{(\alpha)} \\ \varepsilon^p = \sum\limits_{\alpha=1}^{N} P^{(\alpha)} \gamma^{(\alpha)} \left(= (\beta^p)_{sym}\right) \end{cases} \text{with } \gamma^{(\alpha)} = \int \dot{\gamma}^{(\alpha)} dt \tag{15}$$

which are used to explicitly compute the dislocation density and the incompatibility tensors, as defined in Equation (3).

3.3. *Constitutive Equation for CP-FEM*

We employ the general constitutive equation proposed in [57,62] based on dislocation dynamics, applicable both to FCC and BCC metals in a wide range of strain rate and temperature including impact loading conditions. The explicit form is given by

$$\begin{cases} \dot{\gamma}^{(\alpha)} = \dot{A}_{SR} \tau^{*(\alpha)} \left[\left| \tau^{*(\alpha)} \right| B_{SR} \exp \left(1 - \left| \dfrac{\left\langle \tau^{(\alpha)} - \tau^{*(\alpha)}_{Peierls} \right\rangle - \Omega^{(\alpha)}}{K^{(\alpha)}} \right|^p \right)^q + C_{SR} \right] \\ \tau^{*(\alpha)} \equiv \left\langle \tau^{(\alpha)} - \tau^{*(\alpha)}_{Peierls} \right\rangle - \Omega^{(\alpha)} \end{cases} \tag{16}$$

with $\dot{A}_{SR} = \rho_m b L v^*$, $B_{SR} = \Delta G_0^{disloc}/kT$, $C_{SR} = BLv^*/b$, where $K^{(\alpha)}$ and $\Omega^{(\alpha)}$ are drag stress and back stress, respectively, responsible for isotropic and kinematic types of hardening. Here, ρ_m, L, v^*, b and B are mobile dislocation density, mean flying distance of dislocations, the modified Debye frequency, the magnitude of Burgers vector, and the damping coefficient due to, e.g., phonon drag, respectively, and $\Delta G_0^{disloc} \equiv \mu(T) b^3 g_0^{disloc}$ stands for the activation energy for dislocation processes at $T = 0$ K, with g_0^{disloc} being the normalized one and $\mu(T)$ the temperature-dependent shear modulus. For BCC metals, we can set $C_{SR} = 0$, whereas for FCC we may normally assume $\tau^{*(\alpha)}_{Peierls} = 0$. The exponents p and q are the parameters specifying the thermal obstacle of interest, provided $0 \le p \le 1$ and $1 \le q \le 2$. In the above case, a pair of values, $p = 1/2, q = 3/2$, is used for representing dislocation processes. Furthermore, $\langle \bigcirc \rangle \equiv (\bigcirc + |\bigcirc|)/2$ represents the Mackauley parenthesis, with $\tau^{*(\alpha)}_{Peierls}$ expressing the effective stress for the Peierls overcoming process given by

$$\tau^{*(\alpha)}_{Peierls} = \hat{\tau}^{*}_{Peierls}\left[1 - \left(B^{p-1}_{SR}\ln\frac{\dot{\gamma}_{0P}}{\dot{\gamma}^{(\alpha)}}\right)^{1/q_P}\right]^{1/p_P} \tag{17}$$

where $B^{p-1}_{SR} = kT/g^{Peierls}_0 \mu b^3$ with $g^{Peierls}_0$, $\dot{\gamma}_{0P}$, p_P, and q_P are parameters for the thermal activation processess via the Peierls overcomming mechanism. The present study neglects the back stress, i.e., $\Omega^{(\alpha)} = 0$, for simplicity. The evolution of the drag stress $K^{(\alpha)}$ is governed by

$$\begin{cases} \dot{K}^{(\alpha)} = Q_{\alpha\beta} H(\gamma)\left|\dot{\gamma}^{(\beta)}\right| \\ Q_{\alpha\beta} = \delta_{\alpha\beta} + f_{\alpha\kappa} S_{\kappa\beta} + F_k(\alpha^{(\beta)}_k; \eta^{(\beta)}_k) \end{cases} \tag{18}$$

where $Q_{\alpha\beta}$ is the hardening ratio that accounts for additional hardening or softening from the straining/loading history, $H(\gamma)$ represents the referential hardening moduli, $H(\gamma) = h_0\{(h_0/n\tau_0)\gamma + 1\}^{n-1}$ with $\gamma = \sum_{\alpha} \gamma^{(\alpha)}$, $f_{\alpha\beta}$ is the interaction matrix for pairwise dislocation reactions, and $S_{\kappa\beta}$ expresses a history matrix, further given as an increasing function of plastic work performed by the effective stress that is responsible for dislocation processes, e.g., $S_{\alpha\beta} = \tanh\left(W^{p*}_{\alpha\beta}/W^{p}_{sat}\right)$ with $W^{(\beta)}_{p*} = \int \left\langle \tau^{(\alpha)} - \tau^{*(\alpha)}_{Peierls} \right\rangle \cdot \dot{\gamma}^{(\alpha)} dt$ and W^{p}_{sat} the saturation value. The FTMP-based strain gradient term $F_k(\alpha^{(\alpha)}_k; \eta^{(\alpha)}_k)$ collectively represents the strain gradient terms for the dislocation density and incompatibility fields; the explicit forms are given as [57–59,62].

$$\begin{cases} F(\alpha^{(\alpha)}) = k_{p_\alpha}\left(\frac{\left|\alpha^{(\alpha)}\right|}{b}\right)^{1/2} \\ F(\eta^{(\alpha)}) = \mathrm{sgn}(\eta^{(\alpha)}) \cdot k_{p_\eta}\left(\frac{l_{defect}}{b}\left|\eta^{(\alpha)}\right|\right)^{1/2} \end{cases} \tag{19}$$

The parameters include l_{defect} that represents the targeted inhomogeneous fields, which coincides with the Burgers vector b in the case of individual dislocations, along with the coefficients k_{p_α} and k_{p_η}. For evaluating the slip system-wise dislocation density and incompatibility fields, the following projections are employed [57,62]:

$$\begin{cases} \alpha^{(\alpha)} = \left(\boldsymbol{t}^{(\alpha)} \otimes \boldsymbol{s}^{(\alpha)} + \boldsymbol{s}^{(\alpha)} \otimes \boldsymbol{s}^{(\alpha)}\right) : \boldsymbol{\alpha} \\ \eta^{(\alpha)} = \left(\boldsymbol{t}^{(\alpha)} \otimes \boldsymbol{s}^{(\alpha)} + \boldsymbol{s}^{(\alpha)} \otimes \boldsymbol{s}^{(\alpha)} + \boldsymbol{s}^{(\alpha)} \otimes \boldsymbol{m}^{(\alpha)}\right) : \boldsymbol{\eta} \end{cases} \tag{20}$$

where $\boldsymbol{t}^{(\alpha)} = \boldsymbol{s}^{(\alpha)} \times \boldsymbol{m}^{(\alpha)}$ represents the line direction in the case of dislocations. Note that the choice of projection directions can affect the evolved patterning, requiring careful attention, especially regarding the incompatibility term, whereas the dislocation density tensor has a clear physical meaning, specifically representing the edge and screw components via $\boldsymbol{t}^{(\alpha)} \otimes \boldsymbol{s}^{(\alpha)}$ and $\boldsymbol{s}^{(\alpha)} \otimes \boldsymbol{s}^{(\alpha)}$ projections, respectively.

4. Model and Analytical Condition

The finite element (FE) model used for the preliminary analyses is illustrated in Figure 4, which also outlines the analytical conditions. This model, oriented along a single slip direction $[1\bar{5}2]$, is discretized into 24 × 80 × 4 crossed-triangular elements. The evaluation radius for the strain gradient calculation is set to $r_{GR} = 1.0$ μm, within which the average slope of the strain values is determined using the least squares method. Cyclic straining is applied under conditions controlled by the plastic strain amplitude of $\Delta\varepsilon^p/2 = 0.6\%$.

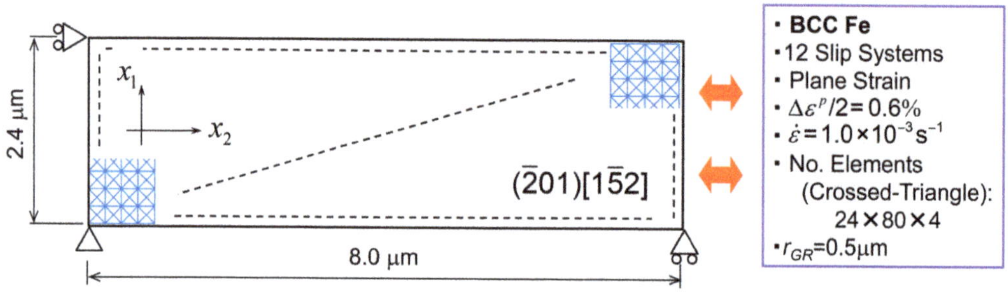

Figure 4. FE model for preliminary analyses: single slip-oriented single crystal. Triangles on the model boundary represent constraint conditions, where displacements are fixed in all directions, except in the rolling direction, which is specified by the pair of rollers at the bottom.

The present study focuses on commercially pure iron (Fe), with the aim of expanding the discussions about Cu-added steels described in conjunction with Figure 1A. In future research, we plan to explore the influence of dislocation substructures on slip band formation and the subsequent transition to crack propagation. All material parameters utilized in the current series of simulations, including those for $F_k(\alpha_k^{(\alpha)}; \eta_k^{(\alpha)})$ in Equation (18), are listed in Table 1, while those for $f_{\alpha\beta}$ in Equation (17) are summarized in Table 2 [57].

Table 1. Material parameters used in CP-FEM simulations, adapted from [57].

Category	Parameter	Value
Shear Strain Rate $\dot{\gamma}^{(\alpha)}$	$\dot{A}_{SR}[/s]$	2.75×10^{10}
	B_{SR}	145.5
	p	0.67
	q	2.0
Peierls Stress $\tau^{*(\alpha)}_{Peierls}$	$\dot{\gamma}_{0P}[/s]$	8.75×10^6
	$\hat{\tau}^*_{Peierls}$[MPa]	356
	B^P_{SR}	3.493
	p_P	0.83
	q_P	2.0
	μ	$(= C_{44})$
Drag Stress $K^{(\alpha)}$	h_0	1.0×10^4
	τ_0[MPa]	116
	W^P_{sat}	6.0
	n	0.13
Shear Gradient Teams $F\left(\alpha^{(\alpha)}; \eta^{(\alpha)}\right)$	k_{p_α}	0.2
	k_{p_η}	0.1
	l_{defect}[μm]	10.0
	b[m]	2.48×10^{-10}
Elastic Stiffness C_{ij}[GPa]	C_{11}	237
	C_{12}	141
	C_{44}	116

Table 2. Interaction matrix for dislocation pairs for BCC metals. adapted from [57].

	1	2	3	4	5	6	7	8	9	10	11	12
1	1.065											
2	0	1.065										
3	0	0	1.065									
4	0	0	0	1.065								
5	1.48	1.343	1.343	0	1.065							
6	1.343	1.48	0	1.343	0	1.065						
7	1.343	0	1.48	1.343	0	0	1.065					
8	0	1.343	1.343	1.48	0	0	0	1.065				
9	1.48	1.343	0	1.343	1.48	0	1.343	1.343	1.065			
10	1.343	1.48	1.343	0	0	1.48	1.343	1.343	0	1.065		
11	0	1.343	1.48	1.343	1.343	1.343	1.48	0	0	0	1.065	
12	1.343	0	1.343	1.48	1.343	1.343	0	1.48	0	0	0	1.065
System Number	1	2	3	4	5	6	7	8	9	10	11	12
Schmid and Boas	B5	C5	D6	A6	B4	C3	D4	A3	B2	C1	D1	A2
Burgers vectors	111	$11\bar{1}$	$1\bar{1}1$	$1\bar{1}\bar{1}$	111	$11\bar{1}$	$1\bar{1}1$	$1\bar{1}\bar{1}$	111	$11\bar{1}$	$1\bar{1}1$	$1\bar{1}\bar{1}$
Slip planes	$1\bar{1}0$	$1\bar{1}0$	110	110	$10\bar{1}$	101	$10\bar{1}$	101	$01\bar{1}$	011	011	$01\bar{1}$

5. Results and Discussion

5.1. General Features

Figure 5 presents a simulated incompatibility contour on a representative slip system (α) along with its projected counterpart on the primary slip system (β), based on the process

$$\left\langle F\left(\eta^{(\alpha)}\right)\right\rangle_{projected} \equiv \sum_{\beta=1}^{n} \left(\boldsymbol{m}_{primary}^{(\alpha)} \cdot \boldsymbol{m}^{(\beta)}\right) F\left(\eta^{(\beta)}\right) \tag{21}$$

where $\boldsymbol{m}^{(\alpha)}$ represents the slip plane normal, and the subscript *"primary"* specifies that for the primary system as $\boldsymbol{m}_{primary}^{(\alpha)}$. As confirmed in Figure 5b, the projection onto the primary slip plane effectively cancels the incompatibility pattern in the matrix region, leaving only the ladder-like pattern along the slip band. This suggests that the incompatibility-based patterning can essentially be categorized as a fluctuation, resulting in a net value of zero overall, in general. In contrast, the dislocation density contour, represented by the edge component α_{31} shown in Figure 5c, naturally exhibits a ladder-like structure, even without this projection. This indicates that the resultant dislocation substructure corresponds to the projected incompatibility distribution.

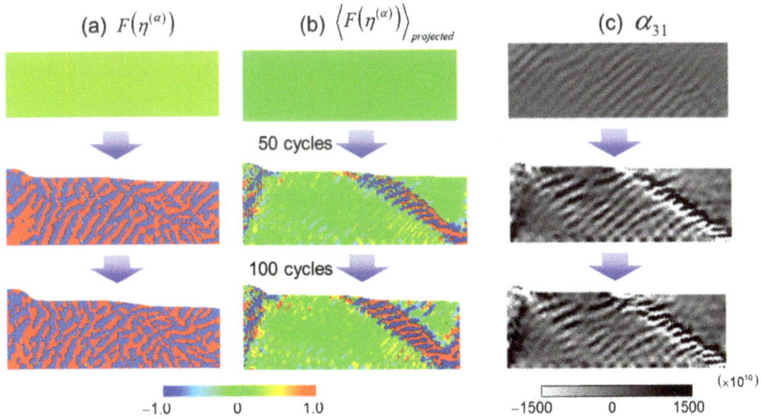

Figure 5. Incompatibility contours before and after projection onto the primary slip plane (via Equation (20)), together with dislocation density contour.

The reproduced PSB accompanied by laddered patterning, as anticipated above (see Section 3.1), underscores the predictive power of the FTMP framework, without any additional ingredients nor specifically-designated intricate mathematical models; moreover, essentially there is no need for elaborate finite element discretization. This reinforces our confidence in modeling seamlessly the fatigue crack initiation processes that such specific dislocation structures can be accurately reproduced without reliance on ad hoc models.

To demonstrate how the ladder-patterned region is softened, the sample is monotonically pulled after interrupting the cyclic straining. Figure 6a compares the samples pulled before the onset of cyclic straining (0 cycles) and after 80 cycles. As expected, the laddered sample exhibits landslide-like localized deformation, while the 0-cycle sample shows relatively uniform elongation. The results of monotonic tension after various straining cycles are presented in Figure 6b, indicating that 15 cycles of straining lead to complete softening of the PSB.

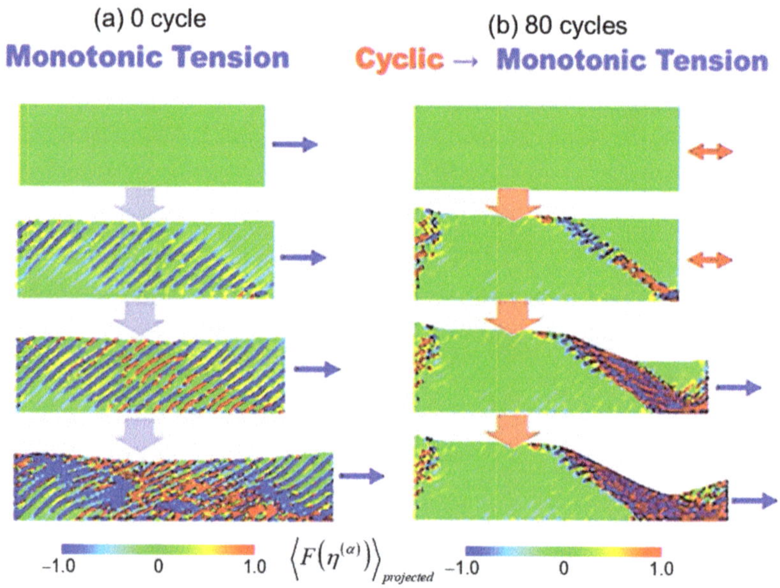

Figure 6. Comparison of sample deformation under monotonic tension (**a**) before (i.e., 0 cycle) and (**b**) after 80 cycles of cyclic straining (see Figure 7i), illustrating that the laddered region has been significantly softened.

Figure 8 compares the hysteresis loops for the matrix and PSB regions. The PSB region exhibits a plastic strain amplitude three times larger, highlighting strain localization along the laddered band, with this region accounting for approximately 14% of the total volume. Notably, the loop for the PSB region tends to shift toward the tension side to balance system-wide deformation, ensuring symmetric push–pull straining. It is worth mentioning that this concentration of deformation is less than what is typically observed in reality, primarily because the present simulation does not distinctly capture vein structures in the matrix region, which can impede deformation. However, reproducing the vein morphology in the matrix is not within the scope of this study at this stage.

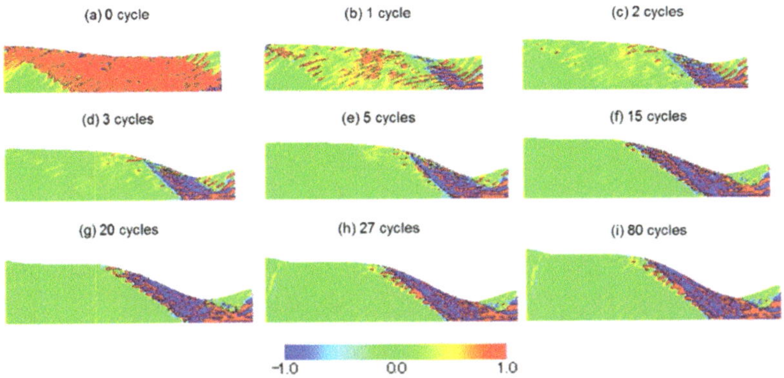

Figure 7. Comparison of sample deformation under monotonic tension for samples interrupted after various cyclic strain cycles. The labels (**a**) through (**i**) indicate the specific interrupting cycles: (**a**) 0, (**b**) 1, (**c**) 2, (**d**) 3, (**e**) 5, (**f**) 15, (**g**) 20, (**h**) 27, and (**i**) 80 cycles, respectively.

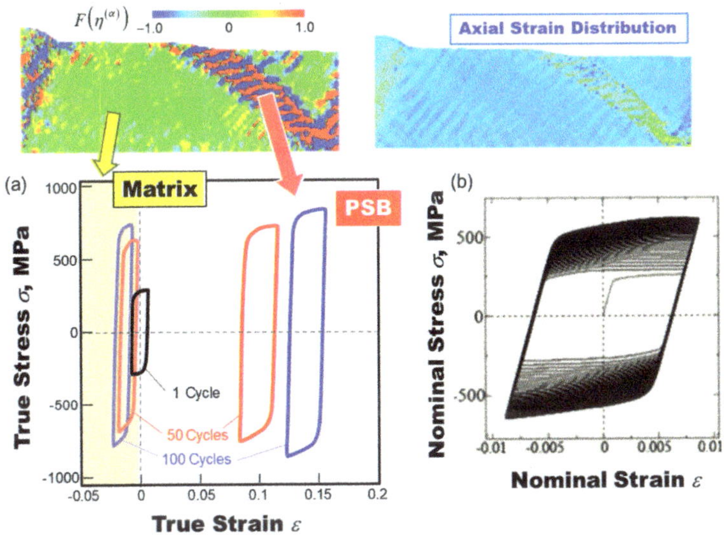

Figure 8. Comparison of simulated hysteresis loops between the matrix and PSB regions, (**a**) the matrix and (**b**) PSB regions.

5.2. Specific Features

Let us examine how the reproduced patterns reflect reality. Figure 9 analyzes the laddered region in terms of component-wise dislocation density distribution, while Figure 10 illustrates the growth of their cross-sectional distributions with increasing straining cycles. Here, we display α_{22} and α_{31} as representatives for the screw and edge components, respectively, while α_{11} and α_{32} show similar trends, although it is not presented. The following points are clearly demonstrated: first, the ladder morphology comprises both screw and edge dislocation components; second, these components are alternately aligned, closely resembling experimental observations, as also confirmed in Figure 10. Furthermore, there is essentially no misorientation developed throughout the sample, with a deviation of less than 1 degree, consistent with experimental observations. To be noted is that incompatibility-induced substructures are accompanied by misorientation developments in general.

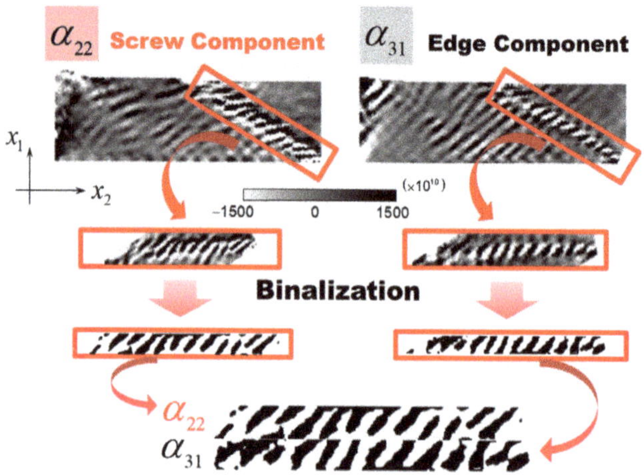

Figure 9. Comparison of dislocation density contours between edge and screw components, where binarized versions are rearranged vertically in the bottom for making comparison easier.

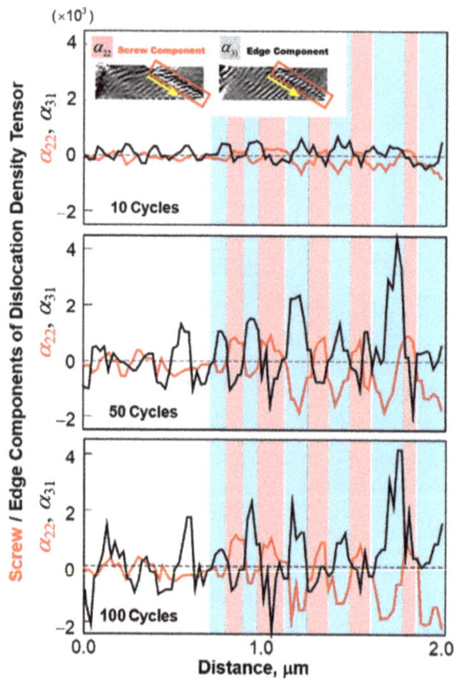

Figure 10. Cross-sectional distributions of dislocation density along PSBs (x-axis) comparing two components of dislocation density tensor, i.e., screw and edge, at 10, 50, and 100 cycles. Background coloring highlights areas with positive values for both components, making them easier to visualize. The red boxes in the inset show the PSB section where the distribution is measured, and the yellow arrows indicate the measurement direction, starting from the surface, corresponding to the horizontal axis.

On the less desirable side, we observe a slightly larger amplitude for α_{31} than for α_{22} in Figure 10, indicating that the structure is less capable of carrying deformation, as the screw segments play a key role in this capacity. This also contributes to the previously mentioned smaller strain concentration, alongside the less-developed vein-like morphology in the matrix region. Additionally, another difference remains in the width ratio of the ladder wall to channel regions, which ought to be much smaller, indicating potential room for further improvement.

5.3. Surface Roughening and Groove Formation

The strain concentration described earlier promotes surface roughening around the banded region, ultimately leading to grooving in that area. Figure 11 illustrates the evolving surface profile with increasing straining cycles, highlighting surface undulation up to N = 90 cycles. This local deformation creates a sample-wide biased surface slope, including a protruded left edge, as the system seeks overall equilibrium. Notably, the banded region displays a significantly larger slope, indicative of localized strain that contributes to grooving at its right edge. The magnified view on the left captures the early stage of groove formation at the intersection of the PSB and the surface, where a shallow but distinct groove tip is evident at the right edge of the laddered band. These observations clearly demonstrate that spontaneous substructuring alone can effectively reproduce surface roughening and initiate the formation of groove embryos.

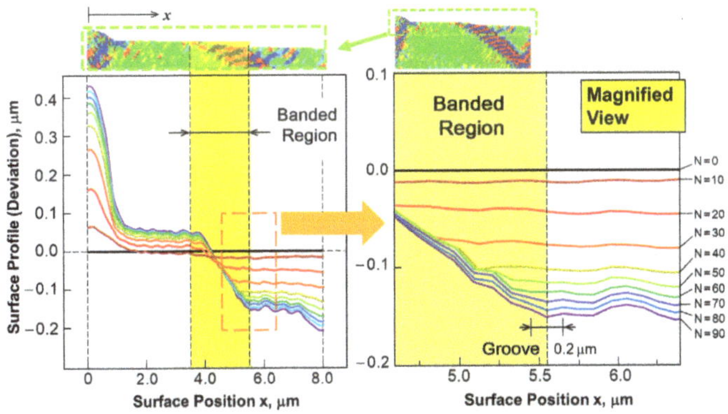

Figure 11. Variation in surface profile with straining cycles, accompanied by a magnified view illustrating the evolving groove at the PSB–matrix interface region. The surface region for measuring the profile is outlined by a green dashed box in the inset, while the yellow highlights denote the banded region.

5.4. Preliminary Analysis with Vacancy Effect Using a Simplified Model

To anticipate the effect of the formed pattern on the subsequent behavior of the PSB and the resulting development of surface undulations, we introduce a simplified model for vacancy formation. This serves as a prelude to the explicit treatments of the vacancy source model coupled with diffusion analysis that will be explored in Part II [69]. Assuming that the gradient of the elastic strain energy attended with the formed laddered pattern can promote vacancy diffusion, combined with the flow-evolutionary law Equation (8), we introduce the following simplified model via the incompatibility term:

$$\eta_{cv} = \kappa_v \delta U^e \tag{22}$$

The idea here is inspired by a systematic series of diffusion analyses described in Appendix D, where the diffusion equation for vacancy concentration is solved against an assumed ladder pattern specified by an a priori elastic strain energy distribution. In these diffusion analyses, diffusion is solely enhanced by the gradient of the elastic strain energy, which is now reflected through the incompatibility term $F(\eta_{cv})$ in the present analysis. We incorporate this term into the original $F(\eta)$ in the hardening ratio $Q_{\alpha\beta}$, anticipating a positive contribution to groove formation and further growth.

Figure 12 presents a comparison of incompatibility contours (top) and the corresponding elastic strain energy fluctuations (bottom) ($F(\eta)$ and δU^e) to assess the effect of the simplified vacancy model on the evolution of laddered patterns in the PSB region. In the absence of the vacancy model (right row), the laddered pattern tends to collapse partially along the righthand side edge of PSBs as straining cycles progress, as indicated by the incompatibility contours. This ultimately results in a relatively scattered distribution of δU^e around the PSB edge on the sample surface. In contrast, incorporating the vacancy model (left row) appears to stabilize the pattern, leading to a concentration of δU^e. We anticipate that this contribution to elastic strain energy fluctuations will foster the healthy development of surface grooves.

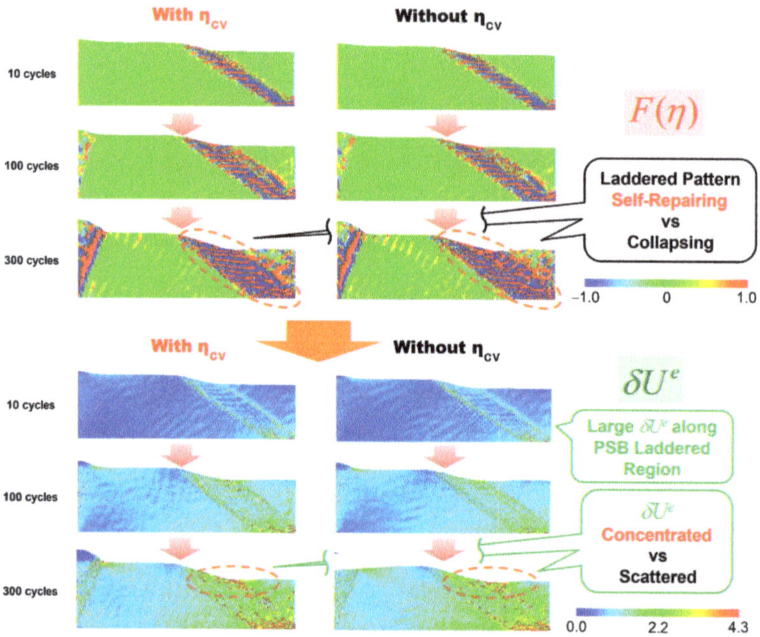

Figure 12. Effect of the simplified vacancy model on the evolution of incompatibility and the associated elastic strain energy fluctuations.

Figure 12 compares the evolution of surface profiles with and without the contribution of η_{cv}. The inclusion of η_{cv} leads to the formation of a single groove, whereas its absence results in a splitting of the groove peaks, becoming evident after 200 straining cycles. This behavior corresponds to the concentrated and diffused strain energy distributions seen in Figure 11.

To further investigate the role of η_{cv} in the final surface grooving, we examine the individual variations of the incompatibility terms $F(\eta)$ and $F(\eta_{cv})$ with respect to straining cycles in Figure 13. The term $F(\eta_{cv})$ itself increases parabolically with the number of cycles while exhibiting periodic oscillations, simply reflecting the applied cyclic straining. Adding $F(\eta_{cv})$ significantly increases the overall variation of $F(\eta)$ in the negative direction (bottom right), while subtracting it leads to a mild decrease in the overall change (bottom left). This upward trend in $F(\eta)$ enhances local deformation, resulting in a converging and sharpened groove, as confirmed by the final surface profile displayed at the top right of Figure 14.

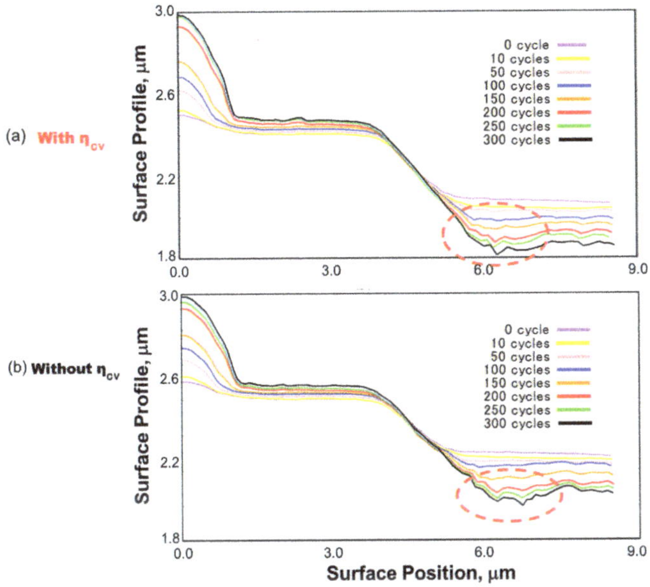

Figure 13. Evolution of surface grooving as a function of straining cycles, comparing scenarios with and without the simplified vacancy model. The influence of the vacancy model on the development of surface features over time is highlighted. Groove regions in the final surface profiles are marked by red dashed circles for clarity.

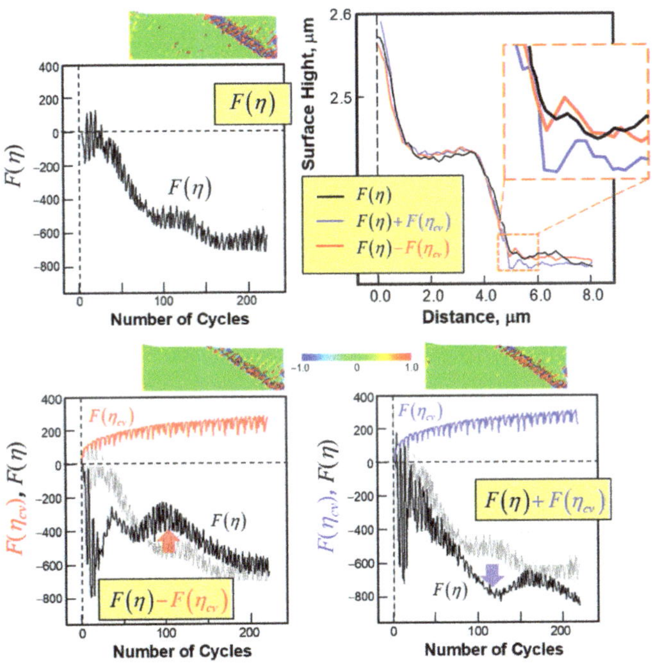

Figure 14. Variation of incompatibility terms with the number of cycles, illustrating the effect of the simplified vacancy model on surface grooving.

These results clearly demonstrate that the addition of $F(\eta_{cv})$ positively promotes grooving by significantly sharpening and deepening the groove profile, as anticipated. This ultimately leads to an accumulation of strain energy at the right edge of the PSB-sample surface intersection, whereas, in the absence of this term, the elastic strain energy tends to diffuse. We will utilize the final surface profile (b) for restart analyses to be conducted in Part II of the present study [69].

5.5. Discussion and Future Scope

In this study, we have successfully demonstrated that the FTMP-incorporated CP-FEM simulation can effectively reproduce the empirically observed laddered morphology and the PSB. Moreover, this simulation adeptly captures the dynamics of surface roughening and the eventual formation of grooves, even in the absence of vacancy formation and diffusion. These findings underscore the significance of the incompatibility tensor-based underlying degrees of freedom, which possess a remarkable capability to reproduce dislocation substructures, features that are often lacking in conventional CP-FEM approaches. This advancement not only helps enhance our understanding of PSB laddered patterning but also opens new avenues for further exploring the complex interplay of microstructural features that dictate the eventual crack initiation process, as will be extended in Part II [69].

The current simulation can be seamlessly extended to multi-grain scenarios, as demonstrated in Figure 15. This figure showcases three preliminary examples featuring single slip-oriented grains located in (a) the upper half, (b) the surface-facing grain, and (c) the central grain. In each case, we observe similar PSB ladder formations exclusively in the single-slip oriented grain, as observed previously, which intriguingly terminate at the grain boundaries. These results underscore the potential for expanding the present study to polycrystalline simulations, highlighting the versatility and applicability of our approach.

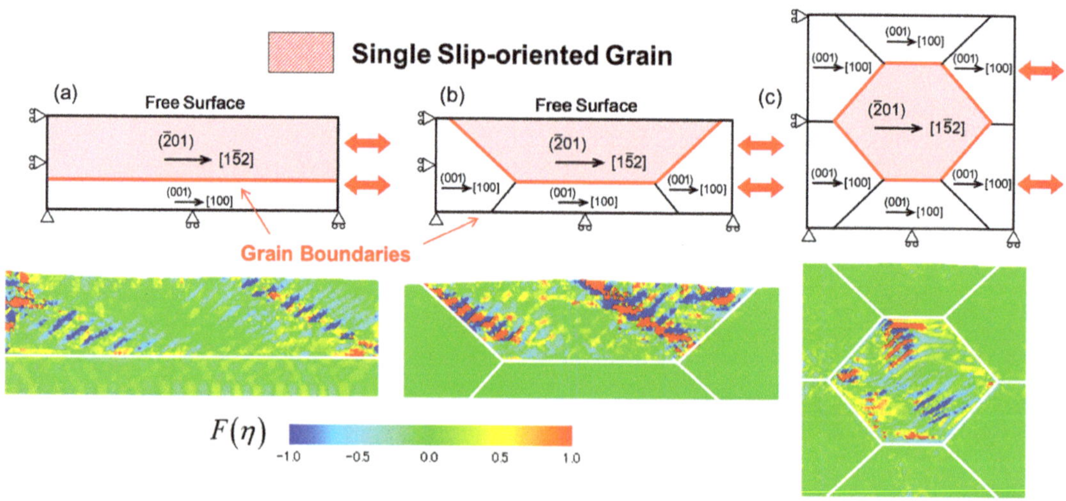

Figure 15. Examples of extended simulation results showing incompatibility contours in multi-grain scenarios featuring grain boundaries, featuring a single slip-oriented grain in (**a**) the upper half, (**b**) the surface-facing grain, and (**c**) the central grain of the samples, respectively.

6. Conclusions

In this study, we first demonstrate that PSB ladder substructuring during cyclic straining can be effectively reproduced through simple deformation analyses, showcasing the predictive capability of the FTMP framework. Notably, this achievement does not require any ad hoc models, intricate mathematical formulations, or elaborate finite element discretization. This reinforces our confidence in the practical feasibility of computational modeling for fatigue crack initiation processes and supports our advancement to the next stage of research without relying on complex modeling approaches.

Furthermore, the FTMP-incorporated CP-FEM simulation not only replicates the empirically observed laddered morphology and PSBs with satisfactory accuracy, but also adeptly simulates surface roughening and eventual grooving, independent of vacancy formation and diffusion. This highlights the crucial role of laddered morphology in these processes and lays the groundwork for further investigations into the effects of vacancy formation through explicit diffusion analyses in Part II [69]. These advancements stem from the incompatibility of tensor-based underlying degrees of freedom, which offer exceptional capabilities in modeling dislocation substructures often overlooked by conventional CP-FEM approaches. Thus, the robustness of the FTMP framework is validated, positioning it as a transformative tool in materials science and paving the way for deeper insights into slip band-fatigue crack transition.

Author Contributions: Conceptualization, T.H.; methodology, T.H.; software, X.Y. and T.H.; validation, T.H.; formal analysis, X.Y.; investigation, X.Y.; resources, X.Y. and T.H.; data curation, X.Y. and T.H.; writing—original draft preparation, X.Y. and T.H.; writing—review and editing, T.H.; visualization, X.Y. and T.H.; supervision, X.Y. and T.H.; project administration, T.H.; funding acquisition, T.H. All authors have read and agreed to the published version of the manuscript.

Funding: This research was partially supported by the Air Force Office of Scientific Research (AFOSR) under grant number FA4869-07-1-4101 and by JSPS KAKENHI for Scientific Research (C) under grant number 21560092.

Data Availability Statement: The original contributions presented in the study are included in the article, further inquiries can be directed to the corresponding author.

Conflicts of Interest: The authors declare no conflict of interest.

Appendix A. Cartesian Coordinate Expressions of Dislocation Density Tensor and Incompatibility Tensor and Extended 4D Spacetime Components

The explicit expressions of the dislocation density tensor and the incompatibility tensor for 3D in terms of plastic distortion tensor β^p_{ij} and the plastic strain tensor ε^p_{ij} are given as follows:

$$\begin{bmatrix} \alpha_{11} & \alpha_{12} & \alpha_{13} \\ \alpha_{21} & \alpha_{22} & \alpha_{23} \\ \alpha_{31} & \alpha_{32} & \alpha_{33} \end{bmatrix} = -\begin{bmatrix} \partial_2\beta^p_{31} - \partial_3\beta^p_{21} & \partial_2\beta^p_{32} - \partial_3\beta^p_{22} & \partial_2\beta^p_{33} - \partial_3\beta^p_{23} \\ \partial_3\beta^p_{11} - \partial_1\beta^p_{31} & \partial_3\beta^p_{12} - \partial_1\beta^p_{32} & \partial_3\beta^p_{13} - \partial_1\beta^p_{33} \\ \partial_1\beta^p_{21} - \partial_2\beta^p_{11} & \partial_1\beta^p_{22} - \partial_2\beta^p_{12} & \partial_1\beta^p_{23} - \partial_2\beta^p_{13} \end{bmatrix} \tag{A1}$$

$$\begin{bmatrix} \eta_{11} & \eta_{12} & \eta_{13} \\ & \eta_{22} & \eta_{23} \\ Sym & & \eta_{33} \end{bmatrix}$$
$$= \begin{bmatrix} (\partial_2^2\varepsilon^p_{33} + \partial_3^2\varepsilon^p_{22}) - 2\partial_2\partial_3\varepsilon^p_{23} & \partial_3\left(\partial_1\varepsilon^p_{23} + \partial_2\varepsilon^p_{31} - \partial_3\varepsilon^p_{12}\right) - \partial_1\partial_2\varepsilon^p_{33} & \partial_2\left(\partial_3\varepsilon^p_{12} + \partial_1\varepsilon^p_{23} - \partial_2\varepsilon^p_{31}\right) - \partial_3\partial_1\varepsilon^p_{22} \\ & (\partial_3^2\varepsilon^p_{11} + \partial_1^2\varepsilon^p_{33}) - 2\partial_3\partial_1\varepsilon^p_{31} & \partial_1\left(\partial_2\varepsilon^p_{31} + \partial_3\varepsilon^p_{12} - \partial_1\varepsilon^p_{23}\right) - \partial_2\partial_3\varepsilon^p_{11} \\ SYM & & (\partial_1^2\varepsilon^p_{22} + \partial_2^2\varepsilon^p_{11}) - 2\partial_1\partial_2\varepsilon^p_{12} \end{bmatrix} \tag{A2}$$

For extended 4D spacetime definition for the incompatibility tensor, we have

$$\eta_{ij} = \in_{iklp}\in_{jmnp}\,\partial_k\partial_m\varepsilon^p_{ln} \tag{A3}$$

The pure temporal component η_{44} is obtained by putting (*i*,*j*) = (4,4) in Equation (A3).

$$\eta_{44} = \in_{4klp}\in_{4mnp}\,\partial_k\partial_m\varepsilon^p_{ln} \tag{A4}$$

While the dummy index p runs from 1 to 3, the alternating symbol is restricted to three cases, as given by

$$\eta_{44} = \in_{4kl1}\in_{4mn1}\,\partial_k\partial_m\varepsilon^p_{ln} + \in_{4kl2}\in_{4mn2}\,\partial_k\partial_m\varepsilon^p_{ln} + \in_{4kl3}\in_{4mn3}\,\partial_k\partial_m\varepsilon^p_{ln} \tag{A5}$$

The first term on the right-hand side, for example, is

$$\begin{aligned}(\text{Eq.}(\text{A4}'))_{1st} &= \in_{4kl1}\in_{4mn1}\,\partial_k\partial_m\varepsilon^p_{ln} \\ &= \partial_2^2\varepsilon^p_{33} + \partial_3^2\varepsilon^p_{22} - 2\partial_2\partial_3\varepsilon^p_{23}\end{aligned} \tag{A6}$$

which coincides with the η_{11} component in Equation (A2). Similarly, the second and third terms correspond to η_{22} and η_{33}, respectively. Therefore, we conclude that the equality is

$$\eta_{44} = \eta_{KK} \tag{A7}$$

A spatio-temporal mixed component, η_{14} for example, by setting $(i\ j)$ = (1,4) in Equation (A3), becomes

$$\eta_{14} = \in_{1klp}\in_{4mnp}\,\partial_k\partial_m\varepsilon^p_{ln} \tag{A8}$$

Similar to the above, while the dummy index p can take values 2 or 3, the alternating symbol is restricted to two cases, as given by

$$\eta_{14} = \in_{1kl2}\in_{4mn2}\,\partial_k\partial_m\varepsilon^p_{ln} + \in_{1kl3}\in_{4mn3}\,\partial_k\partial_m\varepsilon^p_{ln} \tag{A9}$$

In this case, the first term on the right-hand side is further decomposed into four terms as

$$\begin{aligned}(\text{Eq.}(\text{A9}))_{1st} &= \in_{1kl2}\in_{4mn2}\,\partial_k\partial_m\varepsilon^p_{ln} \\ &= \in_{1342}\in_{4132}\,\partial_k\partial_m\varepsilon^p_{ln} + \in_{1342}\in_{4312}\,\partial_k\partial_m\varepsilon^p_{ln} \\ &\quad + \in_{1432}\in_{4312}\,\partial_k\partial_m\varepsilon^p_{ln} + \in_{1432}\in_{4132}\,\partial_k\partial_m\varepsilon^p_{ln} \\ &= \partial_3\partial_1\varepsilon^p_{43} - \partial_3\partial_3\varepsilon^p_{41} + \partial_4\partial_3\varepsilon^p_{31} - \partial_4\partial_1\varepsilon^p_{33} \\ &= \partial_3\left(\partial_1\varepsilon^p_{43} - \partial_3\varepsilon^p_{41}\right) + \partial_3\partial_4\varepsilon^p_{13} - \partial_4\partial_1\varepsilon^p_{33}\end{aligned} \tag{A10}$$

Noticing that $\partial_1\varepsilon^p_{43} - \partial_3\varepsilon^p_{41} = \partial_1\dot{u}^p_3 - \partial_3\dot{u}^p_1 = \dot{\omega}^p_{13}$, $\partial_4\partial_3\varepsilon^p_{13} = \partial_3\dot{\varepsilon}^p_{13}$, and $\partial_4\partial_1\varepsilon^p_{33} = \partial_1\dot{\varepsilon}^p_{33} = \partial_1\dot{\beta}^p_{33}$, along with the definition of α_{23} in Equation (A1), we can rewrite the above as

$$\begin{aligned}(\text{Eq.}(\text{A9}))_{1st} &= \partial_3\left(\dot{\omega}^p_{13} + \dot{\varepsilon}^p_{13}\right) - \partial_1\dot{\beta}^p_{33} \\ &= \partial_3\dot{\beta}^p_{13} - \partial_1\dot{\beta}^p_{33} \\ &= -\dot{\alpha}_{23}\end{aligned} \tag{A11}$$

Similarly, for the second term on the right-hand side, we have

$$\begin{aligned}(\text{Eq.A9}))_{2nd} &= \partial_2\left(\dot{\omega}^p_{12} + \dot{\varepsilon}^p_{12}\right) - \partial_1\dot{\beta}^p_{22} \\ &= -\left(\partial_1\dot{\beta}^p_{22} - \partial_2\dot{\beta}^p_{12}\right) \\ &= \dot{\alpha}_{32}.\end{aligned} \tag{A1}$$

Combining the above results (Equations (A11) and (A12)), we ultimately obtain

$$\eta_{14} = -\left(\dot{\alpha}_{23} - \dot{\alpha}_{32}\right) \tag{A13}$$

This represents a skew-symmetric part of the edge component in the dislocation density tensor from Equation (A1). The other spatio-temporal mixed components are given by

$$\eta_{24} = -\left(\dot{\alpha}_{31} - \dot{\alpha}_{13}\right) \text{and } \eta_{34} = -\left(\dot{\alpha}_{12} - \dot{\alpha}_{21}\right) \tag{A14}$$

In general, we express them collectively as

$$\eta_{A4} = -(\dot{\alpha}_{BC} - \dot{\alpha}_{CB}) = -2(\alpha_{BC})^{\cdot}_{skew} \tag{A15}$$

Appendix B. Historical Context and Derivation of the Flow-Evolutionary Law (FEL)

The current approach in FTMP insists that the flow-evolutionary law (Equation (4)) is based on the idea that "the incompatible displacement (i.e., displacement indeterminacy due to curvature) is driven by a configurational force (e.g., inhomogeneity or singularity), represented as a conserved quantity corresponding to the energy-momentum tensor, based on Noether's theorem" in 4D spacetime [72]. To this end, the incompatibility tensor is extended to 4D spacetime in [57]. Since curvature in crystalline space is geometrical and lacks an inherent driving force, it must be linked to a physical quantity that provides the driving force for its evolution. The flow-evolutionary law, proposed as a working hypothesis within the current FTMP, serves as a candidate for governing this process.

Historically, the incompatibility tensor has been used to derive stress fields related to dislocation density, as discussed by Kröner [73,74]. However, in the current context, we interpret this tensor as a continuum mechanics representation of the curvature tensor of the crystalline space, reflecting the classification of imperfections in terms of torsion and curvature based on differential geometry. This idea, first proposed by Kazuo Kondo in the 1950s as "non-Riemannian plasticity" [75], predates and is independent of the work by Bilby [76], Kröner, and Eshelby [77]. While Bilby acknowledged the relationship between torsion and dislocations, he dismissed the role of curvature, claiming that curvature changes the material vector along parallel displacement, thus no longer representing the crystal. Kondo countered this by stating that if torsion causes closure failure (e.g., in the Burgers circuit), it no longer represents a continuum. This suggests that Bilby's argument is incomplete within the continuum mechanics framework. Additionally, Bilby's work was influenced by E. Cartan's 1923 concept of torsion in spaces with asymmetric connections [78].

The incompatibility tensor coincides with Einstein's tensor in general relativity. Both satisfy the divergence-free condition, a key characteristic shared with the energy-momentum tensor, which appears on the right-hand side of Einstein's field equations for gravity. This connection underscores the mathematical parallels between continuum mechanics and general relativity, reinforcing the idea that the incompatibility tensor is essential for understanding material imperfections in a broader geometric context.

The energy-momentum tensor itself traces back to Emmy Noether's work in classical field theory [72], with its generalized formulation later published by Landau and Lifshitz in The Classical Theory of Fields (1923), Section 33 [79]. For an arbitrary filed variable φ_e^i, the energy momentum tensor T_{ab} is given as

$$T_{ab} = \frac{\partial \mathcal{L}}{\partial(\partial_a \varphi_{ie})}\partial_b \varphi_{ie} - \delta_{ab}\mathcal{L} \tag{A16}$$

where $\mathcal{L}$ represents the Lagrangian density of the system under consideration, and the indices a, b run from 1 to 4, with time as the fourth dimension. For a deformation field, we regard $\varphi_{ij} = \partial_i u_j = \beta_{ij}$ as the distortion tensor. When focussing solely on the spatial components (i.e., a, b, ... = A, B, ...), this expression reduces to that presented by Eshelby [80,81]. Assuming elasticity with C^e_{ABCD} as the elastic stiffness tensor, $\mathcal{L} = \frac{1}{2}\rho_0\beta_{i4}\delta_{ij}\beta_{j4} - \frac{1}{2}C^e_{ABCD}\varepsilon_{AB}\varepsilon_{CD}$, and one obtains

$$T_{AB} = -\sigma_{Ai}\beta_{Bi} + \delta_{AB}W \tag{A17}$$

This is precisely the expression derived by Eshelby in his 1975 paper [81] as P_{AB}. The pure temporal component, i.e., (a,b) = (4,4), on the other hand, corresponds to the total energy of the system

$$T_{44} = \frac{1}{2}\rho_0\beta_{i4}\delta_{ij}\beta_{j4} + \frac{1}{2}C^e_{ABCD}\varepsilon_{AB}\varepsilon_{CD} = \mathcal{H} \tag{A18}$$

where $\mathcal{H} \equiv \mathcal{K} + \mathcal{W}$ represents the Hamiltonian density. In the present study, the potential energy $\mathcal{W}$ is interpreted as the elastic strain energy U^e. Eshelby's work [81,82], particularly his 1975 paper [81], acknowledged Noether's discovery but derived the tensor P_{AB} from a mechanics perspective. While Eshelby's contribution to continuum solid mechanics was significant, his formulation is essentially the spatial part of the energy-momentum tensor defined in 4D spacetime, coinciding with Equation (A17), as shown above.

Parenthetically, a similar attempt is found in Epstein et al. [82], where the energy-momentum tensor (expressed as the Eshelby stress in their context) is interlinked with geometrical quantities. However, they ultimately arrived at a final form (referred to as the "full balance law" in their terminology) [83] involving the torsion tensor, rather than the curvature tensor.

Appendix C. Extended Divergence-Free Condition of the Incompatibility Tensor in 4D Spacetime

In this section, we derive Equation (11) from the divergence-free condition given by Equation (9). When extending the definition of the incompatibility tensor into 4D spacetime, the conservation law represented by the divergence-free condition must also be extended accordingly. The condition is given by

$$\begin{aligned} \frac{\partial \eta_{ab}}{\partial x^a} = \partial_a \eta_{ab} = 0 \Rightarrow \frac{\partial \eta_{44}}{\partial x^4} + \frac{\partial \eta_{A4}}{\partial x^A} = 0 \\ \Leftrightarrow \dot{\eta}_{KK} = -div\eta_{A4} \end{aligned} \tag{A19}$$

where Equation (A7) is used. The physical interpretation of this relationship becomes clear when considered in its integral form as

$$\begin{aligned} \frac{\partial}{\partial t}\int_v \eta_{KK} dv &= -\int_v \frac{\partial \eta_{A4}}{\partial x^A} dv \\ &= -\int_a n_A \eta_{A4} da \end{aligned} \tag{A20}$$

This indicates that the rate of change of incompatibility within the volume element dv is equivalent to the flux of incompatibility η_{A4} across the boundary surface da_A (the x^A-surface). Combining the results from Equation (A15) in Appendix A, we ultimately arrive at the important relationship between the rate of incompatibility and the edge dislocation flux.

$$\dot{\eta}_{KK} = div(\alpha_{BC} - \alpha_{CB})^{\bullet} \tag{A21}$$

Thus, we derive Equation (11).

Appendix D. Preparatory Diffusion Analysis on PSB-Laddered Sample

At the preparatory stage, we conducted pure diffusion analyses, assuming a PSB-laddered strain energy distribution a priori, to examine the effects of laddered morphology on vacancy diffusion and the resulting surface recession, including the wall position relative to the sample surface. This was achieved by solving the diffusion equation employed by Repetto and Ortiz [51] using the finite difference method (FDM). The analytical model and the key results are summarized in Figures A1 and A2. The initial vacancy concentration is assumed to be uniformly distributed across the sample, with an equilibrium value of $c_{veq} = \exp(-\Delta G_v/kT)$. Here, ΔG_v represents the free energy change per vacancy, approximately 0.9eV for α-Fe at room temperature, while k and T are Boltzmann constant and absolute temperature, respectively. The model considers only the contribution of lattice diffusion, represented by $D_{lattice} = 1.5 \times 10^{-15}$ m^2/s, while neglecting pipe diffusion. The findings are as follows:

(i) Insensitivity to the wall position.
(ii) Wherever vacancies are generated—whether in the channel or wall regions—they first flow into the wall, then out of the wall edges into the PSB–matrix interface, subsequently diffusing toward the surface along that interface.

(iii) Ladder walls play a critical role in developing and enhancing surface grooving; without them, surface recession does not occur.

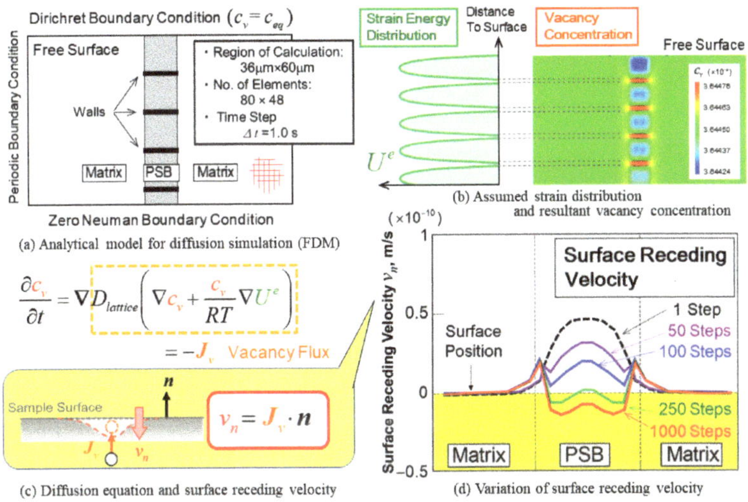

Figure A1. Overview of preparatory diffusion analyses using the finite difference method (FDM): (**a**) analytical model assuming a priori the ladder walls defined by elastic strain energy distribution (**b**), (**c**) diffusion equation employed, and (**d**) resultant surface receding velocity distribution via vacancy flux indicated in (**c**).

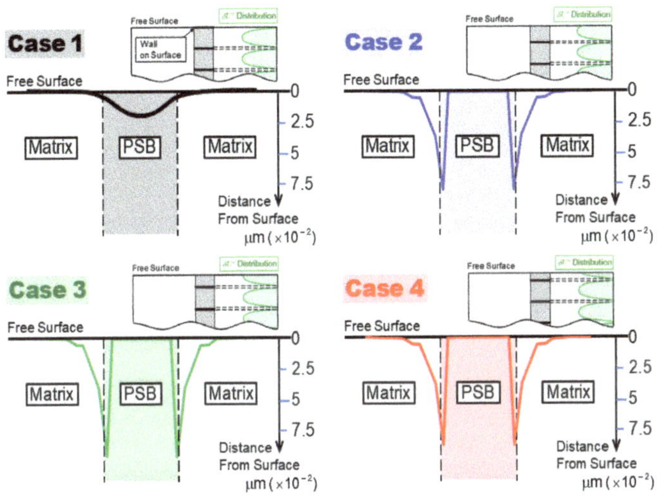

Figure A2. Simulated surface profiles at t = 1000 simulation steps, comparing the effect of ladder wall position relative to the sample surface for Cases 1 through 4.

References

1. Thompson, N.; Wadsworth, N.J.; Louat, N. The Origin of Fatigue Fracture in Copper. *Philos. Mag.* **1956**, *1*, 113–126. [CrossRef]
2. Woods, P.J. Low-amplitude Fatigue of Copper and Copper-5 at.% Aluminium Single Crystals. *Philos. Mag.* **1973**, *28*, 155–191. [CrossRef]
3. Polák, J. On the Role of Point Defects in Fatigue Crack Initiation. *Mater. Sci.* **1987**, *92*, 71–80. [CrossRef]

4. Polák, J.; Petrenec, M.; Man, J. Dislocation Structure and Surface Relief in Fatigued Metals. *Mater. Sci. A* **2005**, *400–401*, 405–408. [CrossRef]
5. Polák, J. Production, Annihilation and Migration of Point Defects in Cyclic Straining. *Materialia* **2020**, *14*, 100938. [CrossRef]
6. Polák, J. Role of Persistent Slip Bands and Persistent Slip Markings in Fatigue Crack Initiation in Polycrystals. *Crystals* **2023**, *13*, 220. [CrossRef]
7. Essmann, U.; Mughrabi, H. Annihilation of Dislocations during Tensile and Cyclic Deformation and Limits of Dislocation Densities. *Philos. Mag. A* **1979**, *40*, 731–756. [CrossRef]
8. Essmann, U. Irreversibility of cyclic slip in persistent slip bands of fatigued pure fcc metals. *Philos. Mag. A* **1982**, *45*, 171–190. [CrossRef]
9. Jones, W.J.D.; Dover, W.D. Enhanced Diffusion Rates during the Fatigue of Metals. *Nature* **1966**, *209*, 704. [CrossRef]
10. Robergre, R.; Herman, H. Fatigue—Generation of Vacancies. *Nature* **1966**, *211*, 178–179. [CrossRef]
11. Zhang, X.; Stinville, J.-C.; Pollock, T.M.; Dunne, F.P.E. Crystallography and Elastic Anisotropy in Fatigue Crack Nucleation at Nickel Alloy Twin Boundaries. *J. Mech. Phys. Solids* **2021**, *155*, 104538. [CrossRef]
12. Li, L.; Zhang, Z.; Zhang, P.; Zhang, Z. A Review on the Fatigue Cracking of Twin Boundaries: Crystallographic Orientation and Stacking Fault Energy. *Prog. Mater. Sci.* **2023**, *131*, 101011. [CrossRef]
13. Zhu, G.; Kang, B.; Zhu, M.-L.; Xuan, F.-Z. Quantitative Assessment of Microstructural Damage for Interior Crack Initiation and High Cycle Fatigue Life Modeling. *Mater. Sci. Eng. A* **2024**, *904*, 146707. [CrossRef]
14. Chen, B.; Jiang, J.; Dunne, F.P.E. Is Stored Energy Density the Primary Meso-Scale Mechanistic Driver for Fatigue Crack Nucleation? *Int. J. Plast.* **2018**, *101*, 213–229. [CrossRef]
15. Pandey, V.B.; Singh, I.V.; Mishra, B.K.; Ahmad, S.; Venugopal Rao, A.; Kumar, V. A New Framework Based on Continuum Damage Mechanics and XFEM for High Cycle Fatigue Crack Growth Simulations. *Eng. Fract. Mech.* **2019**, *206*, 172–200. [CrossRef]
16. Borges, M.F.; Neto, D.M.; Antunes, F.V. Numerical Simulation of Fatigue Crack Growth Based on Accumulated Plastic Strain. *Theor. Appl. Fract. Mech.* **2020**, *108*, 102676. [CrossRef]
17. Pan, Y.; Wu, P.; Fan, S.; Peng, X.; Chen, Z. Peridynamic Simulation of Fatigue Crack Growth in Porous Materials. *Eng. Fract. Mech.* **2024**, *300*, 109984. [CrossRef]
18. Zhou, H.; Suzuki, Y.; Kinefuchi, M.; Schmauder, S.; Dogahe, K.; Shibanuma, K. Bridging Strategy between Microscopic and Macroscopic Crack Growth Simulations to Predict Fatigue Strength of Steels. *Int. J. Fatigue* **2023**, *168*, 107386. [CrossRef]
19. Babu Rao, V.; Phung, B.R.; Johnsson, B.T.; Spear, A.D. Statistical Analysis of Microstructurally Small Fatigue Crack Growth in Three-Dimensional Polycrystals Based on High-Fidelity Numerical Simulations. *Eng. Fract. Mech.* **2024**, *307*, 110282. [CrossRef]
20. Sun, B.; Xu, Y.-L.; Li, Z. Multi-Scale Fatigue Model and Image-Based Simulation of Collective Short Cracks Evolution Process. *Comput. Mater. Sci.* **2016**, *117*, 24–32. [CrossRef]
21. Zhao, Z.; Ramesh, M.; Raabe, D.; Cuitiño, A.M.; Radovitzky, R. Investigation of Three-Dimensional Aspects of Grain-Scale Plastic Surface Deformation of an Aluminum Oligocrystal. *Int. J. Plast.* **2008**, *24*, 2278–2297. [CrossRef]
22. Grilli, N.; Janssens, K.G.F.; Nellessen, J.; Sandlöbes, S.; Raabe, D. Multiple Slip Dislocation Patterning in a Dislocation-Based Crystal Plasticity Finite Element Method. *Int. J. Plast.* **2018**, *100*, 104–121. [CrossRef]
23. Radi, A.; Sajadifar, S.; Seyedmohammadi, S.V.; Krochmal, M.; Bolender, A.; Wegener, T.; Niendorf, T.; Yapici, G.G. On the Low-Cycle Fatigue Behavior of a Multi-Phase High Entropy Alloy with Enhanced Plasticity. *Int. J. Fatigue* **2023**, *173*, 107678. [CrossRef]
24. Prastiti, N.G.; Xu, Y.; Balint, D.S.; Dunne, F.P.E. Discrete Dislocation, Crystal Plasticity and Experimental Studies of Fatigue Crack Nucleation in Single-Crystal Nickel. *Int. J. Plast.* **2020**, *126*, 102615. [CrossRef]
25. Irastorza-Landa, A.; Van Swygenhoven, H.; Van Petegem, S.; Grilli, N.; Bollhalder, A.; Brandstetter, S.; Grolimund, D. Following Dislocation Patterning during Fatigue. *Acta Mater.* **2016**, *112*, 184–193. [CrossRef]
26. Zhai, H.; Jiang, J.; Zhang, W.; Zhang, Q.; Ma, X.; Wang, S.; Li, Z.; Qiu, W.; Xiao, C.; Lin, H. Microstructure Sensitivity of the Low Cycle Fatigue Crack Initiation Mechanisms for the $Al_{0.3}$CoCrFeNi High Entropy Alloy: In-Situ SEM Study and Crystal Plasticity Simulation. *Int. J. Fatigue* **2023**, *176*, 107871. [CrossRef]
27. Kubin, L.; Sauzay, M. Persistent Slip Bands: Similitude and Its Consequences. *Acta Mater.* **2016**, *104*, 295–302. [CrossRef]
28. Hussein, A.M.; El-Awady, J.A. Quantifying Dislocation Microstructure Evolution and Cyclic Hardening in Fatigued Face-Centered Cubic Single Crystals. *J. Mech. Phys. Solids* **2016**, *91*, 126–144. [CrossRef]
29. Heinrich, C.; Sundararaghavan, V. A method to predict fatigue crack initiation in metals using dislocation dynamics. *Corros. Rev.* **2017**, *35*, 325–341. [CrossRef]
30. Wang, Y.; Li, J. Phase Field Modeling of Defects and Deformation. *Acta Mater.* **2010**, *58*, 1212–1235. [CrossRef]
31. Xue, F.; Cheng, T.-L.; Lei, Y.; Wen, Y.-H. Phase-Field Modeling of Crack Growth under Coupled Creep-Fatigue. *Int. J. Fatigue* **2024**, *189*, 108577. [CrossRef]
32. Sangid, M.D. The Physics of Fatigue Crack Initiation. *Int. J. Fatigue* **2013**, *57*, 58–72. [CrossRef]
33. Lukas, P.; Klesnil, M.; Fiedler, R. Plastic Zone Around the Propagating Fatigue Crack. *Philos. Mag.* **1969**, *18*, 799–805. [CrossRef]
34. Grosskreutz, J.C.; Shaw, G.G. Microstructures at the Tips of Growing Fatigue Cracks in Aluminum Alloys. In *Fatigue Crack Propagation*; ASTM International: West Conshohocken, PA, USA, 1967; Volume 415, pp. 226–243. [CrossRef]
35. Wilkins, M.A.; Smith, G.C. Dislocation Structures Near a Propagating Fatigue Crack in an Al 12% mg Alloy. *Acta Metall.* **1970**, *18*, 1035–1043. [CrossRef]

36. Mughrabi, H. Dislocation Wall and Cell Structures and Long-Range Internal Stresses in Deformed Metal Crystals. *Acta Metall.* **1983**, *31*, 1367–1379. [CrossRef]
37. Mughrabi, H.; Ungar, T.; Kienle, W.; Wilkens, M. Long-Range Internal Stress and Asymmetric X-Ray Line Broadening in Tensile Deformed [001]-Oriented Copper Single Crystals. *Philos. Mag. A* **1986**, *53*, 793–813. [CrossRef]
38. Mughrabi, H.; Ackerman, F.; Herz, K. Persistent Slip Bands in Fatigued Face-Centered and Body-Centered Cubic Metals. In *Fatigue Mechanics*; Special Technical Publication 675; Long, J., Ed.; American Society for Testing and Materials (ASTM): West Conshohocken, PA, USA, 1979; pp. 69–105. [CrossRef]
39. Bao-Tong, M.; Laird, C. Overview of Fatigue Behavior in Copper Single Crystals—I. Surface Morphology and Stage I Crack Initiation Sites for Tests at Constant Strain Amplitude. *Acta Metall.* **1989**, *37*, 325–336. [CrossRef]
40. Essmann, U.; Gosele, U.; Mughrabi, H. A Model of Extrusions and Intrusions in Fatigued Metals. *Philos. Mag. A* **1981**, *44*, 405–426. [CrossRef]
41. Suresh, S. *Fatigue of Materials*, 2nd ed.; Cambridge University Press: Cambridge, UK, 1998; pp. 53–68. [CrossRef]
42. Bretschneider, J.; Holste, C.; Kleinert, W. Mechanical Behaviour and Development of Dislocation Arrangements of fcc Single Crystals Fatigued at 77 K. *Mater. Sci. Eng. A* **1995**, *191*, 61–72. [CrossRef]
43. Li, P.; Li, S.X.; Wang, Z.G.; Zhang, Z.F. Fundamental Factors on Formation Mechanism of Dislocation Arrangements in Cyclically Deformed Fcc Single Crystals. *Prog. Mater. Sci.* **2011**, *56*, 328–377. [CrossRef]
44. Ghoniem, N.M.; Walgraef, D. *Instabilities and Self-Organization in Materials*; Oxford University Press: Oxford, UK, 2008.
45. Walgraef, D.; Aifantis, E.C. Dislocation Patterning in Fatigued Metals as a Result of Dynamical Instabilities. *J. Appl. Phys.* **1985**, *58*, 688–691. [CrossRef]
46. Ghoniem, N.M.; Tong, S.-H.; Sun, L.Z. Parametric Dislocation Dynamics: A Thermodynamics-Based Approach to Investigations of Mesoscopic Plastic Deformation. *Phys. Rev. B* **1999**, *61*, 913–927. [CrossRef]
47. Kubin, L.; Sauzay, M. Persistent slip bands: The bowing and passing model revisited. *Acta Mater.* **2017**, *132*, 517–524. [CrossRef]
48. Dodaran, M.; Khonsari, M.M.; Shao, S. Critical operating stress of persistent slip bands in Cu. *Comput. Mater. Sci.* **2019**, *165*, 114–120. [CrossRef]
49. Brown, L.M. Dislocation bowing and passing in persistent slip bands. *Philos. Mag.* **2006**, *86*, 4055–4068. [CrossRef]
50. El-Awady, J.A.; Ghoniem, N.M.; Mughrabi, H. Dislocation modelling of localized plasticity in persistent slip bands. In Proceedings of the 136th TMS annual meeting and exhibition, Orlando, FL, USA, 25 February–1 March 2007; pp. 23–35. Available online: https://www.researchgate.net/publication/253348453_Dislocation_modelling_of_localized_plasticity_in_persistent_slip_bands (accessed on 1 January 2007).
51. Repetto, E.; Ortiz, M. A Micromechanical Model of Cyclic Deformation and Fatigue-Crack Nucleation in fcc Single Crystals. *Acta Mater.* **1997**, *45*, 2577–2595. [CrossRef]
52. Nakai, Y. Evaluation of Fatigue Damage and Fatigue Crack Initiation Process by Means of Atomic-Force Microscopy. *J. Soc. Mater. Sci. Jpn.* **2001**, *50*, 73–81. [CrossRef]
53. Nakai, Y.; Kusukawa, T. Quantitative Evaluation of Slip-Band Growth and Crack Initiation in Fatigue of 70-30 Brass by Means of Atomic-Force Microscopy. *Trans. Jpn. Soc. Mech. Eng.* **2001**, *67*, 476–482. [CrossRef]
54. Nakai, Y.; Maeda, K. Observations of fatigue Slip-Band Growth and Crack Initiation in α-Brass under Cyclic Shear Stresses by Means of Atomic-Force Microscopy. *J. Mater. Sci. Soc. Jpn.* **2003**, *52*, 625–630. [CrossRef]
55. Hasebe, T. Interaction Fields Based on Incompatibility Tensor in Field Theory of Plasticity-Part I: Theory. *Interact. Multiscale Mech.* **2009**, *2*, 1–14. [CrossRef]
56. Hasebe, T. Interaction Fields Based on Incompatibility Tensor in Field Theory of Plasticity-Part II: Application. *Interact. Multiscale Mech.* **2009**, *2*, 15–30. [CrossRef]
57. Hasebe, T.; Sugiyama, M.; Adachi, H.; Fukutani, S.; Iida, M. Modeling and Simulations of Experimentally-Observed Dislocation Substructures Based on Field Theory of Multiscale Plasticity (FTMP) Combined with TEM and EBSD-Wilkinson Method for FCC and BCC Poly/Single Crystals. *Mater. Trans.* **2014**, *55*, 779–787. [CrossRef]
58. Hasebe, T. Field Theoretical Multiscale Modeling of Polycrystal Plasticity. *Trans. MRS-J* **2004**, *29*, 3619–3624.
59. Hasebe, T. Multiscale Crystal Plasticity Modeling based on Field Theory. *Comput. Model. Eng. Sci.* **2006**, *11*, 145–156. [CrossRef]
60. Mizutani, K.; Nawa, Y.; Hasebe, T. Kink Modeling and Simulations Based on Field Theory of Multiscale Plasticity (FTMP) Part I: Explicit Kink Model and Double Compression Test. *Mater. Trans.* **2023**, *64*, 785–794. [CrossRef]
61. Hasebe, T.; Mizutani, K. Kink Modeling and Simulations Based on Field Theory of Multiscale Plasticity (FTMP) Part II: Implicit Kink Model and Scale-Free Treatment. *Mater. Trans.* **2023**, *64*, 795–804. [CrossRef]
62. Hasebe, T. *Field Theory of Multiscale Plasticity*, 1st ed.; Cambridge University Press: Cambridge, UK, 2024. [CrossRef]
63. Yokoi, T.; Takahashi, M.; Maruyama, N.; Sugiyama, M. Application of Controlled Cu Nano-Precipitation for Improvement in Fatigue Properties of Steels. *Nippon Steel Tech. Rep.* **2005**, *91*, 49–55.
64. Yokoi, T.; Takahshi, M.; Maruyama, N.; Sugiyama, M. Cyclic Stress Response and Fatigue Behavior of Cu Added Ferritic Steels. *J. Maters. Sci.* **2001**, *36*, 5757–5765. [CrossRef]
65. Chen, Z.; Kioussis, N.; Ghoniem, N.; Hasebe, T. Lubricant Effect of Copper Nanoclusters on the Dislocation Core in α-Fe. *Phys. Rev. B* **2008**, *77*, 014103. [CrossRef]

66. Kondo, K. Non-Riemannian Geometry of Imperfect Crystals from a Macroscopic Viewpoint. In *Memoirs of the Unifying Study of Basic Problems in Engineering and Physical Sciences by Means of Geometry*; Gakujutsu Bunken Fukyu-kai: Tokyo, Japan, 1955; Volume 1, pp. 458–469.
67. Kondo, K. Derivation of the Differential Equations of Yielding from General Variational Criteria in Analogy with General Relativity Theory and an Extension Thereof to Problems involving Critical Temperature. In *Memoirs of the Unifying Study of Basic Problems in Engineering and Physical Sciences by Means of Geometry*; Gakujutsu Bunken Fukyu-kai: Tokyo, Japan, 1962; Volume 3, pp. 215–227.
68. Ashby, B.F. The Deformation of Plastically Non-Homogeneous Materials. *Philos. Mag.* **1970**, *21*, 399–424. [CrossRef]
69. You, X.; Hasebe, T. Modeling and Simulation of Fatigue Crack Initiation Process Based on Field Theory of Multiscale Plasticity (FTMP) Part II: Modeling Vacancy Formation and Coupling with Diffusion Analysis. *Metals* 2024, *in press*.
70. Khan, A.S.; Huang, S. *Continuum Theory of Plasticity*; Wiley Interscience: Hoboken, NJ, USA, 1995.
71. Nemat-Nasser, S.; Plasticity, A. *Treatise on Finite Deformation of Heterogeneous Inelastic Materials*; Cambridge University Press: Cambridge, UK, 2004.
72. Noether, E. Invariant Variations Problems. *Math.—Phys. Kl.* **1918**, *2*, 235–257. (English Translation in Transp. Theory Stat. Phys. 1991, 1, 186–207). [CrossRef]
73. Kröner, E. *Kontinuumstheorie der Versetzungen und Eigenspannungen*; Ergebnisse der angewandten Mathematik; Springer: Berlin/Heidelberg, Germany, 1958; Volume 5, pp. 1–179.
74. Kröner, E. Allgemeine Kontinuumstheorie der Versetzungen und Eigenspannungen. *Arch. Ration. Mech. Anal.* **1959**, *4*, 273–334. [CrossRef]
75. Kondo, K. Non-holonomic Geometry of Plasticity and Yielding. In *Memoirs of Unifying Study of Basic Problems in Engineering and Physical Science by Means of Geometry*; Gakujutsu Bunken Fukyu-kai: Tokyo, Japan, 1955; pp. 453–572.
76. Bilby, B.A.; Bollough, R.; Smith, E. Continuous Distributions of Dislocations: A New Application of the Methods of Non-Riemannian Geometry. *Proc. Roy. Soc. A* **1955**, *231*, 263–273. [CrossRef]
77. de Wit', R. The Continuum Theory of Stationary Dislocations. *Solid State Phys.* **1960**, *10*, 249–292. [CrossRef]
78. Cartan, E. Sur les variétés à connexion affine et la théorie de la relativité généralisée (première partie). *Ann. Sci. L'école Norm. Supérieure* **1923**, *40*, 325–412. [CrossRef]
79. Lifshitz, E.M.; Landau, L.D. *The Classical Theory of Fields*, 3rd ed.; Pergamon Press: Oxford, UK, 1971; Volume 2, ISBN 0750627689. First published 1923. (In Russian).
80. Eshelby, J.D. The Force on an Elastic Singularity. *Philos. Trans. R. Soc. Lond. Math. Phys. Sci.* **1951**, *244*, 87–112. [CrossRef]
81. Eshelby, J.D. The Elastic Energy-Momentum Tensor. *J. Elast.* **1975**, *5*, 321–335. [CrossRef]
82. Epstein, M.; Elzanowski, M. *Material Inhomogeneities and Their Evolutions*; Springer: Berlin/Heidelberg, Germany, 2007. [CrossRef]
83. Epstein, M.; Maugin, G.A. The energy—Momentum tensor and material uniformity in finite elasticity. *Acta Mech.* **1990**, *83*, 127–133. [CrossRef]

metals MDPI

Article

Modeling and Simulation of Fatigue Crack Initiation Process Based on Field Theory of Multiscale Plasticity (FTMP): Part II: Modeling Vacancy Formation and Coupling with Diffusion Analysis

Xinping You [1] and Tadashi Hasebe [2,*]

1 Department of Mechanical Engineering, Graduate School, Kobe University, 1-1, Rokkodai-cho, Nada-ku, Kobe 657-8501, Japan; you.xinping@mail.mm4.scitec.kobe-u.ac.jp
2 Department of Mechanical Engineering, Kobe University, 1-1, Rokkodai-cho, Nada-ku, Kobe 657-8501, Japan
* Correspondence: hasebe@mech.kobe-u.ac.jp; Tel.: +81-07-8803-6113

Abstract: Cyclic straining simulations using incompatibility-incorporated crystal plasticity-FEM, which exhibit PSB ladder structure evolutions as detailed in Part I, are coupled with diffusion analyses of produced vacancies. A new vacancy source model is introduced based on the Field Theory of Multiscale Plasticity (FTMP), interpreting the relationship between the incompatibility rate and the flux of dislocation density as edge dipole annihilation processes. Both direct and indirect coupling diffusion analyses, with and without cyclic straining, demonstrate that varying incompatibility rates tend to further promote vacancy diffusion, leading to surface grooving, enhanced extension rates, and eventual transition to cracks. The findings reveal that (i) the evolved PSB ladder structure serves as a site for vacancy formation, (ii) it provides a diffusion path toward the specimen surface, and (iii) it significantly enhances groove extension rates. These factors effectively facilitate the transition from a "groove" to a "crack", evidenced by the abrupt acceleration of the extension rate, mirroring systematic experimental observations. These achievements validate the FTMP's capability to simulate complex phenomena and significantly deepen our understanding of slip band–fatigue crack transition mechanisms.

Keywords: fatigue; persistent slip band; crack initiation; crystal plasticity; field theory; non-Riemannian plasticity; finite element method; vacancy diffusion; dislocation dynamic

1. Introduction

Modeling the fatigue crack initiation process remains a critical yet challenging problem in both engineering and materials science, particularly within the multiscale modeling community [1–5]. Despite early breakthroughs, such as the pioneering numerical simulations by Repetto et al. in 1997 [6], progress has been slower than anticipated. Continuum simulations, including crystal plasticity-based finite element method (CP-FEM) [7–11], have largely played a supplementary role, often overshadowed by rapid advancements in experimental techniques like micro-beam X-ray Laue diffraction [12], in situ and ex situ SEM-EBSD (electron backscattering diffraction) [9,13], ECCI (electron channeling contrast imaging) [8], HR-DIC (high-resolution digital image correlation) [9,10,12,13], and AFM (atomic force microscopy) [14–17]. In addition, phase field simulations [18,19], along with physics-based microscopic techniques such as DDD (discrete dislocation dynamics) [20–24] and MD (molecular dynamics) [8,25,26], have further advanced the numerical treatment of plasticity, which may be applied to fatigue crack initiation. However, the limited ability of CP-FEM to capture the full complexity of microscale phenomena has constrained its potential to significantly enhance our understanding of this process [9,27].

In contrast, the Field Theory of Multiscale Plasticity (FTMP) [28–31] offers a more powerful and versatile framework. By naturally modeling deformation-induced field evolutions and resulting patterning, FTMP provides a more accurate and insightful representation of the fatigue process, as demonstrated in Part I of this work [32]. Unfortunately, CP-FEM-based research has largely been relegated to a supportive role, primarily reinforcing experimental results rather than driving novel simulation-based insights. This underutilization of CP-FEM in simulation-driven research represents a significant missed opportunity, as its full potential remains largely untapped in advancing our understanding of fatigue crack initiation and material behavior.

In Part I [32], a simulation model was developed that realistically mimics the ladder structure of persistent slip bands (PSBs), which naturally forms and evolves under cyclic plastic straining in a single-slip-oriented single-crystal sample, accompanied by eventual surface undulation and groove embryos. The effect of vacancy formation on surface grooving through incompatibility contributions was tentatively introduced; however, a more explicit model that incorporates both vacancy formation and subsequent diffusion is needed. The critical roles of vacancy formation in relation to the PSB ladder structure concerning fatigue crack initiation have been recognized for some time [33–35]. While some interpretations and visual models have been proposed in conjunction with PSB ladder structures, no explicit investigations combining deformation analyses have been conducted to date. To fully understand the fatigue crack nucleation process, an approach that reproduces the entire process, including vacancy effects, will be indispensable. This study extends our efforts by coupling analyses with vacancy diffusion after introducing a newly proposed vacancy source model. To this end, new samples are first prepared using a cut-and-paste operation based on the Di-CAP concept (deformation-induced context-dependent autonomic pluripotency; see Appendix C for details), and information about the grooved surface obtained from the results in Part I [32] is transferred.

For the detailed process of fatigue cracks evolving from surface grooves, which manifest the underlying PSBs, Nakai et al. [36–39] reported informative empirical observations using atomic force microscopy (AFM) combined with interrupted fatigue tests on various materials, as detailed in Part I [32] (see Appendix A). They focused on the transition from slip bands to cracks, noting that this transition occurs when the groove depth (referred to as slip distance) reaches a critical value. Focusing on the transition from slip bands to cracks once this critical condition is reached may provide a more effective framework, minimizing ambiguities in the modeling process. Moreover, utilizing this critical condition for assessing the transition to cracks can significantly reduce computational costs when integrated with our FTMP-based approach, as the FTMP does not always require intricate fine mesh divisions to accurately reproduce deformation-induced substructural patterns.

This study further utilizes FTMP [28–31] for modeling vacancy formation via the conservation law for the 4D extended incompatibility tensor, exhibiting a relationship between the incompatibility rate and the edge dislocation flux [28]. Combining with the diffusion equation proposed by Reppeto and Ortiz [6], the study performs indirect and direct coupling analyses on the simulated samples. Based on the simulations, an attempt is made to identify the essential roles of the PSB ladder structure for the fatigue crack initiation process.

2. Theory and Model

2.1. Model Preparation

Figure 1 schematically illustrates the processes involved in preparing the analytical model for the current series of simulations based on the Di-CAP concept. Samples are numerically machined from a parent sample that has developed a PSB ladder structure under cyclic straining, incorporating surface information separately derived from the simulation results obtained in Part I [32]. The material considered here is commercially pure Fe, as in Part I, with plans for future investigations involving Cu-added steel (see Appendix B). These prepared samples are subsequently used for restart analyses. The critical information to be transferred is the plastic distortion tensor $\beta^p_{ij} = \beta^p$, which enables us to reproduce the evolved deformation-induced patterns in the new model through the activation of the dislocation density tensor and the incompatibility tensor.

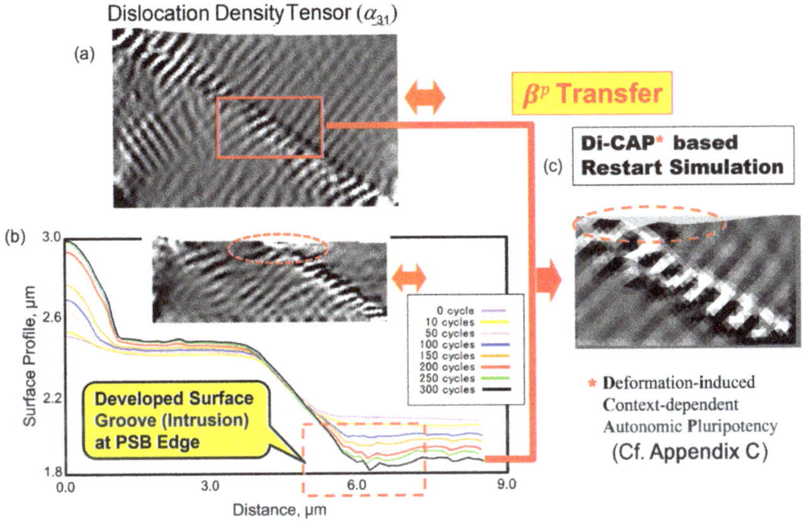

Figure 1. Overview of the present simulation: (**a**) Sampling from the parent model, (**b**) Simulated surface profile to be transferred, and (**c**) Restart simulation, where β^p information from (**a**) and the surface profile from (**b**) are incorporated, based on the Di-CAP concept (see Appendix C).

2.2. *Vacancy Source Model*

In the FTMP framework, the central hypothesis is encapsulated in the "Flow-Evolutionary Law (FEL)", which correlates the incompatibility tensor with fluctuations in elastic strain energy. Unlike the conventional three-dimensional incompatibility tensor, this approach extends the definition into four-dimensional spacetime. As a result, the purely temporal component of the FEL simplifies to the following relationship:

$$\eta_{KK} = \kappa\,\delta\,(U^e + K) \tag{1}$$

Here, the left-hand side originates from η_{44}, while the right-hand side simplifies to δU^e under the static assumptions applied in this study. For the four-dimensional extended incompatibility tensor, the conservation law is expressed as:

$$div^{4D}\boldsymbol{\eta} = 0 \Leftrightarrow \dot{\eta}_{44} + div\eta_{4A} = 0 \tag{2}$$

The spacetime mixed component η_{4A} is given by:

$$\eta_{4A} = -(\alpha_{BC} - \alpha_{CB})^{\cdot} \tag{3}$$

Substituting this expression into Equation (2), one obtains:

$$\dot{\eta}_{KK} = -div(\alpha_{BC} - \alpha_{CB})^{\cdot} \tag{4}$$

Thus, the incompatibility rate $\dot{\eta}_{KK}$ is interpreted equivalently as the edge dislocation flux, providing valuable insights for modeling processes involving edge dislocation pair production and annihilation. In the current study, this framework is specifically applied to model vacancy generation processes.

2.3. Diffusion Equation

The diffusion equation used in the following analysis is consistent with that employed by Repetto et al. [6] and is expressed as:

$$\frac{\partial c_v^{(\alpha)}}{\partial t} = \nabla \cdot \boldsymbol{D}^{(\alpha)} \cdot \left(\nabla c_v^{(\alpha)} + \frac{c_v^{(\alpha)}}{RT} \nabla U^e \right) + s_v^{(\alpha)} \tag{5}$$

In this equation, the diffusion coefficient is decomposed into lattice and pipe components:

$$\boldsymbol{D}^{(\alpha)} = \left\{ D_{lattice} + b^2 D_{pipe} \left(\boldsymbol{s}^{(\alpha)} \otimes \boldsymbol{s}^{(\alpha)} \right) : \boldsymbol{\alpha} \right\} \boldsymbol{I} \tag{6}$$

where D_{pipe} represents the diffusion coefficient for the pipe diffusion process, assuming a ratio against that for lattice diffusion as $D_{pipe}\ /\ D_{lattice} = 800$. The vacancy production rate $s_v^{(\alpha)}$ in Equation (5) is assumed to be driven by the incompatibility rate. Consequently, the vacancy source model is expressed as:

$$s_v^{(\alpha)} = a \times \left\langle \dot{\eta}_{KK}^{(\alpha)} \right\rangle \tag{7}$$

where a represents a conversion constant, and $\langle \bigcirc \rangle \equiv (\bigcirc + |\bigcirc|)/2$ denotes the Macauley bracket, accounting for vacancy production only.

Figure 2 provides an overview of the proposed vacancy source model along with the accompanying diffusion equation (5) and current simulation snapshots. It powerfully illustrates the advantages of the current FTMP-based simulation, emphasizing that all contributions in Equation (5) (and Equation (6)) are directly derived from the simulated patterned structure, specifically through the strain energy fluctuation δU^e, the incompatibility rate $\dot{\eta}_{KK}$ and the screw component of the dislocation density α_{11}, as schematized in Figures 2a, 2b and 2c, respectively. Furthermore, based on the vacancy concentration flux $J_v^{(\alpha)}$ outward from the sample surface, as schematized in Figure 2d, the receding surface velocity $v_v^{(\alpha)}$ is evaluated as follows:

$$v_n^{(\alpha)} = \boldsymbol{J}_v^{(\alpha)} \cdot \boldsymbol{n} \tag{8}$$

where $\boldsymbol{n}$ is the outward normal vector of the sample surface.

A previous application of the incompatibility rate $\dot{\eta}_{KK}$ was demonstrated in a series of dislocation dynamics (DD) simulations by Ihara et al. [40]. In the study, the incompatibility rate was found to correlate well with the number of edge debris loops formed by edge dislocations, which are believed to play a key role in vacancy formation within the PSB ladder structures upon collapse. Figure 3 summarizes these simulation results, with Figure 3a,b showing top and oblique views of snapshots for the edge dipole wall. These images reveal bowing-out dislocation segments, which eventually lead to the formation of debris loops in the channel region. The formation of these debris loops is strongly influenced by the frequency of cross-slip, as seen in Figure 3a. The detailed process for debris loop formation is shown in Figure 3c via a series of snapshots, highlighting several annihilation processes that contribute to it. Crucially, the number of debris loops is found to be well-correlated with the incompatibility rate $\dot{\eta}_{KK}$, as demonstrated in Figure 3d. This correlation provides key insight into the role of the incompatibility rate in vacancy formation within the PSB ladder structures.

The back-and-forth movements of screw dislocations can also lead to the pair annihilation of wall-constructing edge dislocations, as partially depicted in a snapshot in Figure 3b. This effect, however, is not included in the aforementioned correlation in Figure 3d. Since this process also contributes to vacancy formation in the ladder wall regions, it is expected that including it will reduce the data scatter in the correlation. Both of these contributions—

the formation of debris loops and dislocation pair annihilation—are collectively accounted for in Equation (7) in the present simulations.

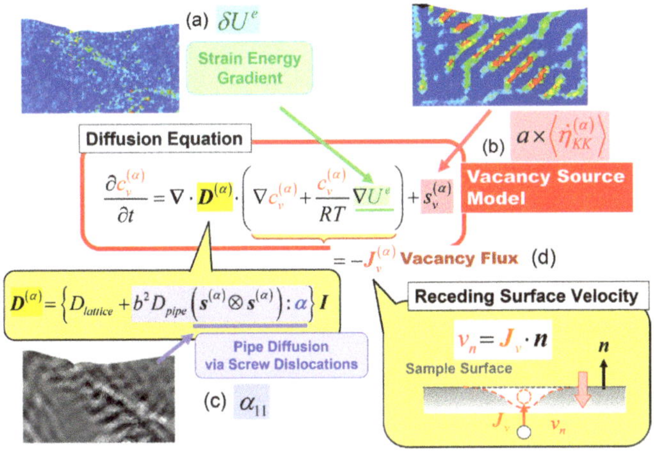

Figure 2. Schematics for diffusion analysis coupled with CP-FEM simulation Diffusion is enhanced by (**a**) the strain energy gradient via δU^e, (**b**) vacancy generation via $\dot{\eta}_{KK}^{(\alpha)}$, and (**c**) pipe diffusion via α_{11}, all of which are driven by the evolved laddered dislocation structure that underlies PSB. These mechanisms work together to promote surface recession via vacancy flux (**d**).

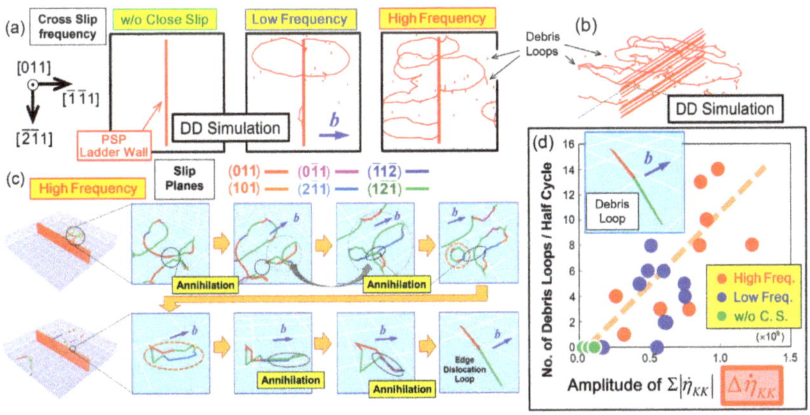

Figure 3. Dislocation dynamics simulations of ladder-wall structure [40]. (**a**) Top and (**b**) oblique views of bowing-out dislocation segments from the edge dipole wall, showing the effect of cross-slip frequency on debris loop formation in the PSB ladder region. (**c**) Process of debris loop formation, including annihilation events, and (**d**) correlation between the incompatibility rate $\dot{\eta}_{KK}$ and the number of debris loops.

Further evaluation of these DD results is provided through the phase-space trajectory (PhS-T) of the incompatibility rate, i.e., $\dot{\eta}_{KK}$ versus $\dot{\eta}_{KK}$, in Figure 4a, while Figure 4b illustrates the variation of $\dot{\eta}_{KK}$ over several stressing cycles. The effectiveness of the current vacancy source model, as described by Equation (7), is confirmed in Figure 4c,d, which compare the FTMP-based results for the PSB-laddered region with the DD-based counterparts, Figure 4a,b, both in terms of the PhS-T and the variation of the incompatibility rate. The comparison demonstrates excellent agreement between the two approaches, with the trends showing close quantitative correlation.

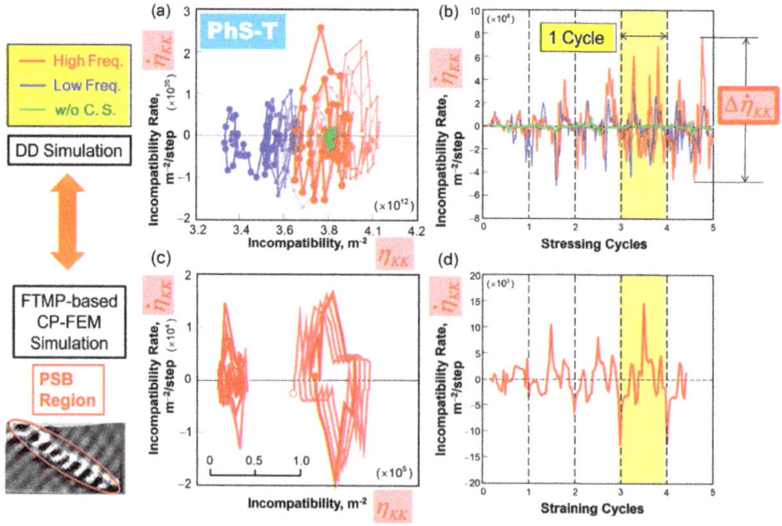

Figure 4. Comparison of phase-space trajectories (PhS-Ts) from discrete dislocation dynamics (DDD) simulations [40] and the current finite element (FE) analysis. (**a**,**b**) DDD results; (**c**,**d**) FE results. The DDD simulation demonstrates that the incompatibility rate $\dot{\eta}_{KK}$ effectively correlates with the number of debris loops (vacancy type with edge character) formed in the channels during screw dislocation motion (see Figure 3d).

2.4. Coupling Analyses

Coupling simulations are conducted by incorporating the proposed vacancy source model into the diffusion analysis in two steps: (i) indirect coupling analyses and (ii) direct coupling analyses. First, an indirect coupling simulation (i) is performed using a stationary snapshot that includes an already-developed PSB ladder structure. In the second step, a direct coupling simulation (ii) is carried out, solving the diffusion and cyclic straining analyses simultaneously. The flowchart of the direct and indirect coupling analyses is displayed in Figure 5, which clarifies the distinction between them by indicating whether cyclically varying data for δU^e, $\dot{\eta}_{KK}$ and α_{11} are utilized, as exemplified in Figures 2a, 2b and 2c, respectively.

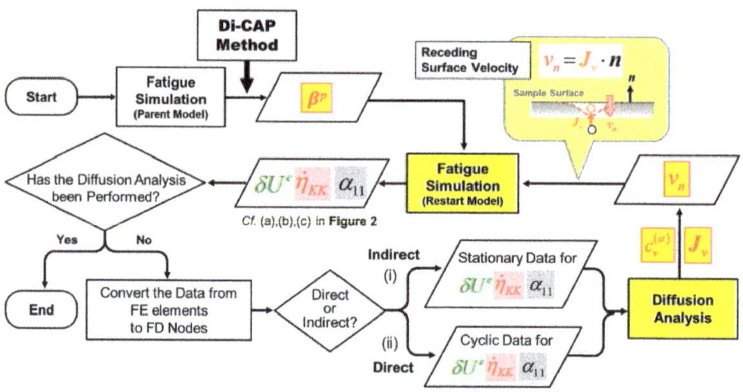

Figure 5. Flowchart of CP-FEM—diffusion coupling analyses for (i) indirect and (ii) direct versions through stationary or cyclically varying data for δU^e, $\dot{\eta}_{KK}$ and α_{11} ((a), (b) and (c) in Figure 2, respectively). The conversion of information between the two analyses is ultimately facilitated through v_n.

For the diffusion analyses, the finite difference method (FDM) is employed to solve Equation (5). The outcome of these analyses includes the vacancy concentration $c_v^{(\alpha)}$ and the attendant flux J_n, from which the surface receding velocity v_n is evaluated based on Equation (8) to update the surface profile. Figure 6 illustrates the analytical models for the CP-FEM and the associated diffusion analyses, highlighting how information about $c_v^{(\alpha)}$ and v_n is converted, specifically from elements to nodes between the FEM and FDM.

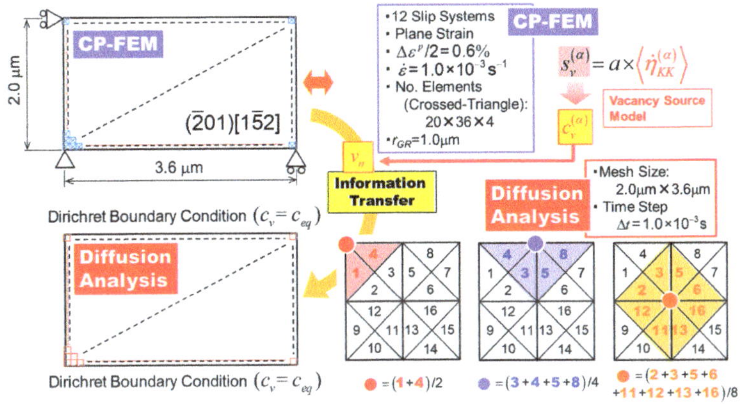

Figure 6. Analytical models for CP-FEM and diffusion analyses, accompanied by schematics that illustrate the information transfer process between the two methods.

3. Analytical Results

3.1. Indirect Coupling Analyses

Figure 7 summarizes the results of the indirect coupling simulations (i), comparing three representative conditions: with and without a threshold for applying the vacancy source model ($|\dot{\eta}_{KK}|$-threshold), and with and without β^p information transfer for the restart analyses, classified as Figures 7a, 7b and 7c, respectively. The effect of the conversion rate a in Equation (7) is also examined in each case.

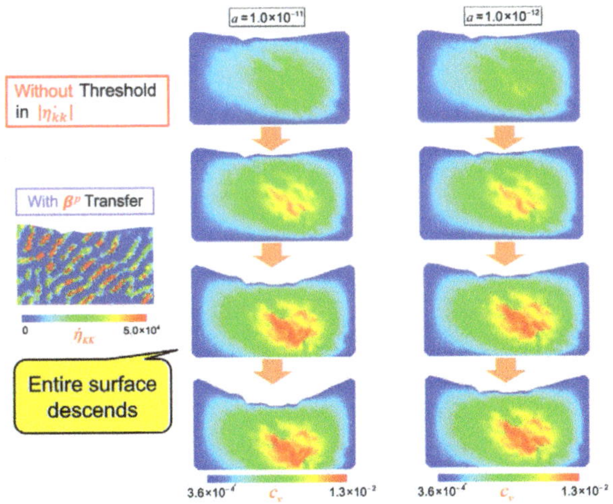

Figure 7. *Cont.*

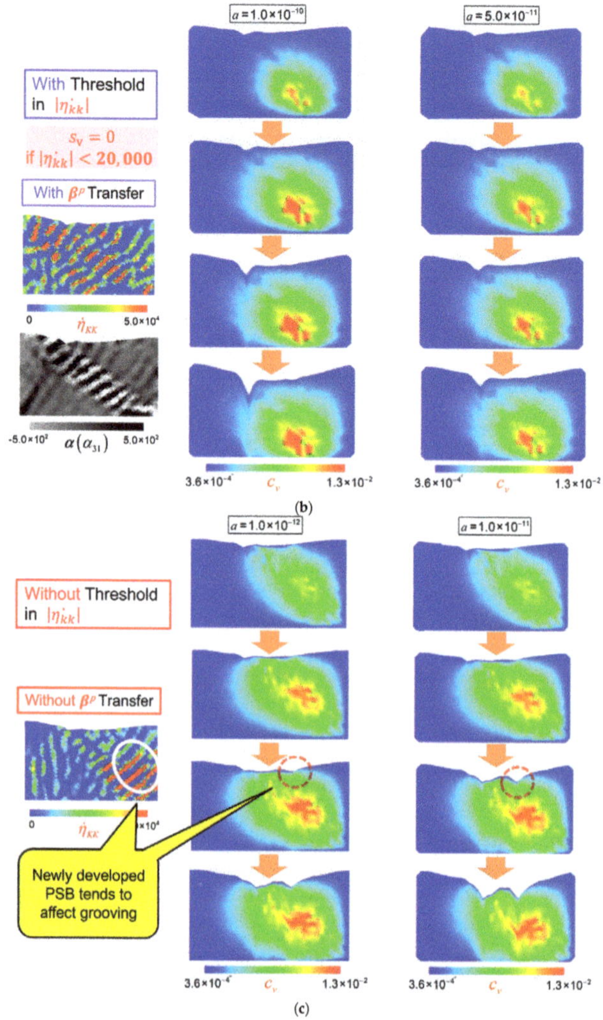

Figure 7. Results of indirect CP-FEM-diffusion coupling analyses for the groove–crack transition process using the proposed vacancy production model. The figure compares three conditions with and without $|\dot{\eta}_{KK}|$-threshold and β^p-information transfer, along with the effect of the conversion rate a: (**a**) without a threshold in $|\dot{\eta}_{KK}|$ and with β^p information transfer, (**b**) with a threshold in $|\dot{\eta}_{KK}|$ and β^p information transfer, and (**c**) without a threshold in $|\dot{\eta}_{KK}|$ and β^p information transfer.

When no threshold is set, as shown in Figure 7a, a sample-wide surface descent occurs regardless of the conversion rate, due to broader vacancy diffusion. This highlights the necessity of introducing a threshold for vacancy formation. When a threshold is applied, as depicted in Figure 7b, surface grooving becomes localized at the edge of the formed PSB, resulting from confined vacancy diffusion in the PSB region, as expected. Overall, a larger conversion rate a leads to a higher extension rate of the grooving.

Regarding the effect of β^p information transfer for the restart analyses, as examined in Figure 7c, it can significantly influence subsequent vacancy production behavior and the following surface grooving, as new PSBs are likely to emerge in response to the new boundary conditions without this transfer. In the present example cases, a new PSB emerges from the bottom-right of the sample, giving ultimately rise to another grooving on the other side of the PSB.

Since this study does not prioritize the further growth of the PSB structure and instead focuses on the extension of a single groove, the indirect coupling analyses employ the condition in Figure 7b. The physical rationale for setting the threshold for vacancy formation is related to the frequency of cross slip, based on DD simulation results by Ihara and Hasebe [40], which indicated no debris loop formation under this condition (see Figure 3). Alternatively, other factors related to vacancy formation may also be relevant and warrant further examination.

The resultant variation of the surface receding velocity distribution, calculated using Equation (8), is shown in Figure 8a as a function of straining cycles. It exhibits multiple peaks that ultimately converge into a single peak as the cycles progress. This leads to surface recession, resulting in groove growth, as demonstrated on the right. For comparison, the result without considering v_n is overlaid with a broken line. The change in the slip distance is measured as schematized in the inset, with a measurement indicated by double-sided arrows in Figure 8b. Figure 9 displays the variation of slip distance with the number of straining cycles, where the slope corresponds to the extension rate of the groove. It is observed that the groove extension rate accelerates sharply when a critical value of slip distance is reached, resembling the experimental observation by Nakai et al. [36–39], briefly described in Appendix A. In this context, it is concluded that the groove, representing a slip band, transitions into a crack, indicating that this transformation is successfully reproduced. In contrast, the scenario without considering vacancy diffusion, depicted by black circular plots, does not exhibit such acceleration; the slip band does not evolve into a crack.

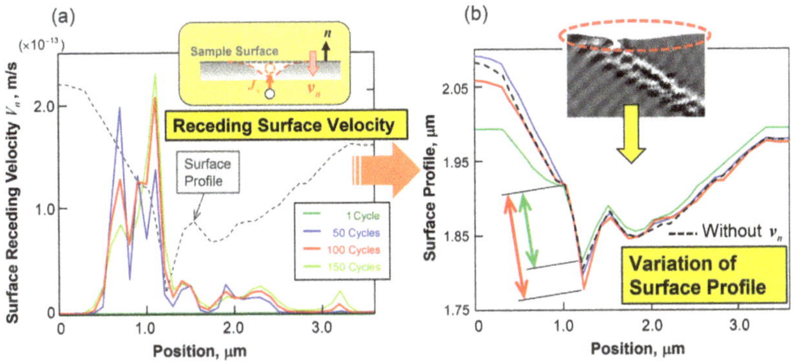

Figure 8. Simulated variations in surface receding velocity (**a**) and the corresponding surface profile with increasing straining cycles (**b**). In (**b**), the surface profiles are compared for the first cycle (green) and after 100 cycles (red), with the double-sided arrows indicating the measured slip distance.

3.2. Direct Coupling Analyses

The series of simulations described above can be readily extended to direct coupling analyses of CP-FEM and vacancy diffusion. In this section, the effect of the sampling site on changes in the surface profile during subsequent restart analyses is extensively examined.

Figure 10 illustrates three sampling sites on a parent specimen subjected to cyclic straining under two conditions: plane stress and plane strain. Markedly different evolutions of the PSB are observed, influenced by both the sampling site and the stress/strain condition. Notably, the plane stress condition promotes the growth of secondary and even tertiary PSBs alongside the original PSB, attributed to the increased freedom in the deformation mode during the restart analyses. This underscores a key feature of the FTMP-based approach, which effectively captures context-dependent evolutionary aspects of inhomogeneous fields, even in simple deformation analyses. Additionally, the figure includes experimental observations via TEM (top), demonstrating similar phenomena for comparison.

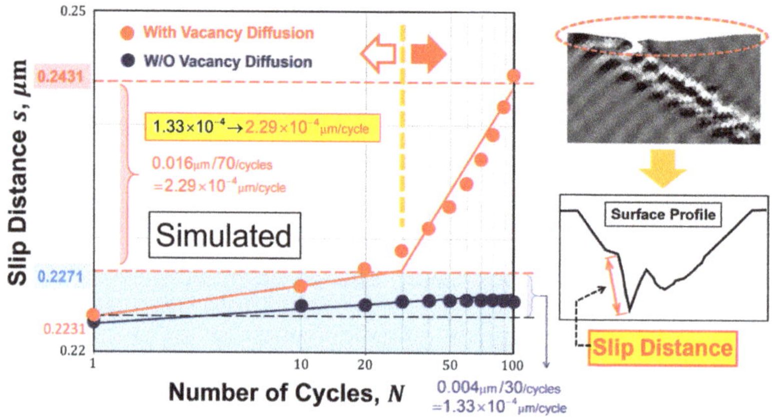

Figure 9. Simulated groove extension rates as they progress into a crack, comparing conditions with and without vacancy diffusion in indirect coupling analyses. The red and black circular plots indicate the simulated variation of slip distance with straining cycles, with and without vacancy diffusion, respectively. Open and solid arrows indicate the points before and after slip distance acceleration due to vacancy diffusion, marked by the orange vertical dashed line.

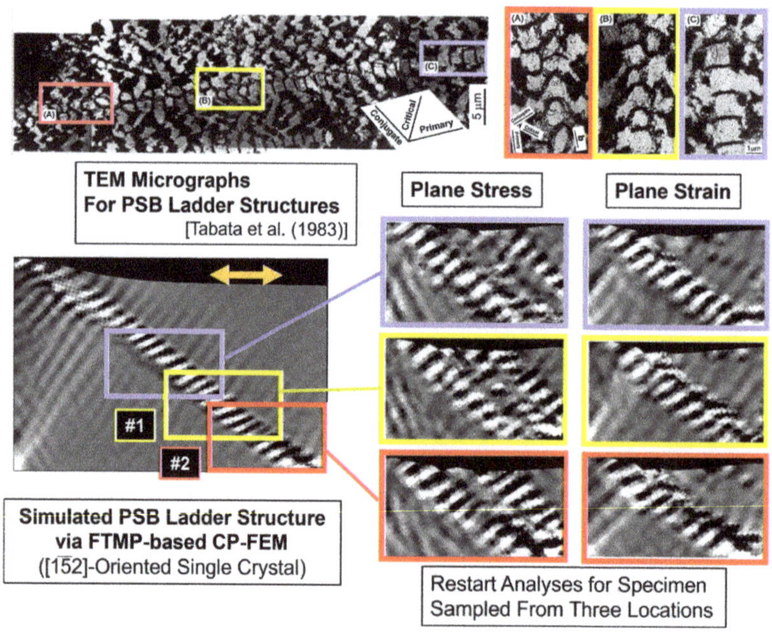

Figure 10. Sampling of restart specimens from a parent model exhibiting a PSB laddered pattern, comparing the results of restarted analyses based on different sampling sites and analytical conditions. Labels #1 and #2 indicate the sites selected for further coupling analyses. For comparison, similar experimental observations from Tabata et al. (1983) [41] are provided above, with labels A, B, and C indicating the sampling portions corresponding to the enlarged micrographs on the right (adapted with permission of the publisher (Taylor & Francis)).

Direct coupling analyses (ii) are performed using the above samples under plane strain conditions, effectively mimicking practical conditions. Two sites—the middle and bottom boxes in Figure 10, labeled #1 and #2—are selected for the coupling analyses, omitting surface information for simplicity. The conversion rate is set to $a = 1.0 \times 10^{-12}$ at%/s, with the threshold applied to the edge components of the dislocation density tensor rather than to $\dot{\eta}_{KK}$.

Figure 11 presents a series of simulated snapshots illustrating the variation in vacancy concentration contours across straining cycles, comparing sites #1 and #2. Notably, site #2 exhibits earlier vacancy production concentrated in the bottom right, followed by rapid growth toward the surface, while site #1 shows a delayed onset and more uniform growth. Contrasting features are observed between the two results: site #2 shows much earlier vacancy production concentrated in the bottom right, followed by rapid growth toward the surface, while site #1 exhibits a delayed onset and subsequent growth that is more uniform.

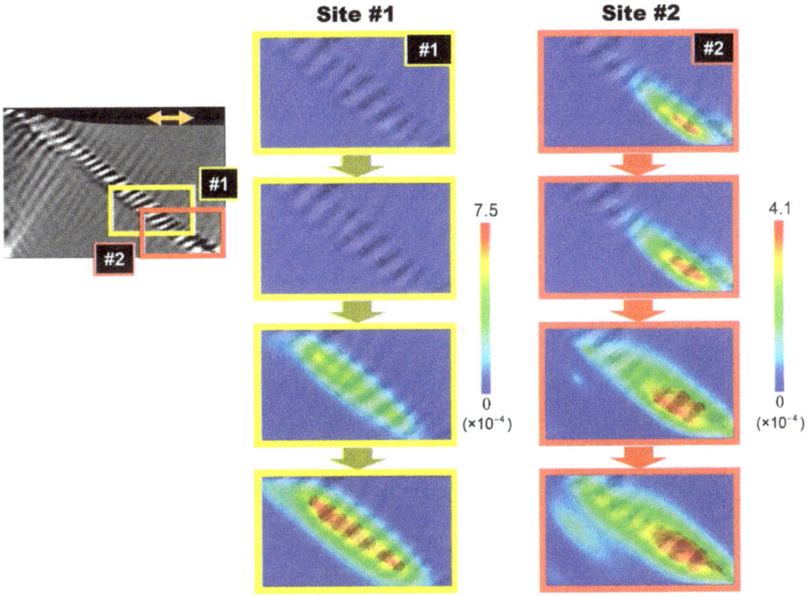

Figure 11. Series of snapshots showing the variation in vacancy concentration contour with straining cycles, comparing two sampling sites #1 and #2.

Figure 12 summarizes the results: variations in the cross-sectional distribution of vacancy concentration (left), surface receding velocity (middle), and surface profile (right) with progressing straining cycles. In response to the contrasting features observed in Figure 11, distinct trends emerge. Site #1 exhibits a relatively uniform evolution, leading to a fluctuating vacancy flux history (accompanied by peak shifts) at the surface, which ultimately results in a relatively blunted surface groove. In contrast, site #2 demonstrates significantly biased growth in the cross-sectional vacancy concentration distribution, resulting in a focused vacancy flux peak history (without peak shifts) that leads to sharp and deep groove formation. Additionally, a small extrusion can be seen forming on the right side of the PSB edge.

3.3. General Discussion

The current indirect/direct coupling analyses incorporate the effects of (i) dislocation density, (ii) elastic strain energy, and (iii) incompatibility rate on the associated vacancy concentration, as schematically illustrated in Figure 2. The key distinction between the indirect and direct simulations is whether these three quantities remain stationary or are subject to cyclic alteration. Therefore, the emphasis here is on how these cyclic updates affect diffusion behavior, surface grooving, and the eventual transition to cracking.

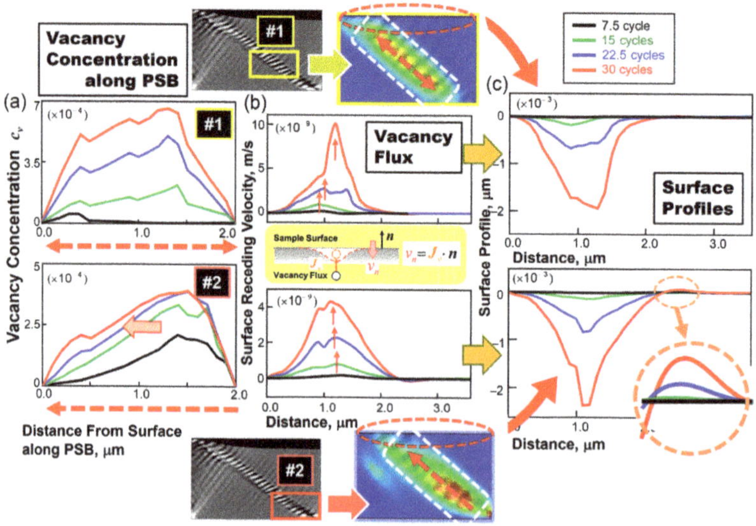

Figure 12. Representative results from direct CP-FEM-diffusion coupling analyses, illustrating variations of vacancy concentration distribution along the PSB (**a**), vacancy flux at the surface (**b**), and evolving surface profiles with straining cycles (**c**). Small arrows in (**b**) indicate the peak positions of vacancy flux, highlighting large and slight shifts for site #1 and site #2, respectively, which ultimately result in broad and narrow groove widths in (**c**).

Comparison is made between the two results regarding groove–crack transition using the slip distance versus number of cycles diagram in Figure 13, where the results for indirect analyses (shown in Figure 9) are overlaid with dashed lines. Additionally, a quantitative comparison—including experimental data—is presented in Table 1, detailing extension rates for grooves and cracks, acceleration, and the critical slip distance for crack nucleation. The numbers 1 through 6 associated with the experimental values correspond to those depicted in Appendix A, illustrating the conditions before and after the acceleration of the extension rate. The following points emerge from this comparison:

(a) The simulation results—both indirect and direct coupling analyses—successfully reproduce the experimental trends, even quantitatively. While no direct comparison is made, the critical slip distance values for the simulations are closely aligned with those for high-strength steel (HSS), lying between the values for α-brass and SUS304 stainless steel.

(b) Cyclic updates of the quantities significantly promote the grooving and subsequent crack nucleation process. The direct coupling analyses demonstrate a smaller critical slip distance and slightly larger acceleration in the extension rate compared to the indirect counterpart.

(c) The slip distance in the direct coupling analyses appears to be insensitive to the differing vacancy diffusion and surface recession trends shown in Figure 12. This observation offers valuable insight into determining the critical slip distance in experiments [36–39], which remains unaffected by the loading conditions.

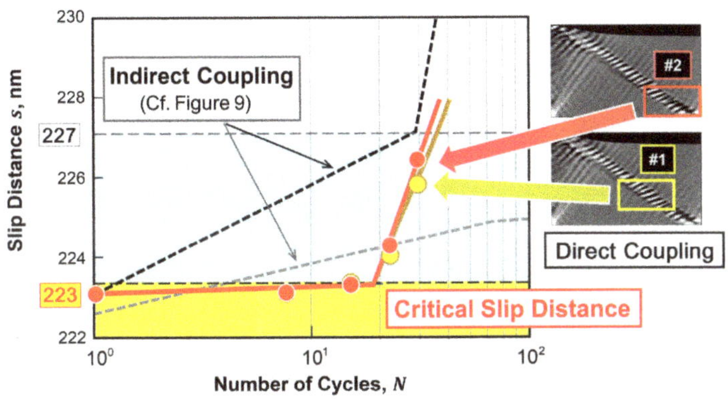

Figure 13. Overall comparison of the simulated groove extension rates (slip distance versus cycle number) for direct (red and yellow plots for sites #1 and #2, respectively) and indirect coupling analyses (black and grey dashed lines for with and without vacancy diffusion, respectively). The original diagram for the indirect coupling analysis is shown in Figure 9.

Table 1. Quantitative comparison of the simulated groove and crack extension rates, their acceleration, and critical slip distances with the experimental counterparts from Nakai et al. [36–39] (Cf. Appendix A). The numbers in circles associated with the experimental extension rates correspond to those specified in Figure A1c. The critical slip distance for high-strength steel was obtained from unpublished work, so detailed data, aside from the value itself, is unavailable.

		Extension Rate [μm/cycle] Groove ⇒	Crack	Acceleration [μm/cycle^2]	Critical Slip Distance [nm]
Analyses					
Indirect Coupling	With η_{cv}	1.33×10^{-4}	2.29×10^{-4}	0.96×10^{-4}	227
	W/o η_{cv}	2.52×10^{-5}	N/A	N/A	N/A
Direct Coupling	Site #1	1.47×10^{-5}	2.89×10^{-4}	2.74×10^{-4}	223
	Site #2	1.78×10^{-5}	2.37×10^{-4}	2.19×10^{-4}	223
Experiments					
SUS304		① 3.70×10^{-6}	② 8.80×10^{-4}	8.76×10^{-4}	100
		③ 6.67×10^{-6}	④ 0.89×10^{-4}	0.82×10^{-4}	
α-Brass		⑤ 5.45×10^{-6}	⑥ 1.11×10^{-5}	0.56×10^{-5}	380
High Strength Steel		Unpublished	Unpublished	Unpublished	168

The indirect case (i), however, requires an unrealistically large vacancy concentration on the order of 10^{-2}, while the direct case (ii) achieves this with a more reasonable concentration on the order of 10^{-4}. This difference is primarily due to the role of $\dot{\eta}_{KK}$. In the indirect case, $\dot{\eta}_{KK}$ remain fixed after the initial vacancy formation, based on its initial value. As a result, a high pre-existing vacancy concentration is necessary to drive the diffusion required for crack transition. In contrast, in the direct case, $\dot{\eta}_{KK}$ is updated cyclically throughout the simulation, ensuring a continuous supply of vacancies (as reflected in the enhanced vacancy formation rate $s_v^{(\alpha)}$ in Equation (5)). This dynamic updating of $\dot{\eta}_{KK}$ allows for a more realistic representation of the crack transition by maintaining vacancy concentrations at realistic levels. This behavior is clearly demonstrated in the snapshots shown in Figure 11, where distinct ladder-like patterns emerge in the vacancy concentration contours from the early stages of the straining cycles and persist throughout the later cycles.

The above results emphasize the critical roles played by the incompatibility tensor in the current FTMP for modeling and simulating the fatigue crack initiation process. It not only generates the PSB laddered pattern but also serves as a source of vacancies through its rate. Furthermore, the reproduced PSB laddered pattern provides a crucial diffusion pathway toward the sample surface, effectively facilitating the groove–crack transition, as previously demonstrated. These aspects also apply to the essential roles of the PSB ladder structure itself in the fatigue crack initiation process.

The current simulation results are insufficient to explain why experimental findings consistently indicate a material-dependent critical slip distance for groove–crack transitions, regardless of straining or loading conditions, aside from the tentative insights provided above in (c). Additionally, they do not yet clarify the factors that dictate this critical distance. However, this series of studies highlights the promising potential of simulation-driven investigations to explore these aspects of fatigue crack nucleation processes effectively far beyond the conventional crystal plasticity framework. Furthermore, the strength of the present vacancy source model based on $\left\langle \dot{\eta}_{KK}^{(\alpha)} \right\rangle$ lies its insensitivity to the specific mechanism of vacancy formation, as long as it accurately represents the pairwise annihilation of edge dislocations, as defined in Equation (4).

The present series of simulation results compellingly highlights the critical role of the incompatibility tensor within the Field Theory of Multiscale Plasticity (FTMP) in accurately capturing the fatigue crack initiation process. This tensor not only enables the reproduction of the PSB laddered pattern but also generates vacancies through its rate. Furthermore, the PSB ladder structure provides a crucial diffusion pathway to the sample surface. Together, these elements are enhanced by the cyclically varying incompatibility rate in the presence of the laddered dislocation structure, underscoring the essential influence of the PSB ladder structure on fatigue crack initiation.

A primary objective of this series of studies, encompassing Part I [32] and Part II (current study), is to elucidate the significant role of the PSB ladder structure in crack initiation, as evidenced by these simulations. Our comprehensive analyses of the transition from PSB to crack reveal three fundamental aspects inherent to the PSB ladder structure: one mechanical and two chemical roles, as summarized in Figure 14. These roles include (1) acting as a flow carrier and a site for stress concentration, (2) serving as a generator of vacancies, and (3) functioning as a diffusion pathway for these vacancies. The incorporation of the incompatibility tensor into the hardening law of the FTMP makes all of these processes possible.

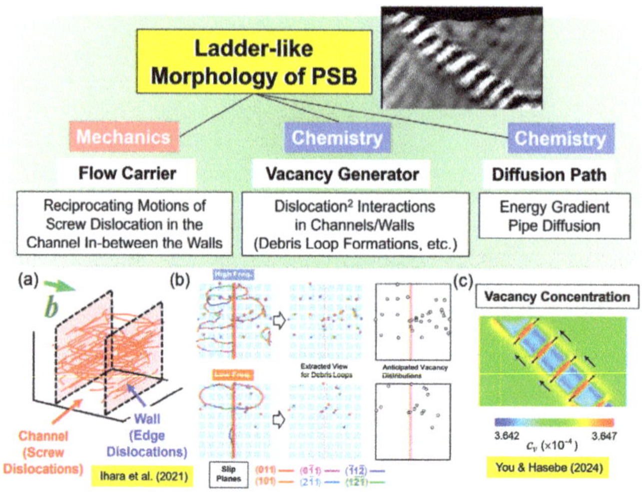

Figure 14. Three fundamental aspects inherent to the PSB ladder structure: one mechanical and two chemical roles. Corresponding representative simulation results are displayed in (a–c): (a) DD results for ladder walls responsible for concentrated strain due to the back-and-forth motions of bowed screw dislocations in the channel region [40], (b) formation of vacancy-type debris loops in the ladder channel regions associated with cross-slips, supposedly leading to vacancy formation [40], and (c) vacancy diffusion along the PSB toward the surface [32].

From this perspective, the formation of a laddered dislocation structure within PSBs plays a multifaceted role in stabilizing the crack initiation process. In contrast, other morphologies, such as cellular structures, may lead to less stable crack nucleation. This observation serves as an indirect yet significant conclusion to the first pivotal research finding presented in Part I [32] (see Appendix B for some details), underscoring the necessity for further investigation as these complexities are explored.

4. Concluding Remarks

This study extends cyclic straining simulations using incompatibility-incorporated CP-FEM, coupled with vacancy diffusion analyses, building on the work presented in Part I [32]. A new vacancy source model based on the Field Theory of Multiscale Plasticity (FTMP) is introduced, linking the incompatibility rate and dislocation density flux to edge dipole annihilation processes. Both direct and indirect coupling diffusion analyses successfully reproduce the transition from groove to crack. Notably, the direct coupling, which accounts for cyclic variations in incompatibility, significantly promotes vacancy diffusion, yielding results quantitatively consistent with experimental observations by Nakai et al. [36–39]. Key findings include (i) the evolved PSB ladder structure serving as a site for vacancy formation, (ii) providing a diffusion path to the surface, and (iii) accelerating groove extension rates, thereby facilitating the transition from groove to crack.

These results validate FTMP's ability to model complex phenomena, providing insights into slip band–fatigue crack transition mechanisms. The framework enhances our understanding of the interplay between the laddered morphology of the PSB and vacancy dynamics in the context of fatigue crack initiation, establishing a new benchmark for modeling dislocation substructures. This lays the groundwork for future advances in multiscale materials science, improving our ability to predict and mitigate fatigue failure.

Author Contributions: Conceptualization, T.H.; Methodology, T.H.; Software, X.Y. and T.H.; Validation, T.H.; Formal analysis, X.Y.; Investigation, X.Y.; Resources, X.Y. and T.H.; Data curation, X.Y. and T.H.; Writing—original draft, X.Y. and T.H.; Writing—review and editing, T.H.; Visualization, X.Y. and T.H.; Supervision, T.H.; Project administration, T.H.; Funding acquisition, T.H. All authors have read and agreed to the published version of the manuscript.

Funding: This research received no external funding.

Data Availability Statement: The original contributions presented in the study are included in the article, further inquiries can be directed to the corresponding author.

Conflicts of Interest: The authors declare no conflict of interest.

Nomenclature

β^p_{ij}; $\boldsymbol{\beta}^p$	Plastic distortion tensor.
α_{ij}; $\boldsymbol{\alpha}$	Dislocation density tensor.
η_{ij}; $\boldsymbol{\eta}$	Incompatibility tensor.
$\dot{\eta}_{KK}$	Incompatibility rate.
κ	Duality coefficient.
δU^e	Fluctuation of elastic strain energy. (J)
K	Kinetic energy. (J)
a	Conversion constant. (at%·s^{-1})
b	Burgers vector. (m)
$\boldsymbol{D}$	Diffusion coefficient tensor. (m^2·s^{-1})
$D_{lattice}$	Diffusion coefficient for the lattice diffusion. (m^2·s^{-1})
D_{pipe}	Diffusion coefficient for the pipe diffusion. (m^2·s^{-1})
$c_v{}^{(\alpha)}$	Diffusion concentration (for slip system α). (m^{-3})
s_v	Vacancy production rate. (m^{-3}·s^{-1})
J_v	Vacancy concentration flux. (m^{-2}·s^{-1})

v_n	Receding surface velocity. ($m \cdot s^{-1}$)
$\boldsymbol{n}$	Outward normal vector to the sample surface.
R	Gas constant. ($J \cdot kg^{-1} \cdot K^{-1}$)
T	Absolute temperatures. (K)

Appendix A. Overview of Experimental Results by Nakai et al. [36–39]

The present study references a significant series of experimental findings by Nakai et al. [36–39], which serve as the basis for the current series of simulations. Their work demonstrated a potential crucial condition for the transition from slip bands to fatigue cracks. The studies utilized semi-continuous atomic force microscopy (AFM) observations of specimen surfaces, combined with interrupted high-cycle fatigue tests on several materials. They concluded that the initiation of fatigue cracks occurs when the critical slip distance is reached, regardless of loading conditions, but depending on the materials tested.

Figure A1 summarizes the experimental results, displaying representative AFM observations that compare two typical cases of evolving surface profiles for intrusion/extrusion—those that have evolved into cracks Figure A1a and those that have not Figure A1b. Additionally, Figure A1c presents the variation of slip distance as a function of the number of stressing cycles. This latter figure primarily includes results for SUS304 austenitic steel, while also annexing results for α-brass and the critical slip distance for high-strength steel (HSS). The numbers attached in Figure A1c correspond to those used in Table 1.

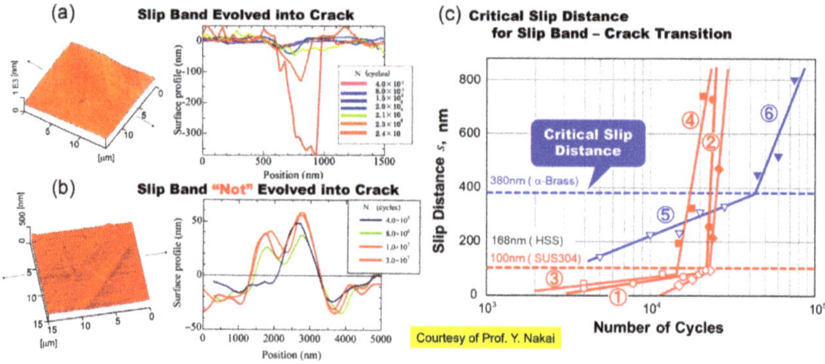

Figure A1. Summary of results by Nakai et al. [36–39] illustrating experimental demonstrations of the critical slip distance for the transition from slip bands to cracks in fatigue. (**a**) Comparison of typical AFM observations showing variations of surface profiles with stressing cycles, differentiating between slip banding that has evolved into cracks and that which has not. (**b**) Corresponding diagrams for SUS304 stainless steel, including the critical slip distances for α-brass and high-strength steel (HSS), where numbers in circles indicate the stage before (odd numbers) and after (even numbers) the transition from grooves (slip bands) to cracks, as referenced in Table 1 (**c**). Courtesy of Prof. Y. Nakai.

Appendix B. Fatigue Contol in Cu-Added Steels

Experimental observations by Yokoi et al. [42,43] show that adding copper (Cu) significantly alters dislocation substructures in steels. Steels without Cu typically exhibit a well-developed 3D dislocation cell structure, leading to coarse intrusions/extrusions and earlier crack initiation. Cu addition transforms the substructure from 3D cells to uniform 2D distributions, resulting in finer intrusions/extrusions, delayed crack initiation, and improved high-cycle fatigue life. Spesifically, solid-solution Cu forms a 2D vein structure, while Cu precipitates create uniformly distributed planar dislocations, shifting cyclic behavior from hardening to softening. These changes are crucial for controlling the fatigue properties of steels, optimizing performance in applications like automotive and power plants.

To understand the effects of Cu, it is essential to explore its quantum-level influence on aspects such as the magnetism of the bcc Fe atomic structure or the core structure of screw dislocations. An approach along these lines was presented by [44], which included one of the current authors, where ab initio simulations demonstrated how Cu addition causes a transition in the Fe screw dislocation core structure, from non-polarized to fully polarized. This transition potentially inhibits cross slip, which is responsible for dynamic recovery. The inhibition of dynamic recovery, in turn, suppresses the formation of 3D dislocation cells.

Appendix C. About Di-CAP Concept for Restart Analyses

A key strength of the FTMP framework is its ability to leverage evolved patterns for further modeling through "cut-and-paste" operations [30,31], maximizing its robust descriptive capabilities. This innovative process involves transferring information about the plastic distortion tensor, denoted as β^p, which enables the accurate reproduction of both the dislocation density tensor and the incompatibility tensor fields. By capturing these critical tensor fields, the methodology allows for a refined and nuanced representation of material behavior under a wide range of loading conditions, enabling simulations to evolve progressively toward more complex and varied scenarios.

The concept underlying this methodology is referred to as Di-CAP, which stands for Deformation-induced Context-dependent Autonomic Pluripotency, as illustrated in Figure A2. The term "pluripotency" signifies the method's ability to adapt and generate multiple potential outcomes based on the context of deformation, thus enhancing the descriptive and predictive capabilities of the model. A typical application of this approach is found in modeling hierarchical martensite lath structures for creep analyses.

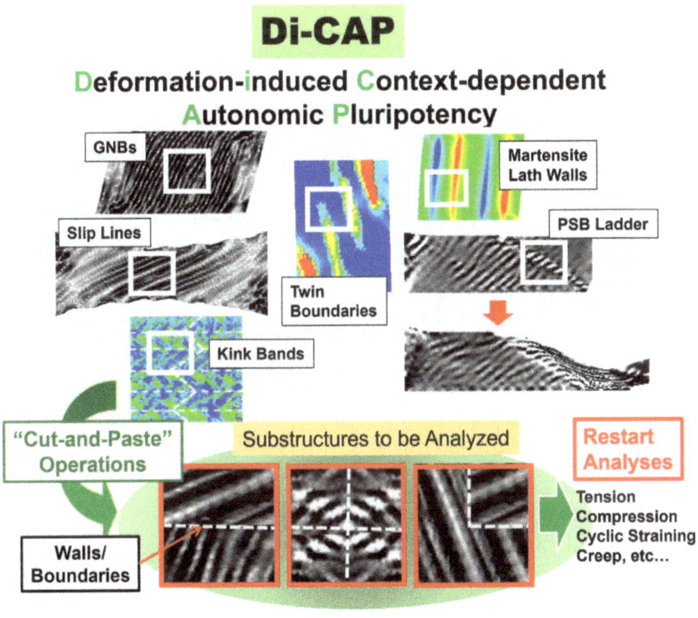

Figure A2. Schematics illustrating the Di-CAP concept and subsequent modeling/analyses based on arbitrary "cut-and-paste" editing operations. White squares on the snapshots above denote "cut" portions, while the red-bordered models at the bottom represent the "paste" constructions for use in restart analyses. White dashed lines indicate the boundaries of the cuts and pastes.

Figure A3 visually illustrates an example application of how unit structures are integrated to create intricate hierarchical lath martensite models, which have been utilized in creep rupture simulations based on CP-FEM. In this framework, complex lath martensite structures are computationally assembled by combining unit lath block structures like building blocks, allowing for a hierarchical representation of the material. Individual lath block models Figure A3a are combined to form single packet models Figure A3b, while these packet models are subsequently embedded into the prior austenite grain aggregate model to create the final model Figure A3c or Figure A3d. It is important to note that the lath block model itself has been developed in advance using FTMP-based CP-FEM, taking into account the initial transformation strain distribution based on the Bain lattice correspondence. This type of hierarchical modeling is undoubtedly crucial for multiscale simulations to accurately capture the mechanical behavior of materials, as different lath arrangements can significantly influence overall creep deformation and rupture processes in the present context.

The present study employs the Di-CAP concept for restart analyses, enabling a seamless transition between different simulation stages. This approach not only enhances computational efficiency but also improves the fidelity of results by ensuring that all relevant substructural information is preserved and accurately modeled. By leveraging the capabilities of Di-CAP, more comprehensive modeling and simulations of material behavior can be conducted, particularly in complex situations and across various loading scenarios.

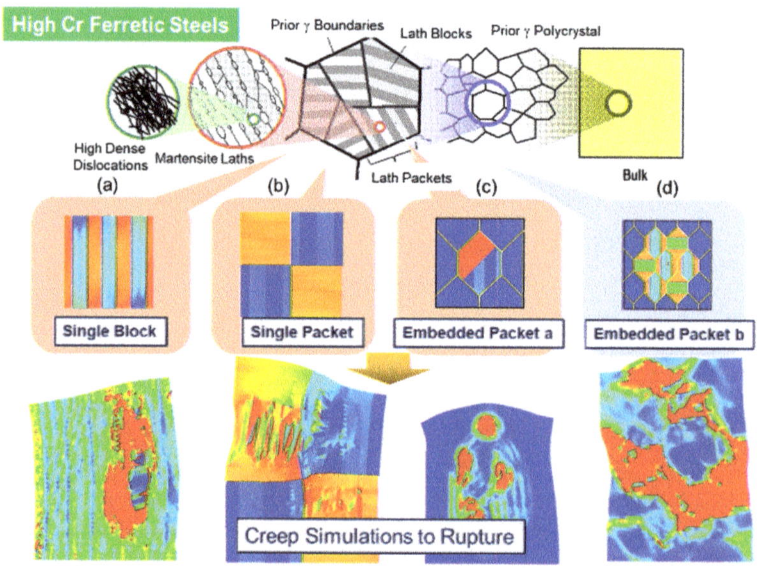

Figure A3. Representative example of the application of the Di-CAP concept from Figure A1, demonstrating the fabrication of hierarchical structures in a building block-like manner to model complex lath martensite-based steels for subsequent creep analyses: (**a**) single block model with laths and variant pairs, (**b**) single packet models with combined block models, (**c**) 7-grained and (**d**) 23-grained embedded packet models, with single packet models embedded in the central regions of both (**c**,**d**).

References

1. Boehme, F.; Hidaka, K.; Weertman, J.R. PSB Observation in Copper Fatigued at One Half the Melting Temperature. *Scr. Metall. Mater.* **1990**, *24*, 2341–2346. [CrossRef]
2. Mughrabi, H. Introduction to the Viewpoint Set on: Surface Effects in Cyclic Deformation and Fatigue. *Scr. Metall. Mater.* **1992**, *26*, 1499–1504. [CrossRef]
3. Mughrabi, H.; Ungar, T.; Kienle, W.; Wilkens, M. Long-Range Internal Stress and Asymmetric X-ray Line Broadening in Tensile Deformed [001]-Oriented Copper Single Crystals. *Philos. Mag. A* **1986**, *53*, 793–813. [CrossRef]
4. Suresh, S. *Fatigue of Materials*, 2nd ed.; Cambridge University Press: Cambridge, UK, 1998; pp. 53–68. [CrossRef]

5. Bao-Tong, M.; Laird, C. Overview of Fatigue Behavior in Copper Single Crystals—I. Surface Morphology and Stage I Crack Initiation Sites for Tests at Constant Strain Amplitude. *Acta Metall.* **1989**, *37*, 325–336. [CrossRef]
6. Repetto, E.; Ortiz, M. A Micromechanical Model of Cyclic Deformation and Fatigue-Crack Nucleation in f.c.c. Single Crystals. *Acta Mater.* **1997**, *45*, 2577–2595. [CrossRef]
7. Zhao, Z.; Ramesh, M.; Raabe, D.; Cuitiño, A.M.; Radovitzky, R. Investigation of Three-Dimensional Aspects of Grain-Scale Plastic Surface Deformation of an Aluminum Oligocrystal. *Int. J. Plast.* **2008**, *24*, 2278–2297. [CrossRef]
8. Grilli, N.; Janssens, K.G.F.; Nellessen, J.; Sandlöbes, S.; Raabe, D. Multiple Slip Dislocation Patterning in a Dislocation-Based Crystal Plasticity Finite Element Method. *Int. J. Plast.* **2018**, *100*, 104–121. [CrossRef]
9. Radi, A.; Sajadifar, S.; Seyedmohammadi, S.V.; Krochmal, M.; Bolender, A.; Wegener, T.; Niendorf, T.; Yapici, G.G. On the Low-Cycle Fatigue Behavior of a Multi-Phase High Entropy Alloy with Enhanced Plasticity. *Int. J. Fatigue* **2023**, *173*, 107678. [CrossRef]
10. Chen, B.; Jiang, J.; Dunne, F.P.E. Is Stored Energy Density the Primary Meso-Scale Mechanistic Driver for Fatigue Crack Nucleation? *Int. J. Plast.* **2018**, *101*, 213–229. [CrossRef]
11. Prastiti, N.G.; Xu, Y.; Balint, D.S.; Dunne, F.P.E. Discrete Dislocation, Crystal Plasticity and Experimental Studies of Fatigue Crack Nucleation in Single-Crystal Nickel. *Int. J. Plast.* **2020**, *126*, 102615. [CrossRef]
12. Irastorza-Landa, A.; Van Swygenhoven, H.; Van Petegem, S.; Grilli, N.; Bollhalder, A.; Brandstetter, S.; Grolimund, D. Following Dislocation Patterning during Fatigue. *Acta Mater.* **2016**, *112*, 184–193. [CrossRef]
13. Zhai, H.; Jiang, J.; Zhang, W.; Zhang, Q.; Ma, X.; Wang, S.; Li, Z.; Qiu, W.; Xiao, C.; Lin, H. Microstructure Sensitivity of the Low Cycle Fatigue Crack Initiation Mechanisms for the Al0.3CoCrFeNi High Entropy Alloy: In-Situ SEM Study and Crystal Plasticity Simulation. *Int. J. Fatigue* **2023**, *176*, 107871. [CrossRef]
14. Man, J.; Valtr, M.; Weidner, A.; Petrenec, M.; Obrtlík, K.; Polák, J. AFM Study of Surface Relief Evolution in 316L Steel Fatigued at Low and High Temperatures. *Procedia Eng.* **2010**, *2*, 1625–1633. [CrossRef]
15. Man, J.; Vystavěl, T.; Weidner, A.; Kuběna, I.; Petrenec, M.; Kruml, T.; Polák, J. Study of Cyclic Strain Localization and Fatigue Crack Initiation Using FIB Technique. *Int. J. Fatigue* **2012**, *39*, 44–53. [CrossRef]
16. Man, J.; Weidner, A.; Klapetek, P.; Polák, J. Slip Activity of Persistent Slip Bands in Early Stages of Fatigue Life of Austenitic 316L Steel. *Key Eng. Mater.* **2013**, *592–593*, 785–788. [CrossRef]
17. Polák, J.; Man, J.; Vystavěl, T.; Petrenec, M. The Shape of Extrusions and Intrusions and Initiation of Stage I Fatigue Cracks. *Mater. Sci. Eng. A* **2009**, *517*, 204–211. [CrossRef]
18. Wang, Y.; Li, J. Phase Field Modeling of Defects and Deformation. *Acta Mater.* **2010**, *58*, 1212–1235. [CrossRef]
19. Xue, F.; Cheng, T.-L.; Lei, Y.; Wen, Y.-H. Phase-Field Modeling of Crack Growth under Coupled Creep-Fatigue. *Int. J. Fatigue* **2024**, *189*, 108577. [CrossRef]
20. Kubin, L.; Sauzay, M. Persistent Slip Bands: Similitude and Its Consequences. *Acta Mater.* **2016**, *104*, 295–302. [CrossRef]
21. Ghoniem, N.M.; Tong, S.-H.; Sun, L.Z. Parametric Dislocation Dynamics: A Thermodynamics-Based Approach to Investigations of Mesoscopic Plastic Deformation. *Phys. Rev. B* **1999**, *61*, 913–927. [CrossRef]
22. Kubin, L.; Sauzay, M. Persistent slip bands: The bowing and passing model revisited. *Acta Mater.* **2017**, *132*, 517–524. [CrossRef]
23. Dodaran, M.; Khonsari, M.M.; Shao, S. Critical operating stress of persistent slip bands in Cu. *Comput. Mater. Sci.* **2019**, *165*, 114–120. [CrossRef]
24. El-Awady, J.A.; Ghoniem, N.M.; Mughrabi, H. Dislocation modelling of localized plasticity in persistent slip bands. In *136th TMS Annual Meeting and Exhibition*; TMS (The Minerals, Metals & Materials Society): Pittsburgh, PA, USA, 2007; pp. 23–35.
25. Brown, L.M. Dislocation bowing and passing in persistent slip bands. *Philos. Mag.* **2006**, *86*, 4055–4068. [CrossRef]
26. Zhang, Y.; Jiang, S.; Zhu, X.; Zhao, Y. Influence of Void Density on Dislocation Mechanisms of Void Shrinkage in Nickel Single Crystal Based on Molecular Dynamics Simulation. *Physica E* **2017**, *90*, 90–97. [CrossRef]
27. Singh, R.; Ajay, V.; Shrivastava, A. Insights into Fatigue Crack Propagation Behaviour of WAAM Produced SS316 Stainless Steel: Molecular Dynamics Simulation and Experimental Analysis. *Eng. Fail. Anal.* **2024**, *165*, 108754. [CrossRef]
28. Hasebe, T. *Field Theory of Multiscale Plasticity*, 1st ed.; Cambridge University Press: Cambridge, UK, 2024. [CrossRef]
29. Hasebe, T.; Sugiyama, M.; Adachi, H.; Fukutani, S.; Iida, M. Modeling and Simulations of Experimentally-Observed Dislocation Substructures Based on Field Theory of Multiscale Plasticity (FTMP) Combined with TEM and EBSD-Wilkinson Method for FCC and BCC Poly/Single Crystals. *Mater. Trans.* **2014**, *55*, 779–787. [CrossRef]
30. Mizutani, K.; Nawa, Y.; Hasebe, T. Kink Modeling and Simulations Based on Field Theory of Multiscale Plasticity (FTMP) Part I: Explicit Kink Model and Double Compression Test. *Mater. Trans.* **2023**, *64*, 785–794. [CrossRef]
31. Hasebe, T.; Mizutani, K. Kink Modeling and Simulations Based on Field Theory of Multiscale Plasticity (FTMP) Part II: Implicit Kink Model and Scale-Free Treatment. *Mater. Trans.* **2023**, *64*, 795–804. [CrossRef]
32. You, X.; Hasebe, T. Modeling and Simulation of Fatigue Crack Initiation Process Based on Field Theory of Multiscale Plasticity (FTMP)Part I: PSB Ladder Formation and Verification. *Metals*, 2024; *in press*. [CrossRef]
33. Polák, J. Role of Persistent Slip Bands and Persistent Slip Markings in Fatigue Crack Initiation in Polycrystals. *Crystals* **2023**, *13*, 220. [CrossRef]
34. Polák, J.; Man, J. Fatigue Crack Initiation—The Role of Point Defects. *Int. J. Fatigue* **2014**, *65*, 18–27. [CrossRef]
35. Essmann, U.; Gösele, U.; Mughrabi, H. A Model of Extrusions and Intrusions in Fatigued Metals I. Point-Defect Production and the Growth of Extrusions. *Philos. Mag. A* **1981**, *44*, 405–426. [CrossRef]

36. Nakai, Y. Evaluation of Fatigue Damage and Fatigue Crack Initiation Process by Means of Atomic-Force Microscopy. *J. Soc. Mater. Sci. Jpn.* **2001**, *50*, 73–81. [CrossRef]
37. Nakai, Y.; Kusukawa, T. Quantitative Evaluation of Slip-Band Growth and Crack Initiation in Fatigue of 70-30 Brass by Means of Atomic-Force Microscopy. *Trans. Jpn. Soc. Mech. Eng.* **2001**, *67*, 476–482. [CrossRef]
38. Nakai, Y.; Maeda, K. Observations of fatigue Slip-Band Growth and Crack Initiation in α-Brass under Cyclic Shear Stresses by Means of Atomic-Force Microscopy. *J. Mater. Sci. Soc. Jpn.* **2003**, *52*, 625–630. [CrossRef]
39. Nakai, Y.; Fukuhara, S.; Ohnishi, K. Observation of Fatigue Damage in Structural Steel by Scanning Atomic Force Microscopy. *Int. J. Fatigue* **1997**, *19*, 223–236. [CrossRef]
40. Ihara, S.; Kojima, N.; Hasebe, T. FTMP-Based Approach toward Dislocation Dynamics of PSB. *Int. J. Multiphys.* **2021**, *15*, 151–168. [CrossRef]
41. Tabata, T.; Fujita, H.; Hiraoka, M.; Onishi, K. Dislocation Behaviour and the Formation of Persistent Slip Bands in Fatigued Copper Single Crystals Observed by High-voltage Electron Microscopy. *Philos. Mag. A* **1983**, *47*, 841–857. [CrossRef]
42. Yokoi, T.; Takahashi, M.; Maruyama, N.; Sugiyama, M. Application of Controlled Cu Nano-Precipitation for Improvement in Fatigue Properties of Steels. *Nippon Steel Tech. Rep.* **2005**, *91*, 49–55.
43. Yokoi, T.; Takahshi, M.; Maruyama, N.; Sugiyama, M. Cyclic Stress Response and Fatigue Behavior of Cu Added Ferritic Steels. *J. Maters. Sci.* **2001**, *36*, 5757–5765. [CrossRef]
44. Chen, Z.; Kioussis, N.; Ghoniem, N.; Hasebe, T. Lubricant Effect of Copper Nanoclusters on the Dislocation Core in α–Fe. *Phys. Rev. B* **2008**, *77*, 014103. [CrossRef]

metals

Communication

Understanding Room-Temperature Ductility of Bcc Refractory Alloys from Their Atomistic-Level Features

Jiayi Yan [1,*] and Cheng Fu [2,*]

[1] Key Laboratory of Advanced Materials (MOE), School of Materials Science and Engineering, Tsinghua University, Beijing 100084, China
[2] CNNC Jianzhong Nuclear Fuel Co., Ltd., Yibin 644000, China
* Correspondence: yanjy2022@tsinghua.edu.cn (J.Y.); fucheng@cjnf.com.cn (C.F.)

Abstract: Many bcc refractory alloys show excellent high-temperature mechanical properties, while their fabricability can be limited by brittleness near room temperature. For the purpose of predicting ductile alloys, a number of ductility metrics based on atomic structures and crystal properties, ranging from mechanistic to empirical, have been proposed. In this work, we propose an "average bond stiffness" as a new ductility metric that is also convenient to obtain from first-principles calculations, in addition to using the average magnitude of static displacements of atoms. The usefulness of average bond stiffness is validated by comparing first-principles calculation results to experimental data on the "rhenium effect" in Mo/W-base and V/Nb/Ta-base binary alloys. The average bond stiffness also correlates well with the room-temperature ductility of refractory high-entropy alloys, with a better performance than some ductility metrics previously reported. While in reality the ductility of an alloy can be influenced by many factors, from processing and microstructure, the average magnitude of static displacements and the average bond stiffness are atomistic-level features useful for design of alloy composition towards a desired level of ductility.

Keywords: refractory alloy; ductility; first-principles calculation

1. Introduction

Body-centered cubic (bcc) refractory alloys, especially high-entropy alloys (HEAs), receive significant research attention for their high-temperature load-bearing capabilities [1,2]. However, an outstanding problem for almost all bcc alloys is their low-temperature brittleness. As temperature decreases, the ductility of bcc alloys can drop abruptly, exhibiting the so-called ductile–brittle transition (DBT) phenomenon [3]. Low-temperature brittleness limits fabricability near room temperature and reduces defect tolerance when the material is in use near or below room temperature. For practical reasons, it is desirable to let the alloy maintain some room-temperature ductility, or let the ductile–brittle transition temperature (DBTT) be sufficiently low [4,5].

However, ductility itself is a quantity involving complex microstructural processes, making it difficult to mechanistically model or predict ductility. There are various methods of measuring ductility, including elongation or reduction in area under uniaxial tension, strain at fracture under uniaxial compression, angle or radius of curvature under bending conditions, etc. [6]. It is difficult to compare or convert results obtained using different methods. For a ductile metallic material, plastic deformation occurs first, and then damage accumulates typically in the form of nucleation, growth, and coalescence of microvoids [7].

 https://doi.org/10.3390/met15080851

Currently, there are already micromechanical models that enable simulations of the processes to reflect deformation and damage until fracture [8], which makes it possible to determine the ductility of the material measured in different types of tests. However, this kind of model usually requires a large number of material parameters, which, in most cases, are determined by fitting to known mechanical behaviors of the material. It means this approach, though mechanistic, is less predictive but more explanatory in linking alloy composition to ductility.

Due to the formidable complexity of the microprocesses behind ductility, researchers have been trying to devise "ductility metrics" for bcc metals and alloys, based on properties that are convenient to calculate or measure. Some commonly used metrics include the valence-electron concentration (VEC), Pugh's ratio (ratio of shear modulus to bulk modulus), local lattice distortion ("LLD"), and the D parameter [9–11]. In some publications, electronic structure [12,13] and enthalpy of mixing [14] are also considered to correlate with ductility. Machine learning models are also implemented to predict ductility [5,15]. Among the aforementioned metrics, except for the D parameter, all the other ductility metrics are based on perfect-crystal properties of pure elements or alloys. Usually, perfect-crystal properties are less costly to compute but also mechanistically less relevant to ductility, due to a lack of consideration of crystal defects (dislocations or microvoids) essential to fracture and ductility. It is of practical interest to exploit the usefulness of perfect-crystal properties with some more physical significance, for the purpose of understanding or even predicting ductility, while maintaining a relatively low computational cost.

In this work, we present our findings on how to relate atomistic features of perfect-crystal solid-solution alloys to ductility. The basic idea is to examine not only the lattice distortion but also a kind of "average bond stiffness" relevant to the energetics associated with the lattice distortion. We then validate our approach using some experimental ductility data of refractory bcc alloys, including some binary alloys and equiatomic high-entropy alloys (HEA). A significant role of bond stiffness in determining ductility is revealed and discussed.

2. Methods

For a good representation of disordered, random bcc solid solutions using supercells, we use a Monte Carlo code "spcm" to generate bcc supercells where atomic configurations are randomized. The generated supercells used in this work include $4 \times 4 \times 4$ bcc unit cells (128 atoms) for binary alloys and quaternary equiatomic HEAs, and $5 \times 5 \times 5$ bcc primitive cells (125 atoms) for quinary equiatomic HEAs. For binary alloys $A_{1-c}B_c$, 128-atom supercells with compositions c = 0.0625, 0.125, 0.25, 0.375, 0.5 (atomic fraction) are generated. The generated atomic configurations exhibit Warren–Cowley short-range order parameters being zero or close to zero within the first eight coordinate shells.

To investigate atomic structures and energetics of the alloys, we use the software VASP (Vienna Ab initio Simulation Package, version 6.4.2) [16,17] for density-functional theory (DFT) [18] calculations. Projector-augmented wave (PAW) potentials [19,20] and PBEsol exchange-correlation [21] are adopted. The plane-wave cutoff energy is set to 500 eV. The k-mesh used for all supercells is a $4 \times 4 \times 4$ Monkhorst–Pack [22]-type mesh. Smearing of partial occupancies is the Methfessel–Paxton type [23] with a width of 0.2 eV. To obtain equilibrium atomic structures and energy, atomic coordinates are relaxed until all force components are below 0.01 eV/Å in magnitude. The relaxations are performed without altering the overall shapes of the supercells.

We compare the atomic positions and total energies before and after relaxation. From the atomic positions in the relaxed, equilibrium configuration and the ideal bcc configuration, we extract the average magnitude of static displacements $\overline{\Delta r}$:

$$\overline{\Delta r} = \frac{1}{N}\sum_{i=1}^{N}\sqrt{\left(x_{i,eq} - x_{i,0}\right)^2 + \left(y_{i,eq} - y_{i,0}\right)^2 + \left(z_{i,eq} - z_{i,0}\right)^2} \quad (1)$$

where N is the number of atoms in the supercell, $\left(x_{i,eq}, y_{i,eq}, z_{i,eq}\right)$ and $\left(x_{i,0}, y_{i,0}, z_{i,0}\right)$ are, respectively, Cartesian coordinates of the relaxed (equilibrium) position and the unrelaxed bcc lattice position of atom i. The average magnitude of static displacements is also called "local lattice distortion" (LLD) by some researchers [24]. We also extract an "average bond stiffness" $\overline{k}$ defined as

$$\overline{k} = \frac{2(E_{bcc} - E_{rel})}{N\overline{\Delta r^2}}$$
$$\overline{\Delta r^2} = \frac{1}{N}\sum_{i=1}^{N}\left[\left(x_{i,eq} - x_{i,0}\right)^2 + \left(y_{i,eq} - y_{i,0}\right)^2 + \left(z_{i,eq} - z_{i,0}\right)^2\right] \quad (2)$$

where E_{bcc} and E_{rel} are, respectively, the supercell total energies before and after relaxation. The average bond stiffness can be regarded as an effective stiffness for a hypothetical process of restoring the ideal bcc structure from the equilibrium configuration. $\overline{\Delta r}$ and $\overline{k}$ are computationally convenient to obtain, because a common procedure to obtain the equilibrium configuration is to relax the atoms from ideal lattice positions.

The roles of $\overline{\Delta r}$ and $\overline{k}$ in affecting alloy ductility can be rationalized as follows: Presumably, the perfect bcc structure has the highest symmetry and the lowest resistance to dislocation motion. If other conditions are kept constant, then larger static displacements (large $\overline{\Delta r}$) are expected to create higher resistance against dislocation motion. However, LLD only contains information about atomic positions, without having any information on force or energy. This is supplemented by the average bond stiffness $\overline{k}$. If interatomic bonds are stiff (high $\overline{k}$), then the energy cost to restore the perfect bcc structure from the relaxed, equilibrium configuration is expected to be high, which is unfavorable for dislocation mobility.

The reasoning can be summarized as two rules:

- Rule 1: Under constant $\overline{k}$, the alloy with lower $\overline{\Delta r}$ should exhibit higher ductility;
- Rule 2: Under constant $\overline{\Delta r}$, the alloy with lower $\overline{k}$ should exhibit higher ductility.

In Section 3, we use some refractory binary alloys and HEAs to showcase the usefulness of $\overline{\Delta r}$ and $\overline{k}$ in analyzing room-temperature ductility.

3. Results

Using the first-principles methods, we have calculated the atomic displacements and the associated total-energy change before and after relaxation, for binary alloys and HEAs. $\overline{\Delta r}$ and $\overline{k}$ are extracted and used to analyze the effects of alloying elements on ductility in some alloy systems.

3.1. The "Rhenium Effect"

It is well known that alloying Re in Mo or W enhances room-temperature ductility and lowers DBTT [6]. The unique ductility-enhancing effect of Re is named "the rhenium effect". However, the rhenium effect is not universal: it is effective in Group VIB elements (Cr, Mo, W), but not in Group VB elements (V, Nb, Ta) [6,25]. By calculating $\overline{\Delta r}$ and $\overline{k}$ of Re-containing alloys using DFT, and then comparing the results from the V/Nb/Ta–Re systems and the Mo/W–Re systems (Figure 1), the selectiveness of the "rhenium effect" can be rationalized by the distinct behaviors of how Re affects the alloy bond stiffness $\overline{k}$:

while Re can significantly reduce bond stiffness in Mo–Re and W–Re alloys, it does the opposite in V–Re, Nb–Re, and Ta–Re alloys, raising the average bond stiffness $\bar{k}$. The effect of Re on $\overline{\Delta r}$ is relatively weak, but a low-level $\overline{\Delta r}$ is maintained, which is a favorable feature for ductility.

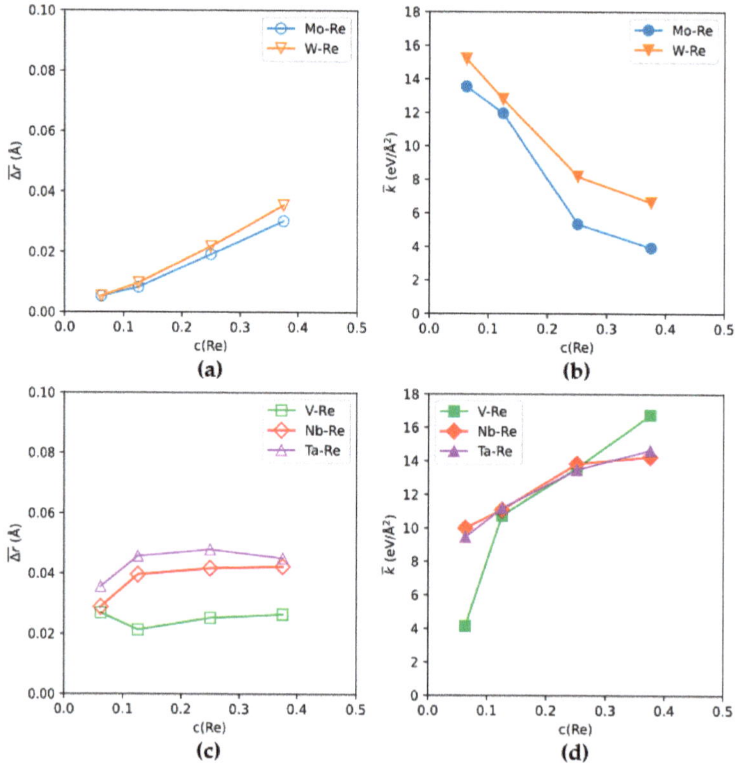

Figure 1. (**a**,**b**) DFT-calculated $\overline{\Delta r}$ and $\bar{k}$ of the Mo–Re, W–Re systems as compared to (**c**,**d**) those of the V–Re, Nb–Re, and Ta–Re systems, as functions of atomic fraction of Re in these alloys.

For example, Buckman [25] has reported (citing Begley [26]) how alloying elements in Nb-alloys raise DBTT. Their efficiency in raising DBTT is ranked as Re > W > Mo > Zr > Ti. The rank in $\bar{k}$ from DFT calculations almost coincides with the rank in the experimentally measured efficiency in raising DBTT (Re > W > Mo > (Zr, Ti)). Re, W, and Mo raise the average bond stiffness $\bar{k}$, while Zr and Ti lower it. The $\bar{k}$s for Zr and Ti are very close, so then the $\overline{\Delta r}$ criterion (Rule 1) can be invoked to explain the experimental result that Ti is less embrittling, for Ti causes smaller $\overline{\Delta r}$ than Zr. The results for Nb-alloys are shown in Figure 2.

The analysis of experimental and DFT results shows that the rhenium effect can be regarded as a bond-stiffness effect to a large extent: In Mo and W, Re addition softens the bonds and promotes ductility, whereas in Nb, Re addition stiffens the bonds and reduces ductility. The role of $\bar{k}$ appears to be dominant over that of $\overline{\Delta r}$ in affecting alloy ductility.

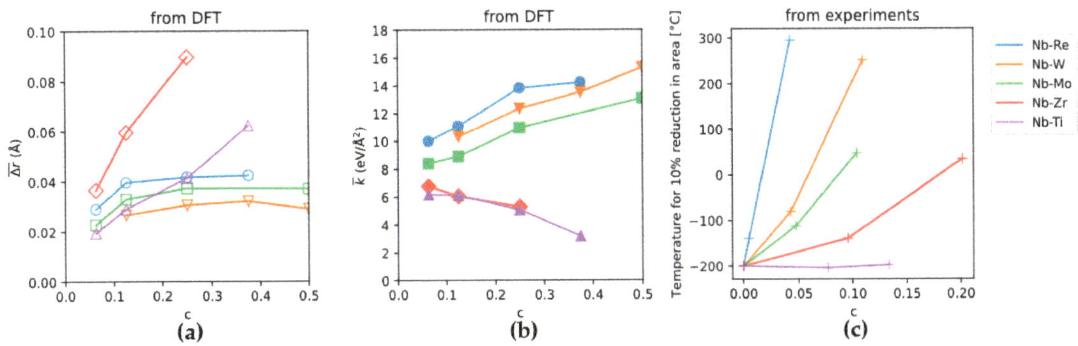

Figure 2. (**a**) $\overline{\Delta r}$ and (**b**) $\bar{k}$ of the Nb–X (X = Re, W, Mo, Zr, Ti) alloys from DFT, compared to (**c**) how experimentally measured DBTT changes with alloying elements (DBTT data from Refs. [25,26]).

3.2. Bcc Refractory HEAs

Bcc refractory HEAs attract significant research interest, one of the reasons being that they can exhibit high-temperature strength superior to the mainstream superalloys. However, usually, an outstanding problem is the insufficient room-temperature ductility or high DBTT. As discussed in the Introduction, researchers have developed and tested various kinds of "ductility metrics" (for example, [10,27,28]), and attempted to use them to guide alloy design in the vast space of chemical composition of HEAs. Singh et al. [9] compiled ductility data of 56 HEAs, and examined four ductility metrics in terms of the goodness of linear fitting to compression fracture strain (ε_f). Here, we also use the calculated $\bar{k}$ from the DFT and available experimental fracture strain data to analyze the usefulness of $\bar{k}$ being a ductility metric for RHEA.

Using the DFT, we have calculated $\overline{\Delta r}$ and $\bar{k}$ for 21 equiatomic HEAs whose data on compression fracture strain ε_f have been reported in the literature, as tabulated in Table 1 and plotted in Figure 3. Figure 3 shows a generally clear linear relationship between $\log_{10}\varepsilon_f$ and $\bar{k}$, apart from a few outliers. Linear fitting to $\log_{10}\varepsilon_f$ instead of ε_f can also avoid the possibility of predicting negative ε_f.

Table 1. Ductility of HEAs (in compressive fracture strain) from the literature, and their $\bar{k}$ and $\overline{\Delta r}$ from DFT. All alloys appear in Figure 3 while only those with an asterisk (*) appear in Figure 4.

Alloy	ε_f (%) Experimental	$\bar{k}$ (eV/Å^2) from DFT	$\overline{\Delta r}$ (Å) from DFT
TiZrVNb	>50 [29], 50 [30]	3.53	0.189
TiZrNbMo *	33 [31]	5.50	0.136
TiVNbTa	>50 [32], 28.3 [33]	5.06	0.101
TiVNbMo *	25.6 [34]	6.30	0.085
ZrHfNbTa *	34 [35]	3.91	0.147
VNbTaW *	12 [32]	9.97	0.061
NbTaMoW *	2.1 [36], 2.6 [37], 1.9 [38], 6.2 [39]	14.58	0.036
TaMoWRe	5.1 [40]	16.00	0.029
TiZrHfNbTa	>50 [41,42]	3.01	0.162
TiZrHfNbMo *	10.12 [43], 10.2 [44], 20 [45]	5.07	0.155
TiZrHfTaMo	4 [45]	4.91	0.159

Table 1. *Cont.*

Alloy	ε_f (%) Experimental	$\overline{k}$ (eV/Å^2) from DFT	$\overline{\Delta r}$ (Å) from DFT
TiZrVNbMo *	26 [31], 32 [30]	5.54	0.139
TiHfNbTaMo	27 [45]	6.25	0.114
TiVNbTaMo *	30 [46]	6.80	0.082
TiVNbTaW *	20 [32], 14.1 [37]	7.14	0.075
TiNbTaMoW *	8.4 [38]	9.33	0.052
ZrHfNbTaMo	15 [45]	6.54	0.131
ZrNbTaMoW	15.9 [45]	8.93	0.087
HfNbTaMoW	5.7 [47]	9.37	0.084
VNbTaMoW *	1.7 [36], 8.8 [37], 1.7 [47]	11.80	0.052
NbTaMoWRe	4.2 [39], 1.7 [48]	15.53	0.035

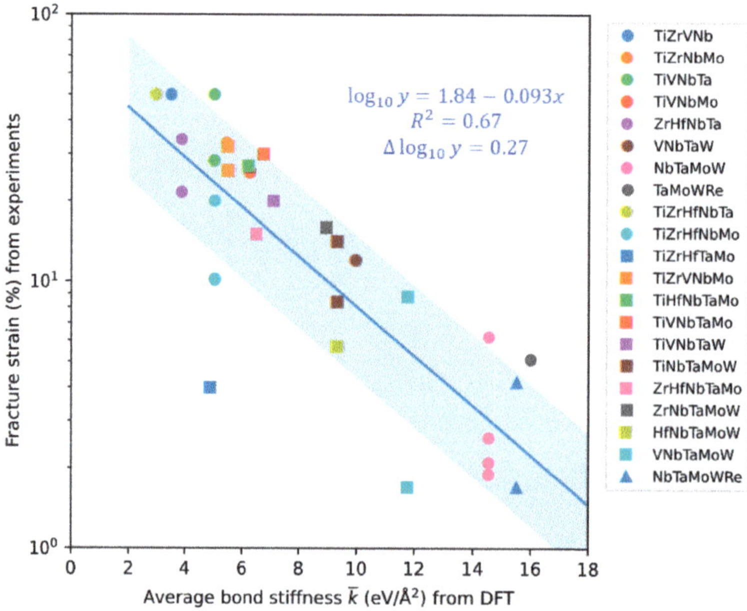

Figure 3. Compression fracture strain ε_f (experimental) versus average bond stiffness $\overline{k}$ (from DFT) of the 21 equiatomic HEAs listed in Table 1. Also given in the figure are the best-fit linear function (the blue line), the R^2 of the linear fit, and the root-mean-square error in $\log_{10} \varepsilon_f$[%] (half-width of the light blue band).

To compare $\overline{k}$ with other ductility metrics, we use the data of 11 equiatomic HEAs (marked with an asterisk in Table 1), of which $\log_{10} \varepsilon_f$, $\overline{k}$, valence-electron concentration (VEC), Cauchy pressure, Pugh's ratio, and the D parameter are all available. Linear regression results (Figure 4) show that $\overline{k}$ fits to $\log_{10} \varepsilon_f$ better (in terms of R^2) than the other four metrics.

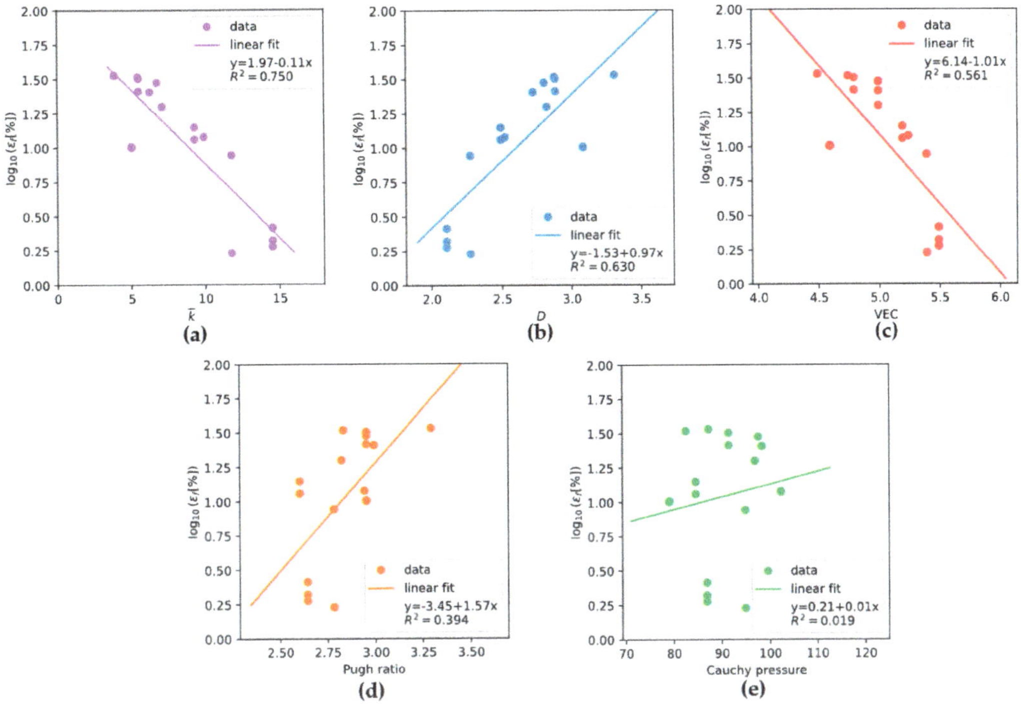

Figure 4. $\log_{10}\varepsilon_f$ vs. five ductility metrics and their linear regression of 11 equiatomic HEAs. The metrics are: (**a**) $\bar{k}$ from this work; (**b**) the *D* parameter; (**c**) valence-electron concentration (VEC); (**d**) Pugh ratio; and (**e**) Cauchy pressure. Data in subplots (**b–e**) are taken from Ref. [9]. Note that there can be several experimental ε_f for the same alloy.

For the 21 HEAs, the $\overline{\Delta r} - \bar{k}$ correlation is shown in Figure 5. Generally, $\bar{k}$ and $\overline{\Delta r}$ data tend to cluster in a band showing a genuine trade-off between the two variables. However, as we have shown, for ductility, the average bond stiffness $\bar{k}$ plays a more dominant role than $\overline{\Delta r}$, which is clear from Figure 5 also.

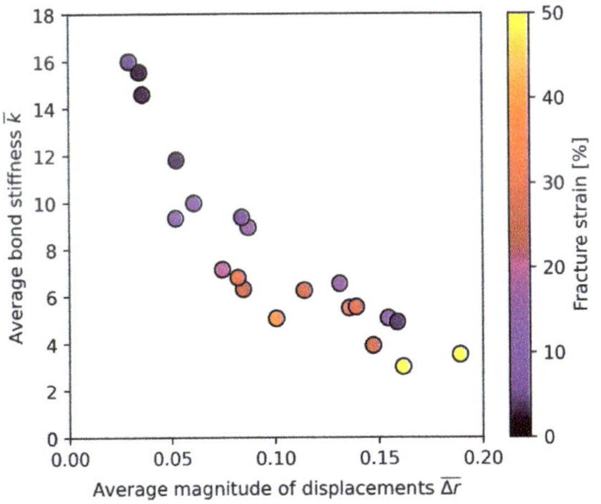

Figure 5. Crossplot of $\bar{k}$ and $\overline{\Delta r}$ from DFT for the alloys in Table 1. Colors represent experimental fracture strain values (averaged if several values are known).

4. Discussion

This work is unique in proposing an "average bond stiffness" that is convenient to obtain and relevant to the structures of solid solutions. The average bond stiffness in this work is representative of the whole supercell of solid solution, and is simple because it requires only atomic relaxation calculations with no extra calculations needed. Pant and Aidhy [24] have also proposed a "bond stiffness" and correlated it to a number of other alloy properties. The "average bond stiffness" we propose is different from theirs, in that ours requires only the total-energy difference in the supercell and static displacements of individual atoms, without the need to make assumptions and approximations for modes of atomic displacements or number of nearest neighbors, etc.

It is obvious that in this work we do not consider numerous other factors that affect alloy ductility: impurity level, degrees of deformation and recrystallization, grain size, inhomogeneity (due to nonequilibrium solidification or phase separation), etc. In fact, the data we have adopted to support the analysis are not always strictly comparable, because the alloys are often different in terms of the factors listed above. Presumably, this is part of the reason for the scatter in Figures 3 and 4. It may be possible in the future to make the ductility model more comprehensive, and not only based on the average bond stiffness.

Also, in this work, due to limitations on supercell size and randomness, we restricted the calculations and analysis to a few special compositions for binary alloys and equiatomic compositions of HEAs. Extension to general compositions will presumably require theoretical advances in better understanding the structure and energetics of solid solutions, including interstitial solutions [49]. The DFT results are invariably affected by the qualities of pseudopotentials and the exchange-correlation potential. Their influences on the atomic displacements and bond stiffness used in this work will require further investigation in order to remove any significant computational artifacts.

5. Conclusions

In this work, we analyze the ductility of bcc refractory alloys from atomistic-level features, namely, the average magnitude of static displacement $\overline{\Delta r}$ and a newly proposed "average bond stiffness" $\bar{k}$. The motivation is to extract parameters useful for ductility, while keeping the low computing cost of first-principles calculations of defect-free crystal structures. From the perspective of solid solution structure and energetics, we hypothesize that decreasing $\overline{\Delta r}$ and decreasing $\bar{k}$ can be favorable for improving ductility.

For bcc Re-containing binary alloys, we show that the "rhenium effect" (Re enhancing the ductility of Mo or W) can be understood as essentially a bond stiffness effect. In Mo/W–Re alloys, adding Re within its solubility limit reduces $\bar{k}$ significantly. On the contrary, in V/Nb/Ta–Re alloys, adding Re significantly raises $\bar{k}$. The results of the two cases are consistent with the experimental findings that Re enhances the ductility of Mo or W, but embrittles V, Nb, and Ta. In both cases, $\overline{\Delta r}$ is kept to a relatively low level. The rank of embrittling efficiency of Re, W, Mo, Zr, and Ti in Nb-base binary alloys can also be explained by the $\bar{k}$ (primarily) and $\overline{\Delta r}$ from DFT calculations.

For bcc refractory HEAs, we also show that $\bar{k}$ is a useful ductility metric. Using the experimental data of compressive fracture strain from 21 equiatomic HEAs, we show that the logarithm of fracture strain follows a linear relationship to $\bar{k}$ reasonably well. Data support the conclusion that $\bar{k}$ works better than other ductility metrics (the D parameter, Pugh ratio, Cauchy pressure, valence electron concentration) in terms of the goodness of a linear fitting to the logarithm of fracture strain.

In summary, our findings reveal the significance of the "average bond stiffness", together with the average magnitude of static displacements, in understanding the ductility of bcc refractory alloys. In spite of many other factors, from the processing and microstructure that influence ductility, our work provides a new "ductility metric" that is convenient and useful. More work in the future is required to generalize $\overline{\Delta r}$ and $\overline{k}$ for more complex compositions, beyond the special compositions considered in this work.

Author Contributions: Conceptualization, J.Y. and C.F.; Methodology, J.Y. and C.F.; Software, J.Y.; Validation, J.Y. and C.F.; Formal Analysis, J.Y. and C.F.; Investigation, J.Y. and C.F.; Writing—Original Draft Preparation, J.Y.; Writing—Review and Editing, J.Y. and C.F.; Visualization, J.Y.; Supervision, C.F.; Project Administration, C.F.; Funding Acquisition, C.F. All authors have read and agreed to the published version of the manuscript.

Funding: This work was financially supported by the CNNC Science Fund for Talented Young Scholars (CNNC-TYS-0812001).

Data Availability Statement: The datasets presented in this article are not readily available because the data are part of an ongoing study.

Acknowledgments: J.Y. thanks A.V. Ruban for sharing the spcm code.

Conflicts of Interest: Author Cheng Fu was employed by the CNNC Jianzhong Nuclear Fuel Co., Ltd. The remaining authors declare that the research was conducted in the absence of any commercial or financial relationships that could be construed as a potential conflict of interest.

References

1. Zhou, Y.; Xiang, H.; Dai, F. *High-Entropy Materials*; Wiley: New York, NY, USA, 2023.
2. Yu, T.; Qi, L.; Ramprakash, S.; Wang, Y. Computational Studies on BCC Superalloys. *Scr. Mater.* **2025**, *267*, 116804. [CrossRef]
3. Argon, A.S. *Strengthening Mechanisms in Crystal Plasticity*; Oxford University Press: Oxford, UK, 2008.
4. Senkov, O.N.; Gorsse, S.; Miracle, D.B.; Rao, S.I.; Butler, T.M. Correlations to Improve High-Temperature Strength and Room Temperature Ductility of Refractory Complex Concentrated Alloys. *Mater. Des.* **2024**, *239*, 112762. [CrossRef]
5. Wen, C.; Zhang, Y.; Wang, C.; Huang, H.; Wu, Y.; Lookman, T.; Su, Y. Machine-Learning-Assisted Compositional Design of Refractory High-Entropy Alloys with Optimal Strength and Ductility. *Engineering* **2025**, *46*, 214–223. [CrossRef]
6. Savitskii, E.M.; Burkhanov, G.S. *Physical Metallurgy of Refractory Metals and Alloys*; Consultants Bureau: New York, NY, USA, 1970.
7. Anderson, T.L. *Fracture Mechanics*; CRC Press: Boca Raton, FL, USA, 2017.
8. Besson, J. Continuum Models of Ductile Fracture: A Review. *Int. J. Damage Mech.* **2010**, *19*, 3–52. [CrossRef]
9. Singh, P.; Vela, B.; Ouyang, G.; Argibay, N.; Cui, J.; Arroyave, R.; Johnson, D.D. A Ductility Metric for Refractory-Based Multi-Principal-Element Alloys. *Acta. Mater.* **2023**, *257*, 119104. [CrossRef]
10. Senkov, O.N.; Miracle, D.B.; Rao, S.I. Correlations to Improve Room Temperature Ductility of Refractory Complex Concentrated Alloys. *Mater. Sci. Eng. A* **2021**, *820*, 141512. [CrossRef]
11. Hu, Y.-J.; Sundar, A.; Ogata, S.; Qi, L. Screening of Generalized Stacking Fault Energies, Surface Energies and Intrinsic Ductile Potency of Refractory Multicomponent Alloys. *Acta. Mater.* **2021**, *210*, 116800. [CrossRef]
12. Gornostyrev, Y.N.; Katsnelson, M.I.; Peschanskikh, G.V.; Trefilov, A.V. On the Nature of the Rhenium Effect. Peculiarities of the Band Structure and Elastic Moduli of W- and Mo-Based Alloys. *Phys. Status Solidi B* **1991**, *164*, 185–193. [CrossRef]
13. Borges, P.P.P.O.; Ritchie, R.O.; Asta, M. Electronic Descriptors for Dislocation Deformation Behavior and Intrinsic Ductility in Bcc High-Entropy Alloys. *Sci. Adv.* **2024**, *10*, eadp7670. [CrossRef]
14. An, Z.; Li, A.; Mao, S.; Yang, T.; Zhu, L.; Wang, R.; Wu, Z.; Zhang, B.; Shao, R.; Jiang, C.; et al. Negative Mixing Enthalpy Solid Solutions Deliver High Strength and Ductility. *Nature* **2024**, *625*, 697–702. [CrossRef]
15. Huang, X.; Zheng, L.; Xu, H.; Fu, H. Predicting and Understanding the Ductility of BCC High Entropy Alloys via Knowledge-Integrated Machine Learning. *Mater. Des.* **2024**, *239*, 112797. [CrossRef]
16. Kresse, G.; Furthmüller, J. Efficiency of Ab-Initio Total Energy Calculations for Metals and Semiconductors Using a Plane-Wave Basis Set. *Comput. Mater. Sci.* **1996**, *6*, 15–50. [CrossRef]
17. Kresse, G.; Furthmüller, J. Efficient Iterative Schemes for Ab Initio Total-Energy Calculations Using a Plane-Wave Basis Set. *Phys. Rev. B* **1996**, *54*, 11169–11186. [CrossRef]
18. Kohn, W.; Sham, L.J. Self-Consistent Equations Including Exchange and Correlation Effects. *Phys. Rev.* **1965**, *140*, A1133–A1138. [CrossRef]
19. Blöchl, P.E. Projector Augmented-Wave Method. *Phys. Rev. B* **1994**, *50*, 17953–17979. [CrossRef] [PubMed]
20. Kresse, G.; Joubert, D. From Ultrasoft Pseudopotentials to the Projector Augmented-Wave Method. *Phys. Rev. B* **1999**, *59*, 1758–1775. [CrossRef]

21. Perdew, J.P.; Ruzsinszky, A.; Csonka, G.I.; Vydrov, O.A.; Scuseria, G.E.; Constantin, L.A.; Zhou, X.; Burke, K. Restoring the Density-Gradient Expansion for Exchange in Solids and Surfaces. *Phys. Rev. Lett.* **2008**, *100*, 136406. [CrossRef] [PubMed]
22. Monkhorst, H.J.; Pack, J.D. Special Points for Brillouin-Zone Integrations. *Phys. Rev. B* **1976**, *13*, 5188–5192. [CrossRef]
23. Methfessel, M.; Paxton, A.T. High-Precision Sampling for Brillouin-Zone Integration in Metals. *Phys. Rev. B* **1989**, *40*, 3616–3621. [CrossRef]
24. Pant, D.; Aidhy, D.S. Electronic Density of States as the Descriptor of Elastic Bond Strength, Ductility, and Local Lattice Distortion in BCC Refractory Alloys. *Mater. Des.* **2025**, *253*, 113885. [CrossRef]
25. Buckman, R.W. Rhenium as an Alloy Addition to the Group VA Metals. In *Rhenium and Rhenium Alloys: Proceedings of the International Symposium on Rhenium and Rhenium Alloys*; Bryskin, B.D., Ed.; TMS: Warrendale, PA, USA, 1997; pp. 629–638.
26. Begley, R.T. Columbium Alloy Development at Westinghouse. In *Evolution of Refractory Metals and Alloys*; Dalder, E.N.C., Grobstein, T., Olsen, C.S., Eds.; TMS: Warrendale, PA, USA, 1994; pp. 29–48.
27. Mak, E.; Yin, B.; Curtin, W.A. A Ductility Criterion for Bcc High Entropy Alloys. *J. Mech. Phys. Solids* **2021**, *152*, 104389. [CrossRef]
28. Temesi, O.K.; Varga, L.K.; Chinh, N.Q.; Vitos, L. Ductility Index for Refractory High Entropy Alloys. *Crystals* **2024**, *14*, 838. [CrossRef]
29. Senkov, O.N.; Rao, S.; Chaput, K.J.; Woodward, C. Compositional Effect on Microstructure and Properties of NbTiZr-Based Complex Concentrated Alloys. *Acta Mater.* **2018**, *151*, 201–215. [CrossRef]
30. Wu, Y.D.; Cai, Y.H.; Chen, X.H.; Wang, T.; Si, J.J.; Wang, L.; Wang, Y.D.; Hui, X.D. Phase Composition and Solid Solution Strengthening Effect in TiZrNbMoV High-Entropy Alloys. *Mater. Des.* **2015**, *83*, 651–660. [CrossRef]
31. Zhang, Y.; Yang, X.; Liaw, P.K. Alloy Design and Properties Optimization of High-Entropy Alloys. *JOM* **2012**, *64*, 830–838. [CrossRef]
32. Yao, H.W.; Qiao, J.W.; Gao, M.C.; Hawk, J.A.; Ma, S.G.; Zhou, H.F.; Zhang, Y. NbTaV-(Ti,W) Refractory High-Entropy Alloys: Experiments and Modeling. *Mater. Sci. Eng. A* **2016**, *674*, 203–211. [CrossRef]
33. Guo, W.; Liu, B.; Liu, Y.; Li, T.; Fu, A.; Fang, Q.; Nie, Y. Microstructures and Mechanical Properties of Ductile NbTaTiV Refractory High Entropy Alloy Prepared by Powder Metallurgy. *J. Alloys Compd.* **2019**, *776*, 428–436. [CrossRef]
34. Chen, S.; Yang, X.; Dahmen, K.; Liaw, P.; Zhang, Y. Microstructures and Crackling Noise of AlxNbTiMoV High Entropy Alloys. *Entropy* **2014**, *16*, 870–884. [CrossRef]
35. Maiti, S.; Steurer, W. Structural-Disorder and Its Effect on Mechanical Properties in Single-Phase TaNbHfZr High-Entropy Alloy. *Acta Mater.* **2016**, *106*, 87–97. [CrossRef]
36. Senkov, O.N.; Wilks, G.B.; Scott, J.M.; Miracle, D.B. Mechanical Properties of Nb25Mo25Ta25W25 and V20Nb20Mo20Ta20W20 Refractory High Entropy Alloys. *Intermetallics* **2011**, *19*, 698–706. [CrossRef]
37. Han, Z.D.; Chen, N.; Zhao, S.F.; Fan, L.W.; Yang, G.N.; Shao, Y.; Yao, K.F. Effect of Ti Additions on Mechanical Properties of NbMoTaW and VNbMoTaW Refractory High Entropy Alloys. *Intermetallics* **2017**, *84*, 153–157. [CrossRef]
38. Han, Z.D.; Luan, H.W.; Liu, X.; Chen, N.; Li, X.Y.; Shao, Y.; Yao, K.F. Microstructures and Mechanical Properties of TixNbMoTaW Refractory High-Entropy Alloys. *Mater. Sci. Eng. A* **2018**, *712*, 380–385. [CrossRef]
39. Zhang, J.; Hu, Y.; Wei, Q.; Xiao, Y.; Chen, P.; Luo, G.; Shen, Q. Microstructure and Mechanical Properties of RexNbMoTaW High-Entropy Alloys Prepared by Arc Melting Using Metal Powders. *J. Alloys Compd.* **2020**, *827*, 154301. [CrossRef]
40. Wan, Y.; Wang, Q.; Mo, J.; Zhang, Z.; Wang, X.; Liang, X.; Shen, B. WReTaMo Refractory High-Entropy Alloy with High Strength at 1600 °C. *Adv. Eng. Mater.* **2022**, *24*, 2100765. [CrossRef]
41. Senkov, O.N.; Scott, J.M.; Senkova, S.V.; Miracle, D.B.; Woodward, C.F. Microstructure and Room Temperature Properties of a High-Entropy TaNbHfZrTi Alloy. *J. Alloys Compd.* **2011**, *509*, 6043–6048. [CrossRef]
42. Juan, C.-C.; Tseng, K.-K.; Hsu, W.-L.; Tsai, M.-H.; Tsai, C.-W.; Lin, C.-M.; Chen, S.-K.; Lin, S.-J.; Yeh, J.-W. Solution Strengthening of Ductile Refractory HfMo x NbTaTiZr High-Entropy Alloys. *Mater. Lett.* **2016**, *175*, 284–287. [CrossRef]
43. Guo, N.N.; Wang, L.; Luo, L.S.; Li, X.Z.; Chen, R.R.; Su, Y.Q.; Guo, J.J.; Fu, H.Z. Effect of Composing Element on Microstructure and Mechanical Properties in Mo–Nb–Hf–Zr–Ti Multi-Principle Component Alloys. *Intermetallics* **2016**, *69*, 13–20. [CrossRef]
44. Guo, N.N.; Wang, L.; Luo, L.S.; Li, X.Z.; Su, Y.Q.; Guo, J.J.; Fu, H.Z. Microstructure and Mechanical Properties of Refractory MoNbHfZrTi High-Entropy Alloy. *Mater. Des.* **2015**, *81*, 87–94. [CrossRef]
45. Tseng, K.-K.; Juan, C.-C.; Tso, S.; Chen, H.-C.; Tsai, C.-W.; Yeh, J.-W. Effects of Mo, Nb, Ta, Ti, and Zr on Mechanical Properties of Equiatomic Hf-Mo-Nb-Ta-Ti-Zr Alloys. *Entropy* **2018**, *21*, 15. [CrossRef]
46. Yao, H.W.; Qiao, J.W.; Hawk, J.A.; Zhou, H.F.; Chen, M.W.; Gao, M.C. Mechanical Properties of Refractory High-Entropy Alloys: Experiments and Modeling. *J. Alloys Compd.* **2017**, *696*, 1139–1150. [CrossRef]
47. Kang, B.; Lee, J.; Ryu, H.J.; Hong, S.H. Ultra-High Strength WNbMoTaV High-Entropy Alloys with Fine Grain Structure Fabricated by Powder Metallurgical Process. *Mater. Sci. Eng. A* **2018**, *712*, 616–624. [CrossRef]

48. Yan, D.; Song, K.; Sun, H.; Wu, S.; Zhao, K.; Zhang, H.; Yuan, S.; Kim, J.T.; Chawake, N.; Renk, O.; et al. Microstructures, Mechanical Properties, and Corrosion Behaviors of Refractory High-Entropy ReTaWNbMo Alloys. *J. Mater. Eng. Perform.* **2020**, *29*, 399–409. [CrossRef]
49. Belcher, C.H.; MacDonald, B.E.; Apelian, D.; Lavernia, E.J. The Role of Interstitial Constituents in Refractory Complex Concentrated Alloys. *Prog. Mater. Sci.* **2023**, *137*, 101140. [CrossRef]

metals

MDPI

Review

Toward Material Property Extraction from Dynamic Spherical Indentation Experiments on Hardening Polycrystalline Metals

John D. Clayton [1,*], Daniel T. Casem [1], Jeffrey T. Lloyd [1] and Emily H. Retzlaff [2]

[1] Army Research Directorate, DEVCOM ARL, Aberdeen, MD 21005, USA
[2] Mechanical and Nuclear Engineering, United States Naval Academy, Annapolis, MD 21402, USA
* Correspondence: john.d.clayton1.civ@army.mil

Abstract: Static indentation and dynamic indentation are reviewed, with a focus on extraction of material properties of isotropic strain-hardening polycrystalline metals that may be rate- and temperature-sensitive. Static indentation is reviewed first, followed by dynamic indentation, since the former is regarded as a specialization of the latter with inertia, rate dependence, and adiabatic heating excluded. Extending concepts from the literature review, a treatment of dynamic indentation using dimensional analysis is forwarded, and a general framework for extraction of material property information (i.e., constitutive model parameters) from instrumented dynamic spherical indentation experiments is set forth. In an example application of the methodology, experimental data obtained from instrumented spherical indentation in a miniature Kolsky bar apparatus are evaluated via dimensional analysis. The substrate material is aluminum alloy Al 6061-T6. Several definitions of indentation strain proposed for static indentation are assessed for dynamic indentation, as are indentation strain rates. While the fidelity of the experimental method and inertial effects could inhibit extraction of elastic properties, extraction of certain plastic constitutive properties may be feasible. Current data are insufficient to enable determination of a complete and unique set of all physical properties. Motivated by the present review and analysis, new experiments and simulations are proposed that would identify influences of material properties, facilitating their extraction from data.

Keywords: indentation; metals; polycrystals; dynamic plasticity; strain hardening; Kolsky bar

1. Introduction

Indentation experiments, being relatively simple and inexpensive to perform, offer the possibility of high-throughput mechanical testing, i.e., numerous experiments conducted in rapid time. Other advantages include the following: only small samples of material are needed, and response variations in different regions of heterogeneous bodies can be easily probed [1]. Furthermore, numerical simulations of spherical indentation of elastic–plastic solids are now routinely accomplished using conventional finite element (FE) methods with widely available commercial software [2–4]. The disadvantage of the indentation experiment, relative to traditional mechanical testing (e.g., uniaxial tension/compression), is proper analysis of the resulting data [5]. Stress and strain fields are highly nonuniform during indentation. Friction at the interface, pile-up, and sink-in effects complicate analysis [2,3,6]. In dynamic indentation, strain rate is also highly nonuniform, even if the indenter's velocity is constant. Inevitably, the indenter will decelerate prior to unloading, drastically reducing the average strain rate over time.

The present research is focused on issues associated with the extraction of material constitutive properties from dynamic spherical indentation tests. The discussion is geared toward classical ductile elastic–plastic solids (e.g., engineering metals), though some issues are shared among other material classes. Materials with viscoelastic response [7,8], or those undergoing brittle fracture [9], are not addressed explicitly. Also excluded from the present review and analysis of classical elastic–plastic solids are considerations of length scale and size effects [10–13] (e.g., specimen size, grain size, and strain gradient effects, with increased complexity often arising at the nanometer scale) and corresponding advanced constitutive theories [14–16] capable of addressing one or more such phenomena. Constitutive models considered later in this work do not contain any intrinsic length scale(s). Phase-field representations of spherical indentation [17] are thus excluded from the formal review. Furthermore, atomic or hybrid atomic–continuum simulations [18], which necessarily probe phenomena at the nanoscale (e.g., in single-crystal domains) due to sizes resolved, are likewise excluded.

Governing relationships among global variables are cast in dimensionless form in an application of dimensional analysis to dynamic spherical indentation. Relationships among useful dependent and independent variables are analyzed in the context of Buckingham's Pi theorem [7,19,20]. This approach, which enables a systematic reduction in the number of independent quantities entering a physical problem, has been used previously to analyze static spherical indentation [21] and ballistic impact [22]. Here, the treatment of Lee and Komvopoulos [23] is extended to account for thermal effects, and different mathematically admissible (and thought more physically useful) choices are made for independent and dependent variables of interest. Results provide guidance on which parameter(s) should be varied systematically in a design of experiments, real or numerical, to enable a complete understanding of the problem. Outcomes are then applied to experimental data on an aluminum alloy, producing dimensionless indentation force, contact radius, and global temperature rise. Different indentation strains and strain rates are calculated and examined for potential suitability in the context of constitutive model/parameter determination.

This paper is organized as follows. The literature review elaborating fundamental concepts is contained in Sections 2 and 3. Pertinent aspects of static indentation, notably governing equations and constitutive model forms, are reviewed first in Section 2, since many carry over to the dynamic regime. Dynamic indentation is addressed in Section 3, including augmentation of basic elastic–plastic constitutive models to incorporate rate and temperature, as well as several proposed definitions of effective indentation strain rate. Sections 2 and 3 are suitably labeled as "Review" sections. Subsequent parts of this paper apply or extend concepts from the literature review to demonstrate their utility in an application to instrumented dynamic spherical indentation. Dimensional analysis of dynamic spherical indentation is undertaken in Section 4, suitably labeled "Extension", since prior dimensional analysis from the literature is updated to include thermal effects and different choices of dimensionless variables. Techniques are applied to instrumented indentation data for aluminum alloy Al 6061-T6 in Section 5, suitably labeled "Application". Based on the literature review and current analysis, recommendations are given for future experiments and possible numerical simulations that should facilitate constitutive model parameterization using dynamic spherical indentation data. Lastly, conclusions follow in Section 6. A list of symbols with definitions and dimensions is included in a Notation section immediately preceding the References list.

2. Review: Static Indentation

The scientific literature on static indentation, even when focused purely on modeling techniques for material property extraction, is immense. A complete review of the entire subject space of indentation experiments, theoretical analyses, and numerical simulations is thus outside the present scope, which instead reports details needed in the subsequent dimensional analysis of the dynamic case. See the book by Johnson [24] for seminal analytical methods applied to contact problems, as well as two review papers [5,25] for contemporary overviews that include computer simulations.

2.1. *Elastic Indentation*

The linear isotropic elastic solutions of Hertz are conventionally used to analyze static indentation data up to initial yield, as well as elastic unloading from a plastically deformed state. The Hertz analysis [24] assumes frictionless contact of homogeneous elastic bodies whose surfaces are parabolic in shape. In the limit of small indentation depths and contact areas pertinent to the elastic regime, the quadratic surface approximation adequately represents the true spherical geometry of the indenter [26]. Provided that the yield strength of the material is a small fraction of the elastic stiffness, effects of elastic nonlinearity are typically deemed negligible, and thus omitted in nearly all reported analytical and numerical studies of indentation of ductile metals. However, for large *elastic* indentation depths, nonlinear compressibility has been shown to mildly increase indentation force [27].

Hertz's equations for spherical indentation into an initially flat substrate are summarized as follows [1,24]. Denote by P the indentation force, h_e the elastic (i.e., reversible) indentation depth, $\bar{E}$ the effective system modulus, $\bar{R}$ the effective system radius, a the contact radius, and $\bar{k}$ a system stiffness. Denote by $E_i, E_s, \nu_i, \nu_s, R_i, R_s$ the elastic moduli, Poisson's ratios, and radii of the indenter and sample, labeled with respective subscripts $(\cdot)_i$ and $(\cdot)_s$. Then, Hertz's solution encompasses

$$P = \bar{k}h_e^{3/2}, \qquad \bar{k} = \tfrac{4}{3}\bar{E}\bar{R}^{1/2}, \qquad a = (\bar{R}h_e)^{1/2}; \tag{1}$$

$$\bar{E} = [(1-\nu_i^2)/E_i + (1-\nu_s^2)/E_s)]^{-1}, \qquad \bar{R} = (1/R_i + 1/R_s)^{-1}. \tag{2}$$

For a rigid indenter, $\bar{E} = E_s/(1-\nu_s^2)$ and $R_i =$ constant. Prior to deformation $R_s \to \infty \Rightarrow \bar{R} = R_i$. Often, $R_s \to \infty$ (i.e., small deformation theory) is used to approximate the entire elastic loading process into a flat substrate, an assumption which produces a very simple closed-form solution when R_i is effectively constant. This approximation is also often used to analyze elastic unloading [28]; however, the unloading process from a plastic impression has been analyzed elsewhere with a finite R_s [6]. If the indenter is deformable, a typical approximation for its elastic displacement h_i is given by Hertz's solution for indentation into a rigid flat surface [1]:

$$h_i \approx 3(1-\nu_i^2)P/(4E_i a). \tag{3}$$

The effective system modulus $\bar{E}$ encompasses all effects of material constitutive behavior on the load–displacement curves in the linear elastic regime. This modulus has also been widely used to fully encompass *elastic* constitutive effects on load–displacement response in the elastic–plastic regime, e.g., in dimensional analysis [29–31]. Justification for this assumption in early stages of elastic–plastic indentation has been obtained from analytical methods [24,32] and verified for $\nu_s \in [0.01, 0.49]$ in FE simulations [3]. For a very stiff or rigid indenter, the effects of Poisson's ratio of the sample on indentation force are reportedly small relative to the effects of the elastic modulus of the sample over typical ranges of constitutive behaviors of ductile metals [7,33,34]. In other words, sensitivity of the static load–displacement response to Poisson's ratio is reportedly low, though exceptions exist for unusual material property combinations [35] and for large indentation depths [3].

2.2. *Elastic–Plastic Indentation*

In the elastic–plastic regime, the total indentation depth for the system, h, is decomposed into [1]

$$h = h_e + h_r = h_s + h_i; \qquad h_e = (P/\bar{k})^{2/3}. \tag{4}$$

The elastic indentation depth of the system is h_e, and when the indenter deforms only elastically, its indentation displacement h_i can be approximated via (3). The residual indentation depth is h_r; after elastic unloading, $h = h_r$ and $h_e = h_i = 0$.

During elastic unloading, the slope S of the load–indentation curve implied from Hertz's theory at small *elastic* deformation (i.e., $\bar{k}(h)$ independent of h_e but possibly dependent on $h_r = h - h_e$ through finite $R_s(h)$) is the derivative

$$S(h) = \partial P(h(h_e, h_r))/\partial h_e = \tfrac{2}{3}\bar{k}(h) \cdot h_e^{1/2} = 2\bar{E} \cdot a(h). \tag{5}$$

If S is measured at a given elastic–plastic contact depth h [6], then the contact radius a at that depth can be inferred from (5) if $\bar{E}$ is known [36]. The radius of the residual indent (i.e., crater) after unloading, denoted by a_r, of a plastically deformed material has been used in this capacity as a coarse approximation of a at peak load [28], though more sophisticated treatments of unloading of elastic–plastic solids exist [6].

Constitutive behavior of an untextured ductile metallic specimen is typically described in engineering practice via isotropic elasto-plasticity with possible power-law hardening [31,37–40]. Although subtle differences exist among numerous functional forms of hardening laws given in the literature, a reasonably standard model for von Mises equivalent flow stress σ is

$$\sigma(\epsilon^P) = \sigma_0\left[1 + \kappa \cdot (\epsilon^P)^n\right], \qquad n = \mathrm{d}\ln(\sigma - \sigma_0)/\mathrm{d}\ln\epsilon^P. \tag{6}$$

The cumulative scalar plastic strain is ϵ^P, the initial yield stress is σ_0, the strain-hardening exponent is n, and κ is a fitting parameter. For perfect plasticity, $\kappa = 0$, and for linear hardening, $n = 1$ with $\kappa > 0$.

Standard, physically justified, and established methods (i.e., associative flow, normality, and consistency) can be used to implement (6) in the context of finite deformations with incremental plasticity in a numerical setting [41]:

$$\mathrm{d}\boldsymbol{\epsilon} = \mathrm{d}\boldsymbol{\epsilon}^E + \mathrm{d}\boldsymbol{\epsilon}^P, \qquad \mathrm{d}w^P = \boldsymbol{\sigma} : \mathrm{d}\boldsymbol{\epsilon}^P = \sigma \mathrm{d}\epsilon^P. \tag{7}$$

The tensor-valued strain increment $\mathrm{d}\boldsymbol{\epsilon}$ is additively decomposed into elastic $(\cdot)^E$ and plastic $(\cdot)^P$ parts, the stress tensor is $\boldsymbol{\sigma}$, and the scalar plastic strain increment $\mathrm{d}\epsilon^P$ is obtained from the increment of the plastic strain tensor such that its work conjugate entering plastic work per unit volume w^P is the von Mises stress σ. For uniaxial stress conditions, with ϵ denoting the total axial strain, such treatment reduces to

$$\sigma = \begin{cases} E\epsilon & (\epsilon \leq \sigma_0/E), \\ \sigma_0\left[1 + \kappa \cdot (\epsilon^P)^n\right] & (\epsilon \geq \sigma_0/E). \end{cases} \tag{8}$$

Define the projected contact area by A and the mean contact pressure by $\bar{p}$. An effective value of the flow stress σ averaged over the indented region in the sample is $\bar{\sigma}$, and c is the constraint factor that depends on geometry, and to a lesser extent, constitutive behavior [2,6,42,43]. Then, the mean pressure and mean flow stress are related by

$$\bar{p} = \frac{P}{A} = \frac{P}{\pi a^2} = c\bar{\sigma}. \tag{9}$$

Yielding in the sample is initiated at $c \approx 1.1$ [6,42], whereby h_p first becomes nonzero. For sufficiently deep indentation, $h \gg h_e$, such that the plastic response dominates. In this regime, the historical analysis and data of Tabor [44] suggest $c \approx 2.8$ for spherical indentation in ductile metals, later corroborated by numerical methods by Hill et al. [45]. Another typical approximation is $c \approx 3$ for this deep plastic regime [6]. In the intermediate regime between elastic and deep plastic indentation, $1.1 \lesssim c \lesssim 3$. At maximum load, the mean pressure can be identified with Meyer's hardness H in an indentation hardness experiment (i.e., spherical or hard ball indentation into a flat substrate) when the maximum force is substituted for P and the residual imprint radius a_r is substituted for a.

For strain-hardening materials ($n > 0$), c likely depends on both n and the definition used for "indentation strain" $\bar{\epsilon}$ substituted into (8) for ϵ^P to acquire a representative value of $\bar{\sigma}$ [2,6]. Tabor [44] proposed $\epsilon^P \approx \bar{\epsilon} = 0.2a/R_i$, which corresponds to plastic strain at the indentation edge [2]. In that work [44], $a \approx a_r$ was approximated as the radius a_r of the residual indent after load removal, and thus is not consistent with the Hertz definition of the true contact area at the instant load removal begins. Additionally, the approximation $\bar{R} \approx R_i$ is often used in practice rather than the second of (2) (with more realistic finite R_s)

to characterize unloading from a plastically deformed state, an approximation which tends to poorly capture the effective stress–strain behavior and unloading modulus [5].

A potential difficulty with many definitions for indentation strain [6,44] is determination of the contact radius a, which may be challenging to obtain directly from experiments. An estimate proposed by Field and Swain [6] for this purpose is

$$a = [2(h - h_e/2)R_i - (h - h_e/2)^2]^{1/2}. \tag{10}$$

However, noted by Kalidindi and Pathak [26], the definition of a in (10)—based on spherical geometry and assuming that the elastic displacement of a preformed spherical impression is evenly divided above and below the circle of contact [6]—is not fully consistent with Hertz's definition in (1).

An alternative measure of indentation strain $\hat{\epsilon}$, with corresponding renamed stress (i.e., mean pressure) $\hat{\sigma}$, is thus proposed by Kalidindi and Pathak [5,26]:

$$\hat{\epsilon} = \frac{4}{3\pi}\frac{h_s}{a} \approx \frac{h_s}{2.4a}, \qquad \hat{\sigma} = \bar{p} = \frac{P}{\pi a^2} = c\bar{\sigma}. \tag{11}$$

The contact radius incorporated in these works [5,26] obeys the Hertz definition in (1) and can be found from (5) during unloading, provided $\bar{E}$ is known from fitting to data collected in the elastic regime. Note that (11) produces $\hat{\sigma} = \bar{E}\hat{\epsilon}$ for $h = h_e$, i.e., for elastic loading/unloading commensurate with Hertz's solution in (1) and (2). Application of (11) to experimental [1] and numerical [37,38] results produces a constraint factor of $c \approx 2$ for metals with no strain hardening, or for strain-hardening metals at a uniaxial-equivalent offset strain of $\epsilon \approx 0.1$–0.2%. Possible reasons for $c < 2.8$ in some investigations [1,37,38] are shallower indentation depths than those used by Tabor and others [2,44,45] and different definitions or measures of a among the different investigations.

Three indentation strain measures are collected below for reference and comparison:

- Tabor [44]: $\bar{\epsilon} = 0.2 \cdot (a/R_i)$;
- Kalidindi and Pathak [26]: $\hat{\epsilon} = (4/\{3\pi\}) \cdot (h_s/a)$;
- Lee and Komvopoulos [46]: $\check{\epsilon} = (E_s/\sigma_0) \cdot (h/\{2R_i h - h^2\}^{1/2})$.

Notably, constitutive scaling factor E_s/σ_0 is present in $\check{\epsilon}$, while the other two measures are purely geometric. Only $\hat{\epsilon}$ requires knowledge of two transient variables (i.e., a and h_s).

The notation for stress measures is clarified in the Notation section at the end of this paper, immediately preceding the References. Specifically, σ is the local von Mises stress, as specified by constitutive stress–strain models, with an initial value of σ_0 prior to strain hardening and in the absence of rate and temperature effects. The measured mean indentation pressure is $\bar{p}$. The mean indentation flow stress of Tabor [44] is $\bar{\sigma}$; this differs from $\bar{p}$ by the constraint factor c. The indentation stress defined by Pathak and Kalidindi [5] is equal to the mean pressure and differs from the mean indentation flow stress by a factor of c. The same notation is used for $\hat{\sigma}$ and $\bar{p}$, since the latter is measured but the former is assumed for generating indentation stress–strain curves (e.g., Equation (12)).

2.3. *Constitutive Property Extraction*

Numerous works have sought to extract uniaxial stress–strain constitutive behavior from spherical indentation data in the quasi-static regime, claiming various degrees of success. A relatively simple model for loading–unloading cycles [6] incorporating (10) is widely mentioned, though it is not valid for the transition regime $1.1 \lesssim c \lesssim 3$. Early FE simulations [45,47,48] of quasi-static spherical indentation date to the 1980s. The existence, uniqueness, and stability of the inverse solution have been investigated in the context of dimensional analysis [31,35]. In related work [31], a representative strain has been determined that renders the dimensionless indentation force independent of the strain-hardening exponent. An optimal data acquisition location from which to extract field variables from FE solutions for material property evaluation has been proposed [49]. An average representative strain and confidence domain for which property correlations are accurate have been defined [50,51].

Inverse methods incorporating numerically generated databases of spherical indentation response data have been used for property extraction [40,52,53]. More sophisticated data-driven approaches utilize neural networks [39,54,55] or Bayesian inference [38,56]. Simplified analytical fitting functions or surrogate numerical models (i.e., calibrated replacements of full FE models) can be invoked in this context to greatly improve computational efficiency [38,39].

Most often, correlation of the loading portion of a predicted force–displacement curve with test data [2] is used to determine plasticity parameters (e.g., σ_0, κ, n), while unloading is used to determine elastic stiffness $\bar{E}$ with some estimate of the contact radius [6]. However, periodic load–unload cycles can be used to determine the projected contact area $A = \pi a^2$ if $\bar{E}$ is measured from the initial elastic loading phase [26]. The strategy advocated by Pathak, Kalidindi, and their coworkers [5,37,38] involves fitting or comparison of indentation stress–strain curves rather than indentation force–displacement curves to ascertain elastic–plastic properties. Elastic–plastic properties have also been identified by numerically matching residual imprints [52,56], as opposed to force–displacement curves.

Protocols for generating uniaxial stress–strain curves from indentation stress–strain curves, where the latter follow (11), are described by Patel and Kalidindi [37]. The equivalent uniaxial stress–strain behavior of the specimen is of the standard form $\sigma = E_s \cdot \epsilon^E = E_s \cdot (\epsilon - \epsilon^P)$. Then, the following correspondence relations apply among indentation stress–strain curves ($\hat{\sigma}$ vs. $\hat{\epsilon}$) and uniaxial curves (σ vs. ϵ):

$$\hat{\sigma} = \begin{cases} \bar{E} \cdot \hat{\epsilon} & (\hat{\sigma} \leq c\sigma_0), \\ \bar{E} \cdot (\hat{\epsilon} - \hat{\epsilon}^P) & (\hat{\sigma} \geq c\sigma_0); \end{cases} \qquad \hat{\epsilon} = \frac{4}{3\pi} \frac{h_s}{a} = \hat{\epsilon}^E + \hat{\epsilon}^P; \tag{12}$$

$$\epsilon^E = \frac{1}{c} \frac{E_s}{\bar{E}} \cdot \hat{\epsilon}^E = \frac{1}{c} \frac{E_s}{\bar{E}} \cdot \frac{\hat{\sigma}}{\bar{E}} = \frac{1}{c} \frac{E_s}{\bar{E}^2} \frac{P}{\pi a^2}, \qquad \epsilon^P = \frac{\hat{\epsilon} - \hat{\epsilon}^E}{\hat{\beta}}. \tag{13}$$

The constraint factor c is measured at the yield point ($\bar{\sigma} = \sigma_0$) for continuity, and $\hat{\beta}$ is a fitting factor that depends on the material. Demonstrative FE simulations for representative elastic–plastic solids [37] with perfect plasticity ($\kappa = 0$) or linear hardening ($n = 0$) with different strain-hardening coefficients $\kappa > 0$ produced $c \approx 2.0 - 2.2$ and $\hat{\beta} \approx 1.3$. The indentation stress–strain fitting method can be advantageous to fitting of force–displacement responses, since pertinent elastic–plastic properties may be strongly correlated to only relatively small regions of the entire force–displacement curve. Thus, a certain precision of fitting for entire force–displacement curves does not necessarily carry over to similar precision in extracted stress–strain curves, elastic modulus, yield strength, and/or strain-hardening parameters [39]. Another method of reducing fitting errors associated with estimation of contact areas involves correlation of total and recoverable strain energies (i.e., integrals of load–displacement relations) rather than indentation forces [7,21], since integral values are less sensitive to experimental noise.

According to Cao and Lu [31], an inverse problem is ill-posed if one of the following properties is not respected:

- Existence: there exists a solution to the problem.
- Uniqueness: there is, at most, one solution to the problem.
- Stability: the solution continuously depends on the data.

The condition number of an inverse problem measures sensitivity of the identified parameter to small changes in the input data. A problem is ill-conditioned if the condition number is large, and it is ill-posed if the condition number is infinity. A thorough numerical analysis of spherical elasto-plastic indentation [31] found stability to degrade with increasing n and with increasing σ_0/E_s. To the above three criteria, an obvious measure of success of an inverse analysis procedure can be added:

- Accuracy: how closely the inverse solution matches the exact solution.

The exact solution would be procured from independent property specifications or different measurements (e.g., uniaxial stress–strain experiments) in the present application to indentation analysis.

Some challenges mentioned in the literature for static property extraction are now reiterated. In Zhao et al. [33], the error of reverse analysis was less than 10% in most cases, where discrepancies were caused in part by the accuracy of fitting functions and in part by the error of applying a power-law hardening model to real materials. In works by Moussa et al. [50,51], it was found that extraction of stress–strain curves from indentation gives precise results only over a range of strain, which was termed a "confidence domain". Patel and Kalidindi [37] noted the difficulty with pinpointing an initial yield stress σ_0 when yielding occurs at low indentation depths, given the typical resolution of experimental data in this regime. Shallow indentation may not provide sufficient information for discriminating hardening parameters [38]. An opinion stated by Dean and Clyne [53] is that a single indentation run with a spherical indenter should be sufficient for property determination, at least for typical quasi-static isothermal behaviors of ductile metals described by (6).

For indenters with self-similar shapes (e.g., conical, Berkovich or Vickers pyramidal), no inherent length scale is introduced by the geometry. It follows that hardness, representative strain, and curvature of the load–displacement relation are independent of indentation depth, which can be proven using dimensional analysis [7,57,58]. This contributes to the nonuniqueness of property extraction from a single indentation using such an indenter shape. So-called "mystical materials" exist that have different elastic–plastic property combinations but identical indentation curves obtained from self-similar indenters [59,60]. Nonuniqueness can be rectified by using combined data from indenters of different shapes, e.g., different tip apex angles [30,53,61] to determine material properties from an inverse analysis. In contrast, nonuniqueness is less problematic for spherical indentation since the indenter radius introduces a geometric length scale independent of the indentation depth [7,21,53].

According to Liu et al. [60], fundamental plastic constitutive properties such as the yield stress and work-hardening coefficient cannot always be uniquely determined from the force–displacement curves of indentation analyses, including both plural sharp indentation and deep spherical indentation. Mystical material pairs can still exist in such cases, with a difference of their force–displacement curves below the resolution of existing indentation techniques for large ranges of sharp indenter angles or spherical indentation depths. From the indentation force–displacement curve alone, it is generally not possible to precisely specify what plastic-hardening model should be used (e.g., (6) vs. some other constitutive equation) for an arbitrary material, and the whole stress–strain curve of the material cannot be measured due to geometric limitations on maximum depth that bound the indentation strain [60]. A Monte Carlo sensitivity analysis [34] demonstrated, in the case of spherical indentation, that the experimental errors must be very small to enable reliable extraction of material properties.

In summary, the existence, uniqueness, stability, and accuracy of an inverse technique for property extraction from static spherical indentation depend on multiple factors. These factors may include the indentation depth and indenter radius, the assumed constitutive model, and presumptive initial conditions (i.e., initial guesses) and bounds on material parameters in a search algorithm. The aforementioned four aspects of performance appear to generally improve as indentation depth increases to the plasticity-dominated regime, wherein the plastic zone under the indenter has enlarged sufficiently to control the stiffness of the substrate that is registered at the contact surface. Incorporation of more data in any calibration procedure, including residual indentation geometry and a range of indentation load–unload depths and geometries, is expected to improve results. Fitting of indentation stress–strain data and consideration of energy–displacement data rather than explicit force–displacement data have been observed to improve accuracy and stability, respectively.

Of course, when the assumed constitutive form is close to true material behavior, with property ranges typical of those for which the constitutive model is intended, success is more likely. On the other hand, for exotic materials (i.e., those very different from typical ductile metals), or for those with highly uncertain behaviors, success of constitutive model parameterization is not ensured a priori and thus should be verified on a case-by-case basis. For example, if the material exhibits simultaneous viscoelastic and irreversible plastic deformation mechanisms, a unique constitutive model prescription from standard indentation data may be impossible [62].

Anisotropy is inherent in the mechanics of single crystals. The present review focuses on isotropic material behavior pertinent to polycrystals with randomly oriented grains. Isotropic elasticity is assumed in the Hertz solutions of Section 2.1 and the constitutive theory of Section 2.2. For polycrystalline material samples, individual grains are anisotropic, but the overall spherical indentation response (e.g., force versus depth) becomes independent of orientations of individual grains when the size of the indenter relative to the grain size is large enough [1]. In this case, the deformed region of material underneath the indenter encompasses enough single crystals, such that their homogenized behavior controls the global response. If the sampled grains are randomly oriented, the overall response will be isotropic and repeatable for multiple indentations on the same material. If, however, the grains are not randomly oriented (i.e., a textured polycrystal), or if the substrate is a single crystal, the homogenized elastic–plastic response will be anisotropic, and a suitable anisotropic constitutive model [63,64] should be implemented instead. Similarly, anisotropic properties become important if the indenter radius–depth combination probes only the local response of a single crystal at the surface of a polycrystalline sample. Finite element simulations of static spherical indentation have been used to understand anisotropic elastic and plastic properties of crystalline materials [17,64–66]. Continuum FE simulations of *dynamic* spherical indentation of anisotropic crystalline solids do not seem present in the available literature.

Spherical indentation has also been used to study the creep properties of solid materials, usually at elevated initial temperatures. Experiments [67,68] and numerical analyses [69–71] of creep in metals (or representative generic ductile solids) under spherical indentation loading have been reported. Time enters the analysis due to the finite relaxation period for manifestation of creep deformation, which is strain-rate-dependent. In creep investigations, indentation rates are finite yet relatively slow, and long loading times are required to ascertain constitutive properties for creep models [7]. Although time affects results in both cases, experiments and modeling for creep behavior are distinguished in the present review from those for dynamic elastic–plastic response reviewed in Section 3. In the dynamic indentation of crystalline metals, strain rates are much higher (e.g., on the order of hundreds or thousands per second for dynamics versus less than unity for creep), and initial temperatures are generally much lower. Physical behaviors are addressed for dynamic indentation with high-rate plasticity for underlying dislocation glide rather than creep viscoelasticity or creep viscoplasticity for underlying diffusion and dislocation climb. Dimensional analysis of indentation of creeping materials obeying a power-law constitutive relation was undertaken by Cheng et al. [7]. Loading rate and temperature entered the analysis but mass density and thermal properties important for dynamics did not.

The basic constitutive theory in Section 2.2, and its augmentation for dynamic regimes in Section 3.1, does not explicitly address inelastic phenomena that are not readily incorporated into the plastic yield and flow functions, such as aforementioned creep plasticity, anisotropic deformation twinning (e.g., modeled in [17,72–74]) or phase transitions [75,76]. Fractures, as more often arising in brittle materials [9,77–79], are also excluded from the current review.

3. Review: Dynamic Indentation

Substantial research has been directed toward material constitutive characterization using dynamic indentation methods, albeit far less than that for the static regime reviewed in Section 2. Early work [80] analyzed the projectile impact of hard spheres into softer metallic targets of much larger dimensions than the spheres. Dynamic stress–strain curves were extracted from analysis of the results. The flow stress was defined as $\bar{p}/c$ with a static value of $c \approx 3$ depending on material, and Tabor's strain measure $\bar{\epsilon} = 0.2a_r/R_i$ was used, with a_r the residual contact radius.

More recent experiments [81–83] likewise analyzed hardness data extracted from spherical projectile impact. Wen et al. [84] used nanoindentation to characterize hardness in the plastic imprint of a titanium alloy following spherical impact. The above studies are not considered instrumented methods. Rather, such investigations tended to rely on knowledge only of the impactor's size and its initial velocity, in conjunction with postmortem analysis of the impact crater, to infer material response information, since transient forces and velocities were not recorded in situ during the impact event.

Dynamic indentation using the split Hopkinson pressure bar (SHPB) (i.e., Kolsky bar) has been implemented for a variety of indenter geometries and target materials, as pioneered by Subhash and co-workers [85,86]. This technique allows for the interrogation of velocities and mechanical forces during the transient indentation process. More recent approaches of dynamic indentation with the SHPB have used a full sphere sandwiched between two specimens [87] or a striker with variable impedance to achieve load cycling [88]. The strain-rate sensitivity of metals was deduced from dynamic conical indentation experiments by Lu et al. [89], where the indenter was propelled by a light gas gun, and a combination of interferometry and load transducer was used to ascertain a time-resolved material response including strain-rate sensitivity. Force and velocity histories were recorded from spherical indentation at moderate impact velocities in an instrumented drop-weight system [90], again to assess rate sensitivity of flow stress.

Dynamic impact experiments with laser-driven flyers [91] may offer an alternative or supplementary means of high-throughput characterization of mechanical properties, at potentially much higher local strain rates and different stress states (e.g., incurring spall) than dynamic indentation with the SHPB. Description and analysis of such contemporary techniques are outside the scope of this review.

3.1. Dynamic Elastic–Plastic Indentation

Analytical and numerical models, the latter primarily dynamic FE methods, have been used to study dynamic indentation, and often more specifically, extract material property information. In addition to mechanisms pertinent to quasi-static loading (i.e., elasticity, yield, and strain hardening), wave propagation enters the dynamic problem, necessitating involvement of mass density ρ_0. For high loading rates, conditions are nearly adiabatic and temperature rise could be substantial, so specific heat capacity (e.g., c_V) may also affect the response in such situations. Finally, dislocation kinetic processes depend on loading rate to varying degrees depending on the specific metallic material. It is assumed that impact conditions are not severe enough to warrant inclusion of nonlinear elasticity (e.g., pressure-dependent compressibility) or thermoelastic coupling (i.e., thermal expansion), which would be needed for accurate analysis of intense shock waves, for example [92].

Focusing attention again on isotropic ductile polycrystalline metals, the inelastic constitutive model for dynamic loading is usually augmented to account for strain-rate and temperature effects, in addition to initial yield and strain hardening. The flow stress of (6) is often simply extended as a product of terms accounting for each mechanism, for example,

$$\sigma(\epsilon^P, \dot{\epsilon}^P, T) = \sigma_0\Big[1 + \kappa \cdot (\epsilon^P)^n\Big] \cdot f(\dot{\epsilon}^P) \cdot g(T), \tag{14}$$

where T is absolute temperature and $\dot{\epsilon}^P$ is a scalar effective plastic strain rate [41]. Functions f and g account, respectively, for strain-rate and thermal sensitivity. A widely used form is the Johnson–Cook model [93,94]:

$$f = 1 + C \cdot \ln[\dot{\epsilon}^P/\dot{\epsilon}_0], \qquad g = 1 - [(T - T_R)/(T_M - T_R)]^q. \tag{15}$$

In the first of (15), $C = \partial\sigma/\partial \ln \dot{\epsilon}^P$ is a fitting parameter and $\dot{\epsilon}_0$ is a normalization constant. In the second of (15), T_R and T_M are a reference temperature and melt temperature, with q a thermal softening exponent. More often in the literature, notations $\sigma_0 \to A$, $\sigma_0 \cdot \kappa \to B$, and $q \to m$ are conventionally used for Johnson–Cook constants. Other models include power-law forms [4,23,75,89]:

$$f = 1 + (\dot{\epsilon}^P/\dot{\epsilon}_0)^m, \qquad g = (T/T_R)^r. \tag{16}$$

In (16), definitions for material constants are $m = \mathrm{d}\ln(f-1)/\mathrm{d}\ln\dot{\epsilon}^P$ for strain-rate sensitivity and $r = \partial \ln\sigma/\partial \ln T$ for thermal sensitivity of flow stress [63,75]. Usually, $C > 0$, $m > 0$, $q > 0$, and $r < 0$, such that strength increases with increasing strain rate and decreasing temperature. In the context of these two models, the number of inelastic constitutive parameters is increased by at least two from the static case, and from three (σ_0, κ, n) to five $(\sigma_0, \kappa, n; C \text{ or } m; q \text{ or } r)$ if T_R and $\dot{\epsilon}_0$ are interpreted as fixed universal constants. More parameters are required to incorporate strain-rate-history effects, and measurements or extraction procedures are presumably needed for mass density and specific heat, where the latter two properties do not directly influence solutions to the static indentation problem.

The local temperature rate can be obtained from the continuum balance of energy [63], where ρ is deformed mass density and c_V specific heat at constant volume per unit mass:

$$\rho c_V \dot{T} \approx \zeta \cdot \dot{w}^P \approx \zeta \cdot \sigma\dot{\epsilon}^P. \tag{17}$$

Thermoelastic coupling is omitted, and adiabatic conditions pertinent to rapid loading relative to time required for heat conduction are assumed. The Taylor–Quinney factor is $\zeta \in [0,1]$, typically much closer to unity than zero and assumed constant in practice, though the ratio of stored to dissipated energy is more realistically expected to evolve with deformation [95,96]. Later, the approximation $\zeta = 1$ is used, presuming the missing thermoelastic heating in (17) under compression is offset by a maximally high value of ζ. A critical examination of the Taylor–Quinney effect, and a novel treatment of plasticity induced heating based on dynamic microstructure adaptation, has been set forth [97].

The basic models in (14) and (15) are considered here as suitable examples that are very widely used by the applied engineering community. More sophisticated constitutive models with stronger fundamental bases in physics and materials science are abundant in the literature, for example [16,92,98–101], though the relative extent of predictive physics versus phenomenology varies among such theories. Notably, novel ideas set forth by Zubelewicz [16,97,100], motivated by dislocation energetics and thermodynamics, include a marked departure from traditional plasticity theory based on a von Mises yield and flow function, most often implemented with a radial return algorithm [41]. Methods of dimensional analysis developed later in Section 4 can presumably be applied to such other models upon suitable consideration of all resulting independent variables (i.e., all requisite material parameters entering the corresponding constitutive theory).

Frictional interactions at the indenter–specimen interface can affect certain aspects of the response. The most common assumption in FE simulations of dynamic indentation appear to be frictionless contact [23,46,90,102]. As shown in FE simulations of static indentation [3], very strong friction (e.g., sticking) significantly affects the local strain field in the vicinity of the indenter and has a non-negligible effect on the contact radius versus indentation depth relation. Most calculations show that the load–displacement curve is not significantly affected by friction for static spherical indentation [3,33,53], though exceptions exist [31]. The assumption of frictionless contact is often made for convenience, but appears in many cases to produce sound agreement between force–depth data extracted from models and experiments [6,38]. In some numerical studies, nonzero (constant) friction coefficients ranging from 0.1 to 0.25 are used [34,40,52], choices which also seem to enable close agreement with experimental data. For dynamic indentation, contact interactions should consider the roles static and dynamic friction separately, with potentially distinct coefficients. Static friction would dictate stick versus potential slip criteria, and dynamic friction (i.e., kinetic friction) would apply when relative velocities of indenter and indented surface are finite. More elaborate constitutive models for static and dynamic friction exist, where coefficients are not constant but depend on local state, time, and/or slip rate [103,104]. Effects of static versus dynamic friction for high-rate spherical indentation remain to be more clearly elucidated in future numerical or experimental studies. The validity of the frictionless assumption likely improves as the surface roughness of the sample and indenter decrease.

3.2. *Survey of Prior Analytical and Numerical Modeling*

Important findings from previous modeling of dynamic indentation are summarized next. The analytical model of Mok and Duffy [80] was used to extract dynamic stress–strain curves from sphere impact experimental data, and a representative strain rate of $\dot{\epsilon} \approx 1500/\mathrm{s}$ was deduced to match dynamic uniaxial compression stress–strain data. More recently, analytical models have been used to assess dynamic hardness and dynamic rate sensitivity (n) for conical [89] and spherical [105] impacts. An analytical model [81] was used to infer that adiabatic plastic strain localization causes a reduction in hardness for sufficiently high sphere impact velocities, at correspondingly high indentation strains and strain rates [82,83]. The model of Tirupataiah and Sundararjan [81] was used by Kumaraswamy and Rao [106] to analyze dynamic sphere impact, wherein the plastic zones from dynamic and static indentation were of similar depth when normalized by indenter diameter. Therein, it was also determined that inertial effects on indentation response were unimportant due to the short time, relative to the total duration of the impact event, required for stress waves to traverse the contact area.

An analytical expanding cavity model was derived by Ito and Arai [94] for dynamic spherical impact, showing reasonable correlation of local field variables with those from FE solutions. This analytical model was later invoked [107] to determine the strain-rate sensitivity parameter C of the Johnson–Cook model in (15) from knowledge of indentation craters from spherical impacts obtained from experiments at two or more distinct impact velocities.

FE simulations provide more information than approximate analytical solutions, and thus enable more reliable material response parameter extraction. In Lu et al. [89], FE simulations with power-law strain-rate dependence as in (16) (but no thermal effects: $g = 1$) were used to validate the strain-rate sensitivity extracted from dynamic conical indentation experiments on copper. In Calle et al. [90], FE simulations were used in conjunction with drop test data to determine strain-rate sensitivity in terms of increase over static yield strength, without direct prescription of a strain-rate-hardening function f as entering (14). Results were discovered to be comparable to conventional SHPB rate sensitivity data, thereby validating their method for four particular metals of study: steel, copper, brass, and a titanium alloy.

The residual indentation imprint from dynamic FE simulations of spherical indentation has been used to identify elastic–plastic constitutive parameters [52] for materials without rate or temperature dependence, i.e., a constitutive model of the form in (6). Similarly, static yield and strain-hardening parameters have been identified from simulated SHPB-driven dynamic indentation using two conical indenters of different apex angles [102]. Spherical impact data for rate-insensitive, linear-hardening materials were generated in dynamic FE simulations [108] for a range of depths and indenter sizes. Then, support vector machine algorithms, Gaussian process regression, and nonlinear regressions as machine learning techniques were employed to estimate the material's plastic properties given only certain indentation response data.

Thorough parametric studies of dynamic spherical indentation using FEM have been reported [46] for rate-insensitive elastic–plastic solids [$m = 0$ in (16)] and were extended [23] for potentially rate-sensitive elastic–plastic solids (e.g., $m \geq 0$). In these studies, the indenter was a rigid sphere moving at constant velocity v_i, conditions that differ from experimental protocols wherein velocity usually decreases with depth, and wherein wave speeds of an elastic indenter are finite. Temperature effects were also omitted ($g = 0$), and nonreflecting boundary conditions were invoked such that wave interactions with sample boundaries vanished in the simulations. For impact speeds that were large, mean contact pressure $\bar{\sigma} = P/(\pi a^2)$ was found to significantly exceed $3\sigma_0$ at early simulation times when inertial effects dominated.

In Lee and Komvopoulos [23], four possible stages of the dynamic deformation process were identified for spherical indentation at constant rates:

- Elastic–plastic deformation characterized by an elastic core separating the plastic zone from the contact interface;
- Initial surface plastic deformation encountered upon disappearance of the elastic core and occurrence of maximum local plastic strain in the subsurface;
- Transient fully plastic deformation, where maximum local plastic strain shifts close to the contact interface and $c = \bar{\sigma}/\sigma_0$ increases with indentation depth;
- Steady full plastic deformation, wherein $c \approx 3$ represents static material hardness. This state is achieved for rate-insensitive materials at sufficient depths, when inertial effects become negligible.

Lee and Komvopoulos [23] also discovered a dimensionless parameter $\chi = (E_s/\sigma_0)(v_i/C_l) = [(\frac{1}{2}\rho_0 v_i^2)/(\frac{1}{2}E_s\epsilon_0^2)]^{1/2} = \sqrt{U_k/U_0}$ that was reported to delineate elastic–plastic, initial surface plastic, and transient fully plastic regimes achieved in the initial state, respectively, with increasing χ. Here, $C_l = \sqrt{E_s/\rho_0}$ is the longitudinal wave speed for uniaxial stress, U_k is a kinetic energy density, and U_0 is elastic energy density at initial yield strain $\epsilon_0 = \sigma_0/E_s$. For the case of rate-sensitive solids ($m > 0$), mean contact pressure decreases, contact area increases, and plastic work increases with increasing m and increasing static hardening coefficient n. The effects of plastic parameters were more easily delineated when inertial effects decayed, as indentation depth and simulation time increased.

A number of other studies of dynamic spherical indentation of ductile metals are pertinent to this review. An early experimental and analytical plasticity study of static and dynamic loading of a lead block by a hard steel sphere was performed by Yew and Goldsmith [109], around the same time as the aforementioned ball impact study of Mok and Duffy [80]. Experiments and analysis of dynamic spherical indentation in the shock loading regime were undertaken by Rudnitsky and Djakovitch [110]. Spherical impact of an indenter into ductile steel via a pendulum apparatus equipped with piezoelectric diagnostics was performed by Nobre et al. [111]; complementary analytical models with several theoretical approaches were used to evaluate findings. Clough et al. [112] performed dynamic hardness tests on steel using a dropped ball and explained data, including size and rate effects, using a dislocation-based theory. Experiments and FE simulations of steel plates loaded statically and dynamically with hemispherical indenters were more recently described by Liu and Soares [113]. Residual stresses from shot peening (a kind of dynamic spherical impact) were studied by Meguid et al. [114] using FE simulations. The effects of strain-rate dependence on indentation hardness were modeled by Almasri and Voyiadjis [115] using power-law and dislocation density-based constitutive theories. A spherical cavity expansion model for dynamic indentation of porous elastic–plastic materials was derived by dos Santos et al. [116] and validated for a large domain of strain rates, albeit for a conical rather than spherical tip.

The investigation of Nguyen et al. [117] used a different plastic constitutive law and a different set of dimensionless parameters than Lee and Komvopoulos [23] for extraction of material properties from FE simulations of dual sharp and spherical indentation at relatively low but finite rates, also under isothermal conditions. In contrast to the example in Section 5 of the current work, instrumented dynamic indentation data from miniaturized SHPB testing were not considered in prior dimensional analysis or in numerical simulations [23,117].

3.3. Indentation Strain Rate

When correlating dynamic indentation data with dynamic experimental data that have been recorded at a lower fluctuating rate (e.g., recorded from traditional SHPB experiments), a measure of global or effective strain rate for the indentation experiment is often sought. Just as there is no unique definition of global strain for static indentation, no unique definition of indentation strain rate exists. Plastic strain is locally heterogeneous during indentation, and local plastic strain rates are even more so, due to inherent changes in velocity of the indenter (e.g., deceleration after impact), as well as stress wave transients.

Several pragmatic definitions have been proposed elsewhere in order to assign an effective strain rate to a dynamic hardness H or strength $\bar{\sigma}$ measurement. In Lu et al. [89], with t_r the loading duration, the following function was proposed for dynamic conical indentation by an indenter of mass m_i, cone angle θ, and initial velocity v_i:

$$\dot{\epsilon}_r = \frac{\epsilon_R}{t_r}; \qquad \epsilon_R = 0.07, \quad t_r = 1.4\left[\frac{9m_i}{4v_i}\frac{\tan^2\theta}{\pi\sigma_0}\frac{1}{1+\ln\{[E_s/(3\sigma_0)]\tan\theta\}}\right], \tag{18}$$

where ϵ_R is the representative strain from Atkins and Tabor [118], and t_r is obtained from derivations by Johnson [32]. The strain rate in (18) is an average measure, i.e., a constant for a given experiment.

Perhaps most often used for interpretation of spherical indentation is the time derivative of Tabor's indentation strain, allowing for a transient contact radius $a(t)$:

$$\dot{\bar{\epsilon}}(t) = \frac{\mathrm{d}}{\mathrm{d}t}\left[0.2\frac{a(t)}{R_i}\right]. \tag{19}$$

A definition for the average strain rate $(\cdot)_A$ over the duration of the experiment, similar to that of Lu et al. [89] but now applicable to spherical indentation, was proposed by Kren et al. [105]; its value was obtained by dividing Tabor's indentation strain by the measured loading time t_m for which strain increases:

$$(\dot{\bar{\epsilon}})_A = \frac{1}{t_m}\int_0^{t_m}\frac{\mathrm{d}}{\mathrm{d}t}\left[0.2\frac{a(t)}{R_i}\right]\mathrm{d}t = 0.2\frac{a_m}{R_i t_m}. \tag{20}$$

The contact radius at maximum depth is a_m; this could be substituted with the residual contact radius after unloading, a_r, for consistency with Tabor's techniques.

The following expressions are derived for a rigid indenter ($h_s = h$) of potentially transient velocity $v_i(t) = \dot{h}(t)$, using Tabor's strain measure and geometry of the indented surface [90]:

$$\begin{aligned} \bar{\epsilon}(t) &= 0.2\frac{a(t)}{R_i} \approx 0.4\left[\frac{h(t)}{2R_i}\left(1-\frac{h(t)}{2R_i}\right)\right]^{1/2} \Rightarrow \\ \dot{\bar{\epsilon}}(t) &\approx \frac{0.1\dot{h}(t)}{R_i}\left[1-\frac{h(t)}{R_i}\right]\left[\frac{h(t)}{2R_i}\left(1-\frac{h(t)}{2R_i}\right)\right]^{-1/2}. \end{aligned} \tag{21}$$

In the context of (10), as derived by Field and Swain [6], it is assumed in the calculation of the contact radius a in (21) that $h \approx h_r \gg h_e/2$. For another alternative strain-rate definition, the time differentiation of (11) produces an effective indentation strain rate of

$$\dot{\hat{\epsilon}}(t) = \frac{4}{3\pi a(t)}\left[\dot{h}_s(t) - \frac{\dot{a}(t)}{a(t)}h_s(t)\right]. \tag{22}$$

Uniaxial-equivalent elastic and plastic strain rates could likewise be obtained from differentiation of (13).

Finally, a general order-of-magnitude relation applicable to any indenter type was defined by Subhash [86]:

$$\dot{\epsilon}_h = v_i/h, \qquad [\dot{\epsilon}_h = \dot{h}/h \text{ for rigid indenter}]. \tag{23}$$

Here, v_i is a representative velocity of the indenter, and depth h could be substituted with another measure of the size of the impression (e.g., the residual imprint's diagonal or the imprint's radius, depending on indenter shape). The rigid-indenter version of (23) was used in the context of creep and viscoelasticity in Cheng and Cheng [7]. In another example application, Kren et al. [105] used (23), with v_i the initial impact velocity and h replaced by the residual (plastic) contact diameter $2a_r$.

3.4. Summary: Dynamic versus Static Indentation

Fundamental differences between high-rate indentation (dynamic regime) and quasi-static indentation (static regime) for elastic–plastic solids are emphasized as follows:

- A time scale enters the problem for dynamics but not statics;
- The indentation strain rate is finite for dynamics, so the rate sensitivity of the plastic response affects dynamics but not statics;
- Inertial effects (i.e., stress waves) appear for dynamics but not statics;
- Due to inertia, mass density is pertinent for dynamics but not statics;
- Adiabatic heating may arise for dynamics but not for statics under conventional thermal boundary conditions;
- Due to adiabatic heating, specific heat and thermal softening properties may be important for dynamics but not statics;
- Both static and dynamic friction could be important for dynamics but only the former for statics in the limit of zero relative interface velocities;
- Under severe impact, nonlinear elasticity and thermal expansion affecting shock waves would arise for dynamics, but shock waves are irrelevant for static loading.

4. Extension: Dynamic Dimensional Analysis

In the analysis that follows, two major assumptions are invoked regarding the indenter and contact interface. These assumptions reduce the number of independent parameters, simplifying analysis.

Firstly, the indenter is assumed rigid. This is a typical assumption in static and dynamic FE simulations (e.g., [23,38,46]), most valid when $E_i \gg E_s$. In addition to reducing $\bar{E} \to E_s/(1-\nu_s^2), R \to R_i = \text{constant}, h \to h_s$ in the Hertz theory of Section 2.1, this assumption eliminates effects of wave transmission in the indenter that could affect contact under dynamic indentation. In the present dimensional analysis, it is not necessary to assume $R_s \to \infty$; this assumption is used later in Section 5, but only when comparing certain experimental results to the then closed-form elastic solution of Hertz.

Secondly, frictionless contact is assumed, as in prior numerical studies of dynamic indentation [23,46,90,102]. Implicit in this assumption is that surfaces are sufficiently smooth to avoid sticking behavior (statics) and dissipative sliding resistance (dynamics). If, on the other hand, friction is substantial (e.g., relatively rough surfaces), then the list of independent variables in the dimensional analysis can be extended to minimally include a static (i.e., sticking) coefficient and, if different, a dynamic (i.e., kinetic) coefficient, both dimensionless. The absence of friction coefficients in the dimensional analysis does not affect any results calculated by example in Section 5. However, frictional effects would complicate a more general treatment of multiple materials of variable surface finish.

Several assumptions are also made regarding the testing apparatus. It is assumed that the radius of the indenter $R = R_i$ is variable from experiment to experiment but that other aspects of the system geometry remain fixed among experiments. Boundaries may be of infinite extent in simulations, or of finite extent in experiments, where the latter (i.e., specimen size) is fixed such that the absolute domain size need not be treated as an independent variable.

It is also assumed that a system velocity, denoted by v, is a defined, controllable constant for each simulation or experiment analyzed and is not a dependent variable. For example, the system velocity v can be simply assigned as the indenter velocity $\dot{h}$ if prescribed as a constant in a simulation [23,46,102] or can be assigned as the initial (measured) projectile impact velocity for a spherical impact experiment [80–83]. In a dynamic hardness or dynamic indentation experiment using the SHPB, the indenter's tip velocity is generally not constant, even during the loading phase [86,88,119]: a transient period may exist over which the indenter accelerates, and then the indenter always decelerates. In that case, the velocity of the striker bar could be used for v as a measure of the input loading rate; otherwise, the average indenter velocity, if controlled, over some finite time interval of the loading phase could be used [86].

4.1. Variable Identification

The current analysis considers only global, scalar quantities that are either (1) imposed or extracted from indentation experiments or (2) homogeneous and stationary material properties. Local field variables (e.g., transient stress and strain distributions with local values depending on position in the sample) are not addressed.

Application of concepts of dimensional analysis and Buckingham's Pi theorem begins with the identification of all dependent and independent variables in dimensional form. Dependent variables are defined as follows:

- Indentation force P;
- Indentation contact radius a;
- Plastic work of indentation W^P:

$$W^P = \int_0^{h_r} P(h)\,\mathrm{d}h. \tag{24}$$

Note that the quantification of W^P requires loading to a maximum depth h_m, followed by unloading to a residual depth h_r, the latter at which $P = 0$. From these three dependent variables, other quantities of interest can be defined, for example:

- Mean pressure $\bar{p} = P/(\pi a^2)$ and the constraint factor measured relative to the initial static isothermal yield strength, hereafter redefined as $c = \bar{p}/\sigma_0$;
- Average temperature rise in a volume $\bar{V}$ of material assuming adiabatic conditions: $\Delta\bar{T} = W^P/(\rho_0 c_V \bar{V})$, where the plastic zone volume can be estimated as the cylindrical region $\bar{V} \approx \frac{3}{4}\pi^2 a^3$, as in other works [5,26].

The inelastic constitutive model of (14) is assumed a priori, with $\zeta = 1$ in (17) for the adiabatic regime. As discussed in Section 3, linear isotropic elasticity is assumed without thermoelastic coupling, and as discussed in Section 4, frictionless contact is assumed. The material is also presumably homogeneous, meaning local grain-to-grain fluctuations in properties are assumed to negligibly affect the global indentation response. In dimensional form, independent variables are then the following:

- Indentation depth h and maximum depth h_m;
- Effective indentation (system) velocity v;
- Indenter radius R;
- Initial temperature T_0;
- Substrate elastic properties (dropping $(\cdot)_s$ subscripts) E, ν;
- Substrate plastic properties σ_0, κ, m, n, r, $\dot{\epsilon}_0$, T_R;
- Substrate initial mass density ρ_0 and specific heat per unit mass c_V.

Of these independent variables, only h varies with time during an indentation simulation or experiment, given the definition of v as a constant explained already. The maximum depth is needed for determination of W^P, since h is multivalued during a load–unload cycle [7,21]. However, h_m can be excluded from the list if only P and a are sought from a monotonic loading process. Sixteen independent variables are listed, of which eleven are material property constants. Time is not an explicit independent variable, since given the system velocity, indenter radius, initial temperature, and material properties, the time at which a particular depth value h is achieved is determined implicitly [7].

Next, the Buckingham Pi theorem is invoked to reduce the number of independent variables when expressed in dimensionless form [7,19,20,22]. The number of independent dimensions entering the problem is four: mass, length, time, and temperature. Since the stress dimension is recovered from mass, length, and time, the four independent dimensions are more conveniently reassigned into stress, length, time, and temperature.

The following independent variable combinations are then used for normalization:

- Stress: modulus E;
- Length: indenter radius R;
- Time: viscoplastic time scale $t_0 = 1/\dot{\epsilon}_0$;
- Temperature: plastic thermal susceptibility $\tilde{T} = \sigma_0/(\rho_0 c_V)$.

The elastic modulus is a standard prescription for stress normalization [7]. Unlike analysis in prior works [7,21,23,46], h and v are herein excluded as normalization factors, since it is more convenient to work with constant quantities R and t_0, which later serve to define dimensionless strain and strain rate. Note that, as t_0 decreases, plastic stress relaxation is faster and a rate-independent response is approached: $f \rightarrow 1$ as $t_0 \rightarrow 0$, so long as $m \geq 0$ in (16). Note also that, as $\tilde{T}$ decreases, the tendency for adiabatic temperature rise decreases.

Applying the Pi theorem, the number of independent variables is reduced from sixteen to twelve, now defined in dimensionless form as follows:

- Indentation depth h/R and maximum depth h_m/R;
- Indentation rate $(v/R) \cdot t_0$;
- Yield strength σ_0/E;
- Elastic wave speed via $R/(C_l t_0)$, where $C_l = \sqrt{E/\rho_0}$;
- Reference temperature $\tilde{T}/T_R$ and initial temperature $\tilde{T}/T_0$;
- Dimensionless elastic and plastic properties ν, κ, m, n, r.

The following physically appealing results are apparent. Normalized depth h/R is an approximate global strain measure. Normalized rate vt_0/R is an approximate global measure of strain rate times plastic relaxation time, where the larger the value of this dimensionless quantity, the greater the anticipated viscoplastic rate effect. As $vt_0/R \rightarrow 0$, the rate-independent case is recovered. Use of the ratio σ_0/E characterizes plastic to elastic stiffness, a standard choice [7,23,46]. The ratio $R/(C_l t_0)$ is interpreted as the elastic wave relaxation time R/C_l divided by the viscoplastic relaxation time t_0. As $R/(C_l t_0) \rightarrow 0$, inertial effects should become less important, since stress wave equilibrium should be achieved more rapidly relative to viscoplastic rate effects. Typically, in practice, T_R is simply fixed at room temperature (≈293–300 K), but $\tilde{T}/T_R \rightarrow 0$ as the material becomes resistive to temperature change. Ambient temperature T_0 will differ from T_R for indentation at other imposed thermal boundary and initial conditions (e.g., testing of preheated samples).

4.2. Functional Forms

Given the independent dimensionless variables, the sought dependent variables can be expressed as dimensionless functions Π_P, Π_a, Π_W:

$$\frac{P}{ER^2} = \Pi_P\left(\frac{h}{R}, \frac{v\,t_0}{R}, \frac{\tilde{T}}{T_0}; \frac{\sigma_0}{E}, \frac{R/t_0}{\sqrt{E/\rho_0}}, \frac{\tilde{T}}{T_R}, \nu, \kappa, m, n, r\right), \tag{25}$$

$$\frac{a}{R} = \Pi_a\left(\frac{h}{R}, \frac{v\,t_0}{R}, \frac{\tilde{T}}{T_0}; \frac{\sigma_0}{E}, \frac{R/t_0}{\sqrt{E/\rho_0}}, \frac{\tilde{T}}{T_R}, \nu, \kappa, m, n, r\right), \tag{26}$$

$$\frac{W^P}{ER^3} = \Pi_W\left(\frac{h}{R}, \frac{h_m}{R}, \frac{v\,t_0}{R}, \frac{\tilde{T}}{T_0}; \frac{\sigma_0}{E}, \frac{R/t_0}{\sqrt{E/\rho_0}}, \frac{\tilde{T}}{T_R}, \nu, \kappa, m, n, r\right). \tag{27}$$

Arguments preceding the the semicolons on the right sides of (25)–(27) are loading conditions, and arguments following the semicolons are material properties. As noted in Section 4.1, h_m/R is not required for determination of force and contact radius during monotonic loading, so it is excluded from Π_P and Π_a. Parametric experiments and/or FE simulations are needed to fully determine the functions on the right sides of (25)–(27). Results from such studies will enable assessment of the relative importance of loading rate, temperature, and material properties on the global mechanical and thermal response.

The isothermal, quasi-static Hertz solution should be recovered as $\sigma_0/E \to \infty$, $v \to 0$, and $C_l \to \infty$, where for small indentation depths $R_s \to \infty$:

$$\Pi_P\left(\frac{h}{R}, 0, \cdot; \infty, 0, \cdot, \nu, \cdot, \cdot, \cdot\right) = \frac{4}{3(1-\nu^2)}\left(\frac{h}{R}\right)^{3/2} \quad [\bar{R} \to R], \tag{28}$$

$$\Pi_a\left(\frac{h}{R}, 0, \cdot; \infty, 0, \cdot, \nu, \cdot, \cdot, \cdot\right) = \left(\frac{h}{R}\right)^{1/2} \quad [\bar{R} \to R], \tag{29}$$

$$\Pi_W\left(\frac{h}{R}, \frac{h_m}{R}, 0, \cdot; \infty, 0, \cdot, \nu, \cdot, \cdot, \cdot\right) = 0. \tag{30}$$

Analytical functional forms, if they can be determined, should be consistent with the limiting cases in (28)–(30). Given (25)–(27), the mean contact pressure, constraint factor relative to σ_0, and mean transient temperature rise can be reconstructed:

$$\bar{p} = \frac{E}{\pi}\left(\frac{R}{a}\right)^2 \cdot \Pi_P \approx c\sigma_0, \qquad \Delta\bar{T} = \frac{4E}{3\pi^2}\left(\frac{R}{a}\right)^3 \cdot \Pi_W = \frac{4E}{3\pi^2} \cdot \frac{\Pi_W}{(\Pi_a)^3}. \tag{31}$$

Similarly, indentation strains [26,44] can be found as

$$\bar{\epsilon} = 0.2\frac{a}{R} = 0.2 \cdot \Pi_a, \qquad \hat{\epsilon} = \frac{4}{3\pi}\frac{h}{a} = \frac{4}{3\pi}\frac{h/R}{\Pi_a}. \tag{32}$$

If the constitutive model of (15) is used instead of (16), then the subset of two independent dimensionless variables (m, r) is replaced with the set of three dimensionless variables $(C, q, T_M/\tilde{T})$ in (25)–(27). Analogous constructions would apply for other constitutive models, for example, as cited in Section 3.1.

5. Application: Analysis of Instrumented Dynamic Indentation Data

5.1. Experimental Protocols

Data from three dynamic spherical indentation experiments are analyzed using the equations and techniques of Sections 2–4. Experimental methods have been discussed by Casem [88,119] and are summarized in what follows.

A miniature Kolsky bar (i.e., SHPB) [120] is adapted for instrumented indentation, whereby transient force, displacement, and velocity data are acquired in each experiment. The loading history (e.g., indenter's velocity) depends on the velocity of the striker bar and geometric properties of the system (including pulse shaping), as well as indentation resistance afforded by the substrate. Initial clearance between indenter and substrate also affects the velocity history. The loading history is thus not strictly controlled; however, different final indentation depths are generally achieved by increasing the striker velocity, commensurate with an increase in the average loading rate. Experimental data include both the loading and unloading histories for each test.

Relevant properties and parameters are listed in Table 1 with supporting references. Of these values, those comprising the set of six strain-hardening, strain-rate, and temperature-sensitivity parameters $\{\kappa, n; C \text{ or } m; q \text{ or } r; T_R, T_M\}$ are not used explicitly in the forthcoming analysis but are included for context to aid in interpretation of results. All current experiments are performed at standard room temperature: $T_0 = T_R$. Prominent results are summarized in Table 2 for reference and are defined and discussed in detail later. Pertaining to loading conditions, h_m in the leftmost column is the maximum indentation depth prior to unloading, and $\dot{h}_A$ in the rightmost column is the average indenter tip velocity over the loading phase of each experiment.

The substrate material is presumed to be isotropic, both elastically and plastically. Characterization and static indentation experiments on the same as-received Al 6061-T6 [1] suggest that, given the presently sized indenter, the volume of material sampled underneath the indenter at initial yield should contain 40 to 50 crystals of random orientation. Results by Weaver et al. [1] confirm that this is a sufficient number to ensure a repeatable, globally isotropic response. It is assumed here that isotropy carries over to the dynamic regime; similar indentation stress-strain curves presented later among multiple dynamic indentation experiments support this assumption, at least post yielding.

Table 1. Material and geometric parameters.

Parameter (Units)	Value	Definition	Source
E (GPa)	71.0	modulus of Al 6061-T6	Wu et al. [121]
ν (-)	0.33	Poisson's ratio of Al 6061-T6	Wu et al. [121]
ρ_0 (g/cm^3)	2.77	mass density of Al 6061-T6	Wu et al. [121]
c_V (J/kg·K)	896	specific heat of Al 6061-T6	Zhu et al. [122]
σ_0 (GPa)	0.25	initial yield strength of Al 6061-T6	Lesuer et al. [123], Zhu et al. [122]
κ (-)	0.35–1.8	hardening coefficient range of Al 6061-T6	Lesuer et al. [123], Zhu et al. [122]
n (-)	0.38–0.43	hardening exponent range of Al 6061-T6	Lesuer et al. [123], Zhu et al. [122]
C (-) $[\simeq m]$	0.002–0.083	rate sensitivity range of Al 6061-T6	Lesuer et al. [123], Zhu et al. [122], Casem et al. [120]
q (-) $[\simeq -r]$	1.34	thermal softening of Al 6061-T6	Lesuer et al. [123]
$\dot{\epsilon}_0$ (1/s)	1.0	reference strain rate (universal)	Zhu et al. [122]
T_M (K)	925	melt temperature of Al 6061-T6	Zhu et al. [122]
T_R (K)	294	reference temperature (ambient)	Zhu et al. [122]
$\bar{E}$ (GPa)	71.2	system modulus with WC indenter	Weaver et al. [1]
R (mm)	3.175	indenter radius	this work

Table 2. Loading conditions and results at max depth $(\cdot)_m$ or averaged over the loading phase $(\cdot)_A$.

Experiment	$\bar{v}$ (m/s)	h_m (μm)	a_m (μm)	W_m^P (mJ)	$\Delta\bar{T}_m$ (K)	$\bar{\epsilon}_m$	$\hat{\epsilon}_m$	$\dot{\bar{\epsilon}}_A$ (1/s)	$\dot{\hat{\epsilon}}_A$ (1/s)	$\dot{h}_A$ (m/s)
1	0.61	17.2	421	1.57	1.15	0.0265	0.0174	1386	908	0.90
2	1.06	25.0	508	3.82	1.59	0.0320	0.0209	2217	1452	1.74
3	1.36	35.7	606	8.43	2.06	0.0382	0.0250	2750	1801	2.57

The miniature SHPB system equipped for indentation testing is sketched in Figure 1. This image only shows the upper half of the system, which is axially symmetric. The indenter is tungsten carbide (WC) with a radius $R = R_i$ of 3.175 mm. The spherical tip of the indenter is machined directly into the input bar, on its right side in Figure 1. The substrate is the aluminum alloy Al 6061-T6 of cylindrical geometry, with $L_s/D_s = \frac{3}{4}$ and $D_s = \frac{3}{2}R_i = \frac{3}{2}R$.

The velocity history $v(t)$ at the right side of the input bar in Figure 1 is acquired from analysis of transient data from a strain gauge. The average of this input velocity over the duration of each test is denoted by $\bar{v}$. As shown in Table 2, $\bar{v}$ increases with test number, as do all other observed or extracted quantities such as maximum depth, maximum indentation strains, and average indentation strain rates. In many subsequent figures, $\bar{v}$ is thus used to further distinguish experiments 1, 2, and 3. A normal displacement interferometer (NDI) focused at each of the indented faces of the sample at its radial edge (left end) and the left end of the output bar records displacement history for each location.

Force and displacement histories for the tip of the indenter are computed from a linear elastic wave analysis with the strain gauge data and the NDI data.

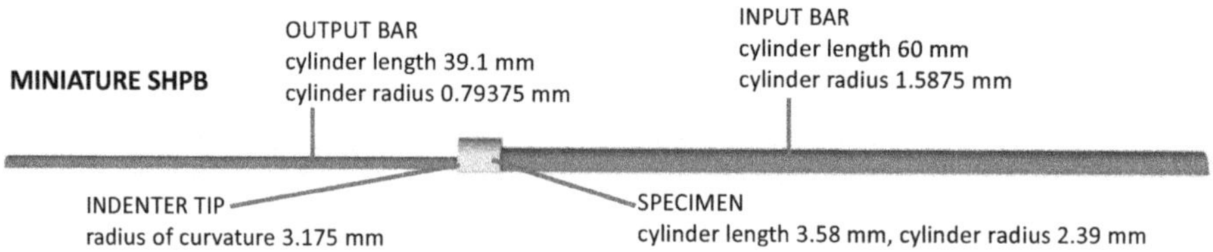

Figure 1. Experimental setup of miniature SHPB with spherical indenter. Spherical tip is machined directly into right end of WC output bar, and input velocity history is recorded at right end of truncated steel input bar. Specimen material is Al 6061-T6.

Static spherical indentation experiments on this material have been reported elsewhere [38], albeit with an indenter of larger radius $2R$. Numerical simulations of the static problem [38] loaded to comparable indentation strain levels suggest that the current dimensions of the substrate are sufficiently large to mitigate boundary edge effects. However, the effects of stress wave interactions with finite boundaries cannot be ruled out in dynamic experiments.

The elastic stiffness of WC ($E_i = 640$ GPa) is an order of magnitude larger than that of aluminum, and $\bar{E}$ in Table 1 accounts for the true elastic modulus of both the indenter and substrate, assumed to be known a priori. If a rigid indenter is assumed instead, then $\bar{E}$ increases by $\approx$10%. However, since the indenter material is held fixed among experiments, elastic properties of the indenter can be excluded from the list of independent variables in the forthcoming dimensional analysis, as assumed in the general framework of Section 4.

Indentation depth and tip velocity (i.e., depth rate) are inferred with respect to the far-field displacement of the surface of the sample and correctly account for rigid body motion. The reported indentation depth and depth rate are the respective true indentation depth h_s and depth rate $\dot{h}_s$ in the substrate, relative to those measured for the far-field surface away from any pile-up or sink-in effects. In the remainder of Section 5, notation is simplified such that $h_s \to h$ and $\dot{h}_s \to \dot{h}$. Cursory calculations with the approximation in (3) confirm that the contribution of deformation of the spherical end of the indenter, h_i, to h should be negligible for loads and contact radii reported in what follows. Summarizing, the indenter can be considered rigid for purposes of setting $h = h_s$ and R_i =constant, but it should have a finite modulus E_i for accurately quantifying $\bar{E}$.

5.2. Data Analysis: Global Response

Experimental force–depth, depth–time, and depth rate–time histories are shown in Figure 2, labeled "exp" (for experiment) 1, 2, and 3. The time t at a given depth h is confirmed by the integral $t = \int (h/\dot{h})\mathrm{d}h$, where $\dot{h}$ is known as a function of h and initially $t = 0$. Shown for reference in Figure 2a is the Hertz elastic solution of (1) obtained assuming $\bar{R} = R$ in (2).

For $h \gtrsim 10\,\mu$m, the data are more compliant than the elastic solution, as expected for an elastic–plastic material post yielding. For $h \lesssim 10\,\mu$m, experiments are similar to the elastic solution. Notably, some data appear slightly stiffer than the elastic solution over small intervals of h. These unusual features could be due to inertial effects and/or imprecision of experimental measurements at very low indentation depths. Depth–time histories shown in Figure 2b indicate total load–unload durations ranging from around 20 to 30 µs. Maximum depth and depth rate increase with experiment number. Velocity histories are drastically different among experiments. For example, velocity ramps up quickly with time for experiment 2, while it is initially near maximal for experiment 3. In the latter case, the indenter clearly accelerates prior to contact with the substrate. Note $\dot{h} < 0$

during the unloading phase of each experiment. Peak loading and unloading velocities are notably smaller for experiment 1 than experiments 2 and 3.

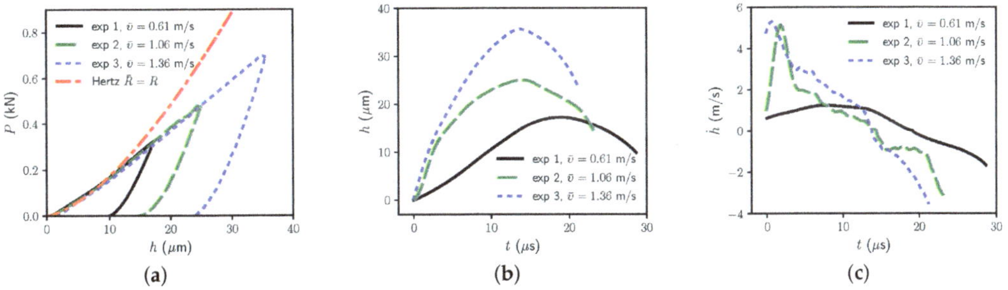

Figure 2. Experimental data: (**a**) force vs. depth (with Hertz analytical solution) (**b**) depth vs. time (**c**) depth rate vs. time.

Force–depth data are delineated for each experiment in Figure 3, wherein the slope S upon initial unloading is extracted from the tangent for each case. The maximum indentation depth is denoted by h_m. The unloading slope S at $h = h_m$ is then used to obtain the Hertz contact radius at maximum depth $a_m = a(h = h_m)$, assuming quasi-static elastic unloading, via (5) [37,38]. Elastic modulus $\bar{E}$ is assumed, a priori, to have the value listed in Table 1. No attempt is made to extract elastic properties from the present indentation data, as has been performed in some investigations elsewhere [6].

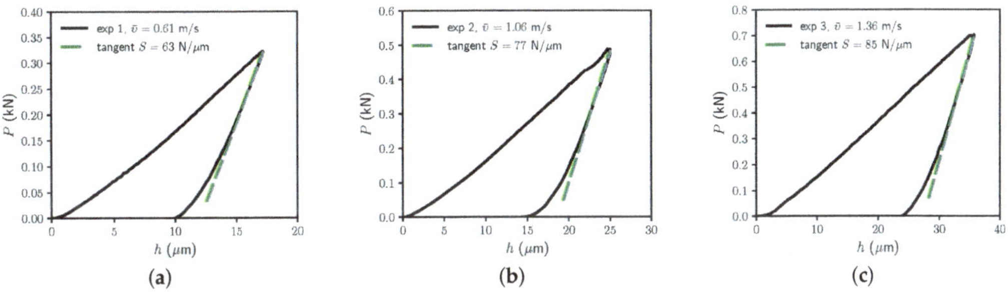

Figure 3. Data and unloading tangent: (**a**) experiment 1, (**b**) experiment 2, and (**c**) experiment 3.

Dimensionless dependent variables (Π-terms) introduced in the analysis of Section 4 are reported in Figure 4. General functional forms are (25)–(27). In each figure, the independent variable resolved on the abscissa is dimensionless depth h/R (or h_m/R), which is well-defined and fully known from the test data. Among different experiments, the normalized indentation velocity vt_0/R also varies, as does the maximum normalized depth h_m/R. Since the substrate material and initial temperature (room temperature) are identical among experiments, the other independent variables on the right sides of (25)–(27) are fixed among the present results. Hence, potential influences of the latter (fixed) properties cannot be fully discerned or quantified among the presently available data. However, qualitative deductions on yielding, strain hardening, and rate sensitivity are still possible, as will be discussed later in the context of indentation stress–strain curves. A candidate definition for effective velocity is the average loading rate $v = \dot{h}_A = h_m/t_m$, where t_m is the time instant at which $h = h_m$. As shown in the rightmost column of Table 2, v_A increases from approximately 0.9 to 1.7 to 2.6 m/s over respective experiment numbers 1 to 2 to 3.

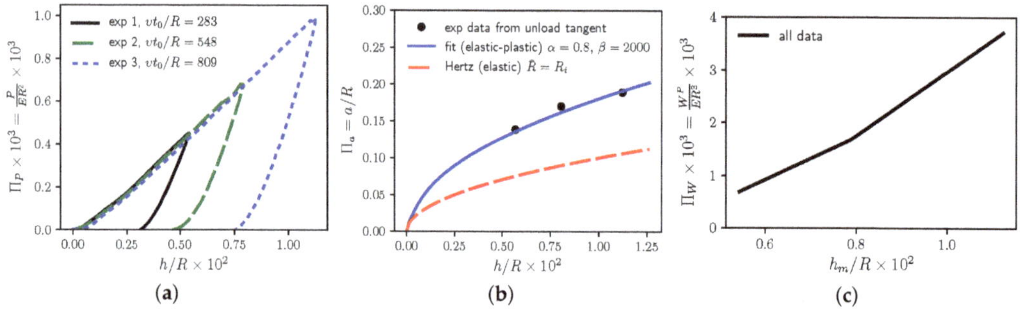

Figure 4. Dimensionless variables: (**a**) normalized force, (**b**) contact radius (loading phase only), and (**c**) plastic work (final).

Normalized force Π_P versus normalized depth h/R in Figure 4a provides the same information as in Figure 2a, since E and R are identical among experiments. Assuming $v = \dot{h}_A$, the results in Figure 4a show that the dependence of dimensionless indentation force Π_P on dimensionless loading rate vt_0/R is low for Al 6061-T6 over the current domain of loading rates.

Since the present data do not contain intermittent unload–reload cycles or continuous stiffness measurements (CSM) to obtain unloading slope $S(h)$ for $h \neq h_m$, approximations are used to compute contact radius $a(h)$ for $h \neq h_m$. First, for the loading phase, $t \leq t_m$, it is assumed that $a(t)$ for each experiment depends only on h/R and not v. As shown in Figure 4b, the normalized radius Π_a is then estimated by the following function, where $\alpha \geq 0$ and $\beta \gg 1$ are dimensionless fitting constants:

$$\Pi_a = a/R = (1 + \alpha\{1 - \exp(-\beta h/R)\})\sqrt{h/R} \qquad [t \leq t_m]. \tag{33}$$

The Hertz solution is recovered in the limit of infinitesimal depth, whereby $a \to \sqrt{hR}$.

Available data do not enable fitting of $a(t)$ for $t > t_m$, i.e., throughout the entire unloading process. Thus, during unloading, it is assumed that the indentation stress–strain data (specifically, $\hat{\sigma} = \bar{p}$ versus $\hat{\epsilon}$) demonstrate a fixed slope of $\bar{E}$, consistent with quasi-static unloading [5,26]. To this end, the following quadratic equation is solved at each time increment for $a(h(t))$ during the unloading phase:

$$\frac{4\bar{E}}{3\pi}\left(\frac{h}{a} - \frac{h_r}{a_f}\right) = \frac{P}{\pi a^2} \qquad [t \geq t_m]. \tag{34}$$

Here, h_r is the final (residual) indentation depth upon complete unloading, and a_f is the projected (not measured) final contact radius. The latter, which generally can differ from the observed residual imprint radius a_r, can be computed by solving (34) at $h = h_m$, with corresponding peak load $P = P_m$ measured and radius $a = a_m$ obtained from (33).

Finally, normalized plastic work Π_W is shown in Figure 4c. Its computation requires the entire load–unload force–displacement cycle for a given h_m; hence, only three data points are available (one for each experiment). Though not shown in Figure 4c, Π_W logically could depend on h_r/R as well, since h_r varies among experiments, increasing from 9.8 to 14.4 to 23.6 μm over experiments 1, 2, and 3. Such dependency is permissible via inclusion of h/R in addition to h_m/R in the listed arguments of (27). However, the identity $\Pi_P(h = h_r) = 0$ provides an additional constraint equation that can be used to eliminate explicit dependence of Π_W on h_r/R. Normalized plastic work clearly increases with increasing maximum penetration depth.

5.3. Data Analysis: Indentation Stress–Strain and Other Extracted Information

Typically, the global force–depth response is of primary interest from indentation testing, along with possible information on the contact radius. In the dynamic case, complete data should include time histories of these quantities. From such data, supplemental quantities such as indentation stress, indentation strain, indentation strain rate, and mean temperature rise can be computed, as demonstrated next. The reader interested only in the primary response (e.g., indentation force versus depth) emphasized in Section 5.2 can bypass the remainder of Section 5.3.

Given a (or equivalently, Π_a), along with force data $P(h)$, mean pressure $\bar{p} = P/(\pi a^2)$ (e.g., Meyer's hardness H when P is maximum and a is the residual imprint radius) is computed. Results are shown in Figure 5a. Constraint factor c is shown in Figure 5b, recalling that the initial yield stress σ_0 from Table 1 is used for normalization in Section 5. Note that this value of σ_0 is considered to be known a priori, rather than extracted from the present indentation data. Mean pressure closely tracks the Hertz solution at a very small h/R for experiments 1 and 2, whereas $\bar{p}$ for experiment 3 suggests anomalously low compliance in the limit $h/R \to 0$. As displayed here in Figures 5 and 6, the Hertz solution invokes the usual assumption $R_s \to \infty$ in (2). Mean pressures are similar among the loading phases of all three experiments for $h/R \gtrsim 0.0025$, in the regime where plastic compliance is expected to overtake elastic compliance. Such similarities suggest strain rate and strain rate–history effects on $\bar{p}$ are small for these experiments on Al 6061-T6, given the very different velocity histories among tests in Figure 2c.

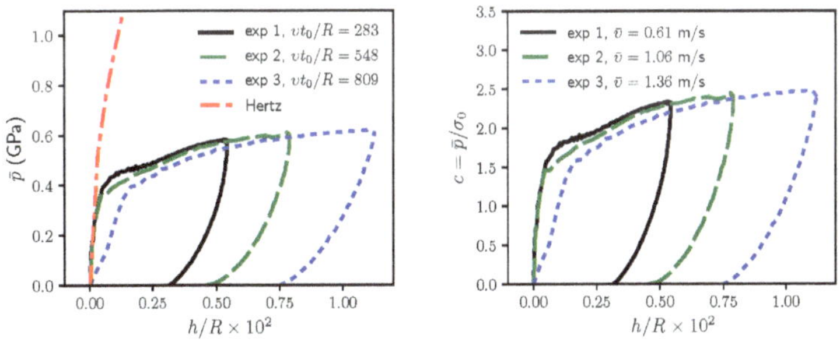

(**a**) (**b**)

Figure 5. Stress vs. depth: (**a**) mean indentation pressure and (**b**) constraint factor.

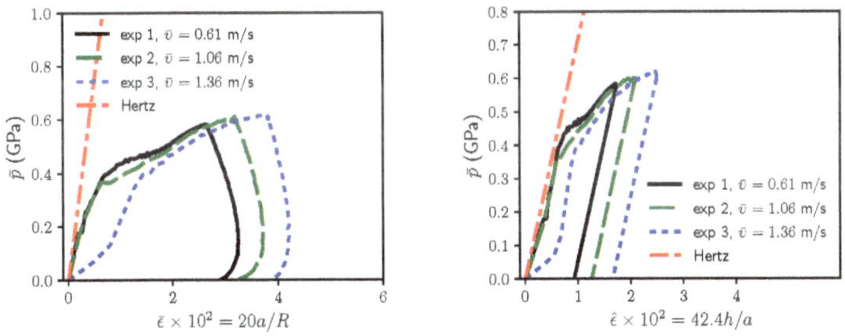

(**a**) (**b**)

Figure 6. Indentation stress vs. (**a**) Tabor's strain [44] and (**b**) Kalidindi and Pathak's strain [26].

Taking $v = \dot{h}_A$, the results in Figure 5a imply that dependence of mean stress on dimensionless loading rate vt_0/R is low. In fact, the mean stress and constraint factor appear to decrease slightly with increasing loading rate, though such decrease may be due to imprecision of the experimental measurements and uncertainty inherent in (33) that neglects possible rate dependence of the transient contact radius.

Recall from the review in Section 2.2 that initial yielding corresponds to $c \approx 1.1$ in the quasi-static Hertz theory [6]. After yielding, the calculated constraint factor increases from $c \approx 1.5$ to $c \approx 2.5$ over the full domain of indentation depths in the experiments. Strain hardening likely influences c to increase with increasing h/R in the fully plastic regime. The computed range of c is within bounds observed elsewhere in quasi-static spherical indentation experiments and simulations [1,2,6,37,42–45]. Similar ranges have been observed in dynamic spherical indentation simulations of elastic–perfectly plastic, rate-independent solids for constant rigid indenter velocities that are not too large (e.g., $\dot{h} \lesssim 75$ m/s), albeit for representative metallic substrates with different properties than those of aluminum [46]. Transient increases in c due to inertial effects manifest at velocities on the order of 100 m/s [46]; at such high velocities, the effects of strain rate sensitivity on c also become stronger [23]. During elastic unloading, c is easily computed given P and the current method of estimation of a, but its value does not have any physical significance.

Indentation stress (i.e., mean pressure $\bar{p}$) is reported versus the two different indentation strain measures of (32) in Figure 6. Elastic deformation followed by plastic yielding and mild-to-moderate strain hardening is apparent in each representation. The hardening behavior is qualitatively consistent with the traditional dynamic uniaxial stress–strain behavior of this material [120,122]. The anomalously high initial compliance of experiment 3 is also apparent in Figure 6. Elastic unloading is perfectly represented in Figure 6b, as a result of the implementation of (34). On the other hand, the unloading portions of $\bar{p}$ versus $\bar{\epsilon}$ curves in Figure 6a show slopes inconsistent with $\bar{E}$. Similar inconsistencies have been noted when $\bar{\epsilon}$ is used as the indentation strain measure for static indentation [5,26]. In Figure 6b, yielding is apparent at $\hat{\epsilon} \approx 0.6\%$ for experiments 1 and 2. From (13), with $c \approx 2.5$, the uniaxial-equivalent yield strain is then estimated as $\sigma_0/E \approx 0.25\%$.

Strain rates in (19) and (22) are consistently obtained from numerical differentiation of (32). Results are shown in Figure 7. According to each strain definition, strain rates are initially large over domains wherein both the indentation depth and contact radius increase rapidly with time. Initial values of $\dot{\bar{\epsilon}}$ and $\dot{\hat{\epsilon}}$ range from 7000/s to 28,000/s, with the highest strain rates observed in experiment 3. Strain rates decrease subsequently with increasing time, as the indenter necessarily decelerates. Negative rates persist during unloading. However, oscillations also arise during unloading to accommodate the assumed form of contact radius in the elastic response function of (34).

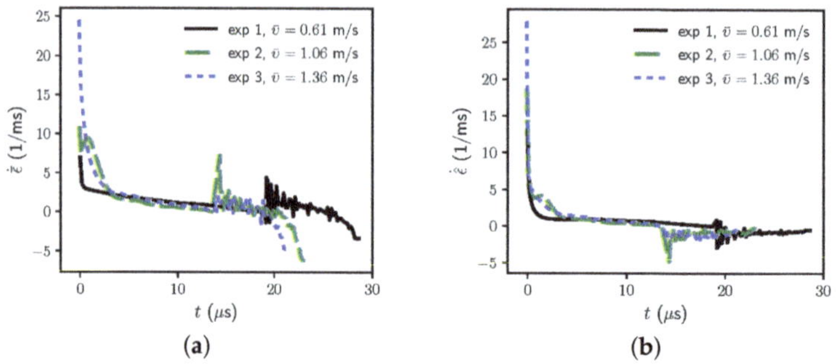

Figure 7. Indentation strain rates from (**a**) Tabor's strain [44] and (**b**) Kalidindi and Pathak's strain [26].

Due to uncertainty in assumed relations (33) and (34), the strain-rate histories shown in Figure 7 should be considered highly approximate. However, average strain rates during the loading phase of each test should be much more accurate, since these do not rely on (33) or (34). Rather, such averages depend only on the unloading slope S in (5) used to obtain $a_m = \frac{1}{2}S/\bar{E}$ at known time t_m at the measured depth $h = h_m$:

$$\dot{\bar{\epsilon}}_A = 0.2\frac{a_m}{Rt_m}, \qquad \dot{\hat{\epsilon}}_A = \frac{4}{3\pi}\frac{h_m}{a_m t_m}. \tag{35}$$

Values of these average strain rates are listed in Table 2. Notably, average rates follow the same trends over all three experiments, with $\dot{\bar{\epsilon}}_A \approx 1.5\dot{\hat{\epsilon}}_A$ in each case. These average strain rates are within ranges that can be obtained in conventional uniaxial SHPB experiments on this material [122,123].

The same trends apply for maximum strains: $\bar{\epsilon}_m \approx 1.5\hat{\epsilon}_m$ in all three experiments. Values of $\hat{\epsilon}_m$ range from 1.8% to 2.5%. Taking $\hat{\beta} \approx 1.3$ in (13), as in earlier simulations of other elastic–plastic materials [37], maximum uniaxial-equivalent strains are then approximated as ranging from 1.4% to 1.9%.

Applying a representative rate sensitivity parameter of $C = 0.01$ from Table 1, the ratio of dynamic flow stress at a strain rate of 2000/s is predicted by the Johnson–Cook model to be only 7.6% higher than that at a rate of 1000/s. Accordingly, any effects of different strain rates witnessed in experiments 1, 2, and 3 may be too small to be discerned in the load–displacement or indentation stress–strain curves. Viewed differently, the relatively low strain-rate sensitivity of the Al 6061-T6 material is corroborated by the similar indentation stress–strain curves among experiments at different transient and average loading rates. However, mild rate sensitivity effects, if they exist, might also be obscured by limited fidelity and the limited strain-rate range of the data.

Other discrete data of interest are included in Table 2. Maximum depth h_m and corresponding contact radius a_m were discussed already in the context of Figure 4a,b. Average velocities $\dot{h}_A$ are used to represent v entering dimensionless rates in the legends of Figures 4a and 5a. Likewise, plastic work W^P_m is used in the construction of Figure 4c. Note that W^P_m is the residual plastic work after complete unloading from depth h_m.

The average adiabatic temperature rise at maximum depth, $\Delta\bar{T}_m$, is computed from the second of (31), with $W^P = W^P_m$ and $a = a_m$. This is the temperature change that would be experienced by a cylindrical column of plastically deformed material of radius a_m and height $\frac{3}{4}\pi a_m$ [5,26]. Given the temperature sensitivity $q \approx 1$ of Al 6061-T6 in Table 1, the predicted heating-induced change in flow stress (i.e., thermal softening) for this volume of material should be negligible. However, localized temperature increases in regions of concentrated plastic strain (e.g., near the edges of the contact surface) could be substantially larger, whereas heat conduction could counteract such increases given sufficient time. Thus, thermal effects cannot be completely ruled out without verification from simulations, for example.

5.4. Summary and Recommendations

The following key points are ascertained from analysis of the dynamic indentation data:

- Maximum uniaxial-equivalent strains are estimated from maximum indentation strains to be on the order of 1% to 2%;
- Average uniaxial-equivalent strain rates are estimated from indentation strain rates to be on the order of 700 to 1500/s;
- Subtle variations in indentation force–depth curves can lead to drastic changes in indentation stress–strain curves, particularly at small indentation depths;
- Dynamic indentation stress–strain curves are qualitatively similar to those given elsewhere [37,38] for static spherical indentation, with constraint factors within ranges observed for static experiments on ductile metals;
- Mean pressure (i.e., indentation stress) shows evidence of yielding and mild-to-moderate strain-hardening characteristic in uniaxial stress–strain data for the aluminum alloy 6061-T6;
- Strain rate and inertial effects are not detected among the experimental datasets, whereby an increase in average indentation strain rate by a factor of 2 produces no apparent increase in indentation stress;

- A negligible effect of average strain rate correlates with the low strain-rate sensitivity of flow stress for the aluminum alloy, as measured in traditional SHPB experiments;
- Plastic work results in a trivially small adiabatic temperature rise ($\lesssim$ 2K) averaged over the entire plastically deformed zone, though magnitudes of localized temperature increases at plastic strain concentrations are unknown.

Consider the inverse problem of determination of material parameters (elastic, thermomechanical, and plastic entering (14) and (16)) from recorded indentation force versus depth, i.e., $P = ER^2 \cdot \Pi_P$ versus h, data for different loading velocities (e.g., average indentation depth rates) v and ambient temperatures T_0. In this context, R is known a priori, T_0 and v are imposed, P and h are measured, and E is presumably unknown and thus to be determined.

It may not be possible to determine E given only dimensionless Π_P versus h/R data. For example, for purely elastic Hertz-type indentation, the dimensionless force–displacement response of (28) is independent of E. Therefore, at least some experimental data should be recorded in dimensional, rather than dimensionless, form to allow for determination of all of the normalization constants of the dimensional analysis: $R, E, t_0, \tilde{T}$. In other words, measurement of P rather than Π_P reintroduces a stress scale into the inverse problem, in principle enabling implicit extraction of E. Similarly, measurement or knowledge of R reintroduces an independent length scale into the problem. Measurement or imposition of v rather than $\frac{vt_0}{R}$ reintroduces an independent time scale (i.e., $\frac{R}{v}$ given R), and measurement or imposition of T_0 rather than $\frac{\tilde{T}}{T_0}$ reintroduces an independent temperature scale.

The inverse problem can be stated as follows:

- Given $\frac{P}{R^2} = E \cdot \Pi_P\left(\frac{h}{R}, \frac{vt_0}{R}, \frac{\tilde{T}}{T_0}; \frac{\sigma_0}{E}, \frac{R/t_0}{\sqrt{E/\rho_0}}, \frac{\tilde{T}}{T_R}, \nu, \kappa, m, n, r\right)$ versus $\frac{h}{R}$ for different effective loading strain rates v/R and initial temperatures T_0, determine the eleven material properties $E, \nu, \rho_0, c_V, \sigma_0, \kappa, \dot{\epsilon}_0 = t_0^{-1}, m, n, r, T_R$. (The list of sought properties is reduced to nine if $\dot{\epsilon}_0$ and T_R are regarded as fixed universal constants.)

In the present dynamic indentation experiments, the dimensionless response function Π_a for contact radius cannot be precisely measured at present over the entire deformation history, and response function $ER^3 \cdot \Pi_W$ ultimately offers no additional information over $ER^2 \cdot \Pi_P$, since the former can be constructed from the history of the latter. Based on dimensional analysis of available results, the following deductions and recommendations are then proposed to facilitate the solution of the inverse problem:

- The relatively small magnitudes of impact velocities and the similarities of static and dynamic indentation curves suggest that inertial effects associated with ρ_0 cannot be discerned in the data. Thus, a standard (e.g., Archimedes) method should be used to measure ρ_0.
- The relatively low changes in average temperature suggest that effects of specific heat capacity c_V cannot be easily discerned in the force response data. Thus, a standard (e.g., calorimetry) method should be used to measure c_V.
- Precision of current experimental methods in the very small-depth regime is likely insufficient to directly ascertain elastic compliance via comparison with Hertz's solution. Elastic (shallow) force–depth data also seem unable to delineate E and ν distinctly, since P depends only on $E/(1-\nu^2)$ and not E and ν independently in the Hertz solution. Thus, a standard (e.g., longitudinal and shear wave speed) method should be used to measure E and ν. Any experimental facility equipped for dynamic instrumented indentation should include the capability for such sound speed measurements, presuming material samples are available.
- The precision of current experimental methods is likely inadequate for determination of an "exact" initial yield stress σ_0. However, an offset yield stress should be measurable, which can provide an approximate value of σ_0 as in static experiments [37].

- For low loading rates or rate-insensitive materials, extraction of static hardening parameters κ and n should be possible from measured increases of Π_P with increasing h/R, though unique determination of both parameters may or may not be difficult.
- Effects of loading rate vt_0/R on Π_P for highly rate-sensitive materials remains unknown. Experiments on other solids with much greater strain-rate sensitivity of flow stress are needed to determine if such rate sensitivity manifests in dynamic indentation force–displacement (and corresponding indentation stress–strain) curves over comparable domains of average indentation strain rates. If differences in Π_P at vastly different vt_0/R do not manifest for such materials, the present experimental method might be unsuitable to extract rate sensitivity parameters (e.g., m, or C if (15) is used).
- Following typical protocols [93,122], the two parameters $\dot{\epsilon}_0$ and T_R can be set universally at 1/s and 294 or 300 K, which reduces the complexity of the inverse problem, as noted already.
- Presumably, systematic matching of experimental $ER^2 \cdot \Pi_P$ data with results of parametric FE simulations on the same geometry (sample size and R), loading rate history, and initial temperature, and covering a sufficient domain of possible material property sets, will produce the sought material property relationships. Similar efforts have been undertaken for static indentation, as reviewed in Section 2.3, though most not invoking dimensional analysis techniques.
- The existence, uniqueness, stability, and accuracy of the inverse method should be verified for multiple materials, with constitutive properties validated by comparison with values obtained from independent, alternative experimental techniques (e.g., standard SHPB compression tests rather than dynamic indentation).

Several other recommendations are in order. Firstly, experimental methods to directly measure the contact radius during dynamic indentation could supply data that would render the assumed rate-independent forms in (33) and (34) unnecessary. Efforts are presently underway to measure residual impressions with confocal microscopy. Techniques involving periodic dynamic unloading in the SHPB apparatus to measure contact stiffness are also under development [88], though their accuracy remains unclear.

While knowledge of $a = \Pi_A \cdot R$ is not needed to solve the stated inverse problem, the contact radius is needed to calculate indentation stress (mean pressure) $\bar{p}$, as well as indentation strains $\bar{\epsilon}$ and $\hat{\epsilon}$. Constitutive property extraction might be more efficient and accurate by matching experimental and simulated indentation stress–strain curves rather than matching $P = ER^2 \cdot \Pi_P$ vs. h or h/R.

Comparison of static and dynamic isothermal FE simulations on the same geometry (i.e., same sample size and R), for a hypothetical rate-independent material, would enable verification of the tentative conclusion that inertial effects are negligible for the present range of loading rates. Lastly, adiabatic FE simulations could be used to provide an upper bound on localized temperature rise in highly strained regions under the indenter, and thus verify the tentative conclusion that effects of temperature rise (but not necessarily initial temperature) on indentation force are negligible over the present range of loading rates.

6. Conclusions

Static and dynamic indentation methods for ductile metals were reviewed. This review focused on spherical indentation, including experiments, analytical fundamentals, and numerical models; the motivation being extraction of information on the constitutive response of the indented material. Analytical foundations for the interpretation of data from instrumented dynamic spherical indentation experiments were established.

To demonstrate and extend concepts covered in the literature review, a framework to guide future experiments and numerical simulations was set forth based on principles of dimensional analysis. The ultimate intent is acquiring constitutive properties, here focused on plastic properties of ductile metals, from dynamic indentation force–depth data at different loading rates and different initial temperatures. In a representative application, the potential utility and limitations of this framework were assessed using data collected on Al 6061-T6, obtained from SHPB experiments equipped for instrumented spherical indentation. The analysis revealed that average strain rates achieved in experiments are on the order of 10^3/s, with maximum equivalent strains on the order of 2%. Indentation stress–strain results verify that the strain-rate sensitivity of the alloy is low.

Author Contributions: Conceptualization, J.D.C., D.T.C., J.T.L. and E.H.R.; methodology, J.D.C., D.T.C., J.T.L. and E.H.R.; formal analysis, J.D.C.; investigation, J.D.C., D.T.C. and E.H.R.; writing—original draft preparation, J.D.C.; writing—review and editing, J.D.C., D.T.C., J.T.L. and E.H.R. All authors have read and agreed to the published version of the manuscript.

Funding: This research received no external funding.

Institutional Review Board Statement: Not applicable.

Informed Consent Statement: Not applicable.

Data Availability Statement: The data presented in this study that are not already included in the tables and figures of this paper may be available upon request from the corresponding author. All such data are not publicly available at this time due to policies of the authors' organization.

Acknowledgments: The authors acknowledge support from the intramural component of the High-Throughput Materials Discovery for Extreme Conditions program at the US Army Combat Capabilities Development Command (DEVCOM) Army Research Laboratory (ARL). Information exchanges with Surya Kalidindi of the Georgia Institute of Technology on spherical indentation experiments and models are acknowledged as well. A preliminary version of this article, not refereed or peer-reviewed, first appeared in 2022 in an internal institutional preprint [124] (i.e., lab report) by the same authors. A few minor errors in the text and equations of that preprint [124], which did not affect calculated results, are corrected in the current work.

Conflicts of Interest: The authors declare no conflict of interest.

Notation

The following symbols (dimensions; force = mass·length/time^2) are used:

a	projected contact radius	[length]
a_r	radius of residual imprint	[length]
A	projected contact area	[length^2]
c	spherical constraint factor	[-]
c_V	specific heat capacity	[force·length/mass·temperature]
C	strain-rate sensitivity of ref. [93]	[-]
C_l	elastic wave speed	[length/time]
E	elastic modulus	[force/length^2]
h	total indentation depth	[length]
h_e	elastic indentation depth	[length]
h_i	indenter indentation depth	[length]
h_m	maximum indentation depth	[length]
h_r	residual indentation depth	[length]
h_s	substrate indentation depth	[length]
H	spherical indentation (Meyer's) hardness	[force/length^2]
$\bar{k}$	indentation stiffness	[force/length$^{3/2}$]
m	strain-rate sensitivity exponent	[-]
n	strain-hardening exponent	[-]
$\bar{p}$	mean indentation pressure	[force/length^2]
P	total indentation force	[force]
q	thermal softening exponent of ref. [93]	[-]
r	thermal softening exponent of ref. [4]	[-]
R	radius of rigid indenter	[length]
R_i	radius of indenter	[length]
R_s	radius of substrate	[length]

$\bar{R}$	indentation system radius	[length]
S	unloading slope	[force/length]
t	time	[time]
t_0	reference time (inverse strain rate)	[time]
T	absolute temperature	[temperature]
T_0	initial temperature	[temperature]
T_M	melt temperature	[temperature]
T_R	reference temperature	[temperature]
$\tilde{T}$	normalization temperature	[temperature]
$\Delta\bar{T}$	mean temperature rise	[temperature]
w^P	local plastic work density	[force/length^2]
W^P	plastic work of indentation	[force·length]
ϵ	total strain	[-]
ϵ^E	elastic strain	[-]
ϵ^P	plastic strain	[-]
$\bar{\epsilon}$	indentation strain of ref. [44]	[-]
$\hat{\epsilon}$	indentation strain of ref. [5]	[-]
$\check{\epsilon}$	indentation strain of ref. [46]	[-]
$\dot{\epsilon}$	total strain rate	[1/time]
$\dot{\epsilon}^E$	elastic strain rate	[1/time]
$\dot{\epsilon}^P$	plastic strain rate	[1/time]
$\dot{\epsilon}_0$	reference strain rate	[1/time]
κ	strain hardening coefficient	[-]
ν	Poisson's ratio	[-]
Π_a	dimensionless contact radius	[-]
Π_P	dimensionless indentation force	[-]
Π_W	dimensionless plastic work	[-]
ρ	current mass density	[mass/length^3]
ρ_0	initial mass density	[mass/length^3]
σ	local von Mises stress	[force/length^2]
$\bar{\sigma}$	mean indentation flow stress of ref. [44]	[force/length^2]
$\hat{\sigma}$	mean indentation flow stress of ref. [5]	[force/length^2]
σ_0	initial athermal static yield strength	[force/length^2]
v	indentation system velocity	[length/time]
$\bar{v}$	average velocity of input bar	[length/time]
ζ	Taylor–Quinney ratio	[-]

References

1. Weaver, J.; Khosravani, A.; Castillo, A.; Kalidindi, S. High throughput exploration of process-property linkages in Al-6061 using instrumented spherical microindentation and microstructurally graded samples. *Integr. Mater. Manuf. Innov.* **2016**, *5*, 192–211. [CrossRef]
2. Taljat, B.; Zacharia, T.; Kosel, F. New analytical procedure to determine stress-strain curve from spherical indentation data. *Int. J. Solids Struct.* **1998**, *35*, 4411–4426. [CrossRef]
3. Mesarovic, S.; Fleck, N. Spherical indentation of elastic–plastic solids. *Proc. R. Soc. Lond. A* **1999**, *455*, 2707–2728. [CrossRef]
4. Clayton, J. *Spherical Indentation in Elastoplastic Materials: Modeling and Simulation;* Technical Report ARL-TR-3516; US Army Research Laboratory, Aberdeen Proving Ground: Adelphi, MD, USA, 2005.
5. Pathak, S.; Kalidindi, S. Spherical nanoindentation stress–strain curves. *Mater. Sci. Eng. R* **2015**, *91*, 1–36. [CrossRef]
6. Field, J.; Swain, M. A simple predictive model for spherical indentation. *J. Mater. Res.* **1993**, *8*, 297–306. [CrossRef]
7. Cheng, Y.T.; Cheng, C.M. Scaling, dimensional analysis, and indentation measurements. *Mater. Sci. Eng. R* **2004**, *44*, 91–149. [CrossRef]
8. Barick, M.; Gaillard, Y.; Lejeune, A.; Amiot, F.; Richard, F. On the uniqueness of intrinsic viscoelastic properties of materials extracted from nanoindentation using FEMU. *Int. J. Solids Struct.* **2020**, *202*, 929–946. [CrossRef]
9. Chen, X.; Hutchinson, J.; Evans, A. The mechanics of indentation induced lateral cracking. *J. Am. Ceram. Soc.* **2005**, *88*, 1233–1238. [CrossRef]
10. Armstrong, R. On size effects in polycrystal plasticity. *J. Mech. Phys. Solids* **1961**, *9*, 196–199. [CrossRef]
11. Voyiadjis, G.; Yaghoobi, M. Review of nanoindentation size effect: Experiments and atomistic simulation. *Crystals* **2017**, *7*, 321. [CrossRef]
12. Voyiadjis, G.; Yaghoobi, M. *Size Effects in Plasticity: From Macro to Nano;* Academic Press: London, UK, 2019.

13. Armstrong, R. Dislocation mechanics pile-up and thermal activation roles in metal plasticity and fracturing. *Metals* **2019**, *9*, 154. [CrossRef]
14. Abu Al-Rub, R.; Voyiadjis, G. Analytical and experimental determination of the material intrinsic length scale of strain gradient plasticity theory from micro-and nano-indentation experiments. *Int. J. Plast.* **2004**, *20*, 1139–1182. [CrossRef]
15. Voyiadjis, G.; Song, Y. Strain gradient continuum plasticity theories: Theoretical, numerical and experimental investigations. *Int. J. Plast.* **2019**, *121*, 21–75. [CrossRef]
16. Zubelewicz, A. Mechanisms-based transitional viscoplasticity. *Crystals* **2020**, *10*, 212. [CrossRef]
17. Clayton, J.; Aydelotte, B.; Becker, R.; Hilton, C.; Knap, J. Continuum modelling and simulation of indentation in transparent single crystalline minerals and energetic solids. In *Applied Nanoindentation in Advanced Materials*; Tiwari, A., Ed.; Wiley: Hoboken, NJ, USA, 2017; Chapter 15, pp. 347–368.
18. Knap, J.; Ortiz, M. Effect of indenter-radius size on Au (001) nanoindentation. *Phys. Rev. Lett.* **2003**, *90*, 226102. [CrossRef] [PubMed]
19. Buckingham, E. On physically similar systems; illustrations of the use of dimensional equations. *Phys. Rev.* **1914**, *4*, 345–376. [CrossRef]
20. Sedov, L. *Similarity and Dimensional Methods in Mechanics*; Academic Press: London, UK, 1959.
21. Ni, W.; Cheng, Y.T.; Cheng, C.M.; Grummon, D. An energy-based method for analyzing instrumented spherical indentation experiments. *J. Mater. Res.* **2004**, *19*, 149–157. [CrossRef]
22. Clayton, J. Penetration resistance of armor ceramics: Dimensional analysis and property correlations. *Int. J. Impact Eng.* **2015**, *85*, 124–131. [CrossRef]
23. Lee, A.; Komvopoulos, K. Dynamic spherical indentation of strain hardening materials with and without strain rate dependent deformation behavior. *Mech. Mater.* **2019**, *133*, 128–137. [CrossRef]
24. Johnson, K. *Contact Mechanics*; Cambridge University Press: Cambridge, UK, 1985.
25. Ghaednia, H.; Wang, X.; Saha, S.; Xu, Y.; Sharma, A.; Jackson, R. A review of elastic–plastic contact mechanics. *Appl. Mech. Rev.* **2017**, *69*, 060804. [CrossRef]
26. Kalidindi, S.; Pathak, S. Determination of the effective zero-point and the extraction of spherical nanoindentation stress–strain curves. *Acta Mater.* **2008**, *56*, 3523–3532. [CrossRef]
27. Clayton, J.; Becker, R. *Modeling Nonlinear Elastic-Plastic Behavior of RDX Single Crystals during Indentation*; Technical Report ARL-TR-5864; US Army Research Laboratory, Aberdeen Proving Ground: Adelphi, MD, USA, 2005.
28. Oliver, W.; Pharr, G. An improved technique for determining hardness and elastic modulus using load and displacement sensing indentation experiments. *J. Mater. Res.* **1992**, *7*, 1564–1583. [CrossRef]
29. Dao, M.; Chollacoop, N.; Van Vliet, K.; Venkatesh, T.; Suresh, S. Computational modeling of the forward and reverse problems in instrumented sharp indentation. *Acta Mater.* **2001**, *49*, 3899–3918. [CrossRef]
30. Chollacoop, N.; Dao, M.; Suresh, S. Depth-sensing instrumented indentation with dual sharp indenters. *Acta Materialia* **2003**, *51*, 3713–3729. [CrossRef]
31. Cao, Y.P.; Lu, J. A new method to extract the plastic properties of metal materials from an instrumented spherical indentation loading curve. *Acta Mater.* **2004**, *52*, 4023–4032. [CrossRef]
32. Johnson, K. The correlation of indentation experiments. *J. Mech. Phys. Solids* **1970**, *18*, 115–126. [CrossRef]
33. Zhao, M.; Ogasawara, N.; Chiba, N.; Chen, X. A new approach to measure the elastic–plastic properties of bulk materials using spherical indentation. *Acta Mater.* **2006**, *54*, 23–32. [CrossRef]
34. Phadikar, J.; Bogetti, T.; Karlsson, A. Aspects of experimental errors and data reduction schemes from spherical indentation of isotropic materials. *J. Eng. Mater. Technol.* **2014**, *136*, 031005. [CrossRef]
35. Le, M.Q. Material characterization by instrumented spherical indentation. *Mech. Mater.* **2012**, *46*, 42–56. [CrossRef]
36. Donohue, B.; Ambrus, A.; Kalidindi, S. Critical evaluation of the indentation data analyses methods for the extraction of isotropic uniaxial mechanical properties using finite element models. *Acta Mater.* **2012**, *60*, 3943–3952. [CrossRef]
37. Patel, D.; Kalidindi, S. Correlation of spherical nanoindentation stress-strain curves to simple compression stress-strain curves for elastic-plastic isotropic materials using finite element models. *Acta Mater.* **2016**, *112*, 295–302. [CrossRef]
38. Fernandez-Zelaia, P.; Joseph, V.; Kalidindi, S.; Melkote, S. Estimating mechanical properties from spherical indentation using Bayesian approaches. *Mater. Des.* **2018**, *147*, 92–105. [CrossRef]
39. Lu, L.; Dao, M.; Kumar, P.; Ramamurty, U.; Karniadakis, G.; Suresh, S. Extraction of mechanical properties of materials through deep learning from instrumented indentation. *Proc. Natl. Acad. Sci. USA* **2020**, *117*, 7052–7062. [CrossRef] [PubMed]
40. Chen, G.; Zhang, X.; Zhong, J.; Shi, J.; Wang, Q.; Guan, K. New inverse method for determining uniaxial flow properties by spherical indentation test. *Chin. J. Mech. Eng.* **2021**, *34*, 94. [CrossRef]
41. Simo, J.; Hughes, T. *Computational Inelasticity*; Springer: New York, NY, USA, 1998.
42. Francis, H. Phenomenological analysis of plastic spherical indentation. *J. Eng. Mater. Technol.* **1976**, *98*, 272–281. [CrossRef]
43. Matthews, J. Indentation hardness and hot pressing. *Acta Metall.* **1980**, *28*, 311–318. [CrossRef]
44. Tabor, D. *The Hardness of Metals*; Oxford University Press: Oxford, UK, 1951.
45. Hill, R.; Storåkers, B.; Zdunek, A. A theoretical study of the Brinell hardness test. *Proc. R. Soc. Lond. A* **1989**, *423*, 301–330.
46. Lee, A.; Komvopoulos, K. Dynamic spherical indentation of elastic-plastic solids. *Int. J. Solids Struct.* **2018**, *146*, 180–191. [CrossRef]

47. Follansbee, P.; Sinclair, G. Quasi-static normal indentation of an elasto-plastic half-space by a rigid sphere–I: Analysis. *Int. J. Solids Struct.* **1984**, *20*, 81–91. [CrossRef]
48. Bhattacharya, A.; Nix, W. Finite element simulation of indentation experiments. *Int. J. Solids Struct.* **1988**, *24*, 881–891. [CrossRef]
49. Lee, H.; Lee, J.H.; Pharr, G. A numerical approach to spherical indentation techniques for material property evaluation. *J. Mech. Phys. Solids* **2005**, *53*, 2037–2069. [CrossRef]
50. Moussa, C.; Hernot, X.; Bartier, O.; Delattre, G.; Mauvoisin, G. Identification of the hardening law of materials with spherical indentation using the average representative strain for several penetration depths. *Mater. Sci. Eng. A* **2014**, *606*, 409–416. [CrossRef]
51. Moussa, C.; Hernot, X.; Bartier, O.; Delattre, G.; Mauvoisin, G. Evaluation of the tensile properties of a material through spherical indentation: Definition of an average representative strain and a confidence domain. *J. Mater. Sci.* **2014**, *49*, 592–603. [CrossRef]
52. Arizzi, F.; Rizzi, E. Elastoplastic parameter identification by simulation of static and dynamic indentation tests. *Model. Simul. Mater. Sci. Eng.* **2014**, *22*, 035017. [CrossRef]
53. Dean, J.; Clyne, T. Extraction of plasticity parameters from a single test using a spherical indenter and FEM modelling. *Mech. Mater.* **2017**, *105*, 112–122. [CrossRef]
54. Huber, N.; Tsakmakis, C. Determination of constitutive properties from spherical indentation data using neural networks. Part I: The case of pure kinematic hardening in plasticity laws. *J. Mech. Phys. Solids* **1999**, *47*, 1569–1588. [CrossRef]
55. Huber, N.; Tsakmakis, C. Determination of constitutive properties from spherical indentation data using neural networks. Part II: Plasticity with nonlinear isotropic and kinematic hardening. *J. Mech. Phys. Solids* **1999**, *47*, 1589–1607. [CrossRef]
56. Wang, M.; Wang, W. An inverse method for measuring elastoplastic properties of metallic materials using Bayesian model and residual imprint from spherical indentation. *Materials* **2021**, *14*, 7105. [CrossRef] [PubMed]
57. Cheng, Y.T.; Cheng, C.M. Relationships between hardness, elastic modulus, and the work of indentation. *Appl. Phys. Lett.* **1998**, *73*, 614–616. [CrossRef]
58. Cheng, Y.T.; Li, Z.; Cheng, C.M. Scaling relationships for indentation measurements. *Philos. Mag. A* **2002**, *82*, 1821–1829. [CrossRef]
59. Chen, X.; Ogasawara, N.; Zhao, M.; Chiba, N. On the uniqueness of measuring elastoplastic properties from indentation: The indistinguishable mystical materials. *J. Mech. Phys. Solids* **2007**, *55*, 1618–1660. [CrossRef]
60. Liu, L.; Ogasawara, N.; Chiba, N.; Chen, X. Can indentation technique measure unique elastoplastic properties. *J. Mater. Res.* **2009**, *24*, 784–800. [CrossRef]
61. Cao, Y.; Lu, J. Depth-sensing instrumented indentation with dual sharp indenters: stability analysis and corresponding regularization schemes. *Acta Mater.* **2004**, *52*, 1143–1153. [CrossRef]
62. Barick, M. Identification of the Viscoelastic-Viscoplastic Properties of Materials by Instrumented Nanoindentation. Ph.D. Thesis, Université Bourgogne Franche-Comté, Besançon, France, 2020.
63. Clayton, J. *Nonlinear Mechanics of Crystals*; Springer: Dordrecht, The Netherlands, 2011.
64. Wu, J.; Wang, M.; Hui, Y.; Zhang, Z.; Fan, H. Identification of anisotropic plasticity properties of materials using spherical indentation imprint mapping. *Mater. Sci. Eng. A* **2018**, *723*, 269–278. [CrossRef]
65. Clayton, J.; Becker, R. Elastic-plastic behavior of cyclotrimethylene trinitramine single crystals under spherical indentation: Modeling and simulation. *J. Appl. Phys.* **2012**, *111*, 063512. [CrossRef]
66. Patel, D.; Al-Harbi, H.; Kalidindi, S. Extracting single-crystal elastic constants from polycrystalline samples using spherical nanoindentation and orientation measurements. *Acta Mater.* **2014**, *79*, 108–116. [CrossRef]
67. Choi, I.C.; Yoo, B.G.; Kim, Y.J.; Jang, J.I. Indentation creep revisited. *J. Mater. Res.* **2012**, *27*, 3–11. [CrossRef]
68. Yang, S.; Lu, W.; Ling, X. Spherical indentation creep characteristics and local deformation analysis of 310S stainless steel. *Eng. Fail. Anal.* **2020**, *118*, 104946. [CrossRef]
69. Ogbonna, N.; Fleck, N.; Cocks, A. Transient creep analysis of ball indentation. *Int. J. Mech. Sci.* **1995**, *37*, 1179–1202. [CrossRef]
70. Huber, N.; Tyulyukovskiy, E. A new loading history for identification of viscoplastic properties by spherical indentation. *J. Mater. Res.* **2004**, *19*, 101–113. [CrossRef]
71. Mohammed, Y.; Stone, D.; Elmustafa, A. Strain rate sensitivity of hardness in indentation creep with conical and spherical indenters taking into consideration elastic deformations. *Int. J. Solids Struct.* **2021**, *212*, 143–151. [CrossRef]
72. Clayton, J. A continuum description of nonlinear elasticity, slip and twinning, with application to sapphire. *Proc. R. Soc. Lond. A* **2009**, *465*, 307–334. [CrossRef]
73. Clayton, J.; Knap, J. A phase field model of deformation twinning: Nonlinear theory and numerical simulations. *Phys. D* **2011**, *240*, 841–858. [CrossRef]
74. Clayton, J.; Knap, J. Phase field modeling of twinning in indentation of transparent single crystals. *Model. Simul. Mater. Sci. Eng.* **2011**, *19*, 085005. [CrossRef]
75. Clayton, J.; Lloyd, J. A dynamic finite-deformation constitutive model for steels undergoing slip, twinning, and phase changes. *J. Dyn. Behav. Mater.* **2021**, *7*, 217–247. [CrossRef]
76. Clayton, J.; Lloyd, J. Finite strain continuum theory for phase transformations in ferromagnetic elastic-plastic solids. *Contin. Mech. Thermodyn.* **2022**, *34*, 1579–1620. [CrossRef]
77. Leavy, R.; Clayton, J.; Strack, O.; Brannon, R.; Strassburger, E. Edge on impact simulations and experiments. *Procedia Eng.* **2013**, *58*, 445–452. [CrossRef]

78. Clayton, J.; Guziewski, M.; Ligda, J.; Leavy, R.; Knap, J. A multi-scale approach for phase field modeling of ultra-hard ceramic composites. *Materials* **2021**, *14*, 1408. [CrossRef]
79. Clayton, J. Modeling deformation and fracture of boron-based ceramics with nonuniform grain and phase boundaries and thermal-residual stress. *Solids* **2022**, *3*, 643–664. [CrossRef]
80. Mok, C.H.; Duffy, J. The dynamic stress-strain relation of metals as determined from impact tests with a hard ball. *Int. J. Mech. Sci.* **1965**, *7*, 355–371. [CrossRef]
81. Tirupataiah, Y.; Sundararajan, G. A dynamic indentation technique for the characterization of the high strain rate plastic flow behaviour of ductile metals and alloys. *J. Mech. Phys. Solids* **1991**, *39*, 243–271. [CrossRef]
82. Sundararajan, G.; Tirupataiah, Y. The localization of plastic flow under dynamic indentation conditions: I. Experimental results. *Acta Mater.* **2006**, *54*, 565–575. [CrossRef]
83. Sundararajan, G.; Tirupataiah, Y. The localization of plastic flow under dynamic indentation conditions: II. Analysis of results. *Acta Mater.* **2006**, *54*, 577–586. [CrossRef]
84. Wen, Y.; Xie, L.; Wang, Z.; Wang, L.; Lu, W.; Zhang, L.C. Nanoindentation characterization on local plastic response of Ti-6Al-4V under high-load spherical indentation. *J. Mater. Res. Technol.* **2019**, *8*, 3434–3442. [CrossRef]
85. Anton, R.; Subhash, G. Dynamic Vickers indentation of brittle materials. *Wear* **2000**, *239*, 27–35. [CrossRef]
86. Subhash, G. Dynamic indentation testing. *ASM Handb. Mech. Test. Eval.* **2000**, *8*, 519–529.
87. Li, S.; Haozhe, L. A new technique of dynamic spherical indentation based on SHPB. In *Dynamic Behavior of Materials*; Song, B., Casem, D., Kimberley, J., Eds.; Springer: Cham, Switzerland, 2014; Volume 1, pp. 81–87.
88. Casem, D. *A Model for a Kolsky Bar Experiment: Application to Experiment Design*; Technical Report ARL-TR-9416; DEVCOM Army Research Laboratory, Aberdeen Proving Ground: Adelphi, MD, USA, 2022.
89. Lu, J.; Suresh, S.; Ravichandran, G. Dynamic indentation for determining the strain rate sensitivity of metals. *J. Mech. Phys. Solids* **2003**, *51*, 1923–1938. [CrossRef]
90. Calle, M.; Mazzariol, L.; Alves, M. Strain rate sensitivity assessment of metallic materials by mechanical indentation tests. *Mater. Sci. Eng.* **2018**, *725*, 274–282. [CrossRef]
91. Mallick, D.; Zhao, M.; Parker, J.; Kannan, V.; Bosworth, B.; Sagapuram, D.; Foster, M.; Ramesh, K. Laser-driven flyers and nanosecond-resolved velocimetry for spall studies in thin metal foils. *Exp. Mech.* **2019**, *59*, 611–628. [CrossRef]
92. Clayton, J. *Nonlinear Elastic and Inelastic Models for Shock Compression of Crystalline Solids*; Springer: Cham, Switzerland, 2019.
93. Johnson, G.; Cook, W. A constitutive model and data for materials subjected to large strains, high strain rates, and high temperatures. In Proceedings of the 7th International Symposium on Ballistics, The Hague, The Netherlands, 19–21 April 1983; pp. 541–547.
94. Ito, K.; Arai, M. Expanding cavity model combined with Johnson-Cook constitutive equation for the dynamic indentation problem. *J. Eng. Mater. Technol.* **2020**, *142*, 021005. [CrossRef]
95. Rosakis, P.; Rosakis, A.; Ravichandran, G.; Hodowany, J. A thermodynamic internal variable model for the partition of plastic work into heat and stored energy in metals. *J. Mech. Phys. Solids* **2000**, *48*, 581–607. [CrossRef]
96. Clayton, J. Dynamic plasticity and fracture in high density polycrystals: constitutive modeling and numerical simulation. *J. Mech. Phys. Solids* **2005**, *53*, 261–301. [CrossRef]
97. Zubelewicz, A. Century-long Taylor-Quinney interpretation of plasticity-induced heating reexamined. *Sci. Rep.* **2019**, *9*, 1–7. [CrossRef] [PubMed]
98. Zerilli, F.; Armstrong, R. Dislocation-mechanics-based constitutive relations for material dynamics calculations. *J. Appl. Phys.* **1987**, *61*, 1816–1825. [CrossRef]
99. Preston, D.; Tonks, D.; Wallace, D. Model of plastic deformation for extreme loading conditions. *J. Appl. Phys.* **2003**, *93*, 211–220. [CrossRef]
100. Zubelewicz, A. Metal behavior at extreme loading rates: Thermodynamics. *Phys. Rev. B* **2008**, *77*, 214111. [CrossRef]
101. Voyiadjis, G.; Almasri, A. A physically based constitutive model for fcc metals with applications to dynamic hardness. *Mech. Mater.* **2008**, *40*, 549–563. [CrossRef]
102. Si, B.; Li, Z.; Xiao, G.; Shu, X. Determination of mechanical properties from sharp dynamic indentation. *J. Strain Anal. Eng. Des.* **2022**, *57*, 607–613. [CrossRef]
103. Berthoud, P.; Baumberger, T.; G'sell, C.; Hiver, J.M. Physical analysis of the state-and rate-dependent friction law: Static friction. *Phys. Rev. B* **1999**, *59*, 14313–14327. [CrossRef]
104. Baumberger, T.; Berthoud, P.; Caroli, C. Physical analysis of the state-and rate-dependent friction law. II. Dynamic friction. *Phys. Rev. B* **1999**, *60*, 3928–3939. [CrossRef]
105. Kren, A.; Rudnitskii, V.; Lantsman, G.; Khudoley, A. Influence of the dynamic indentation parameters on the behavior of metals during the penetration of an indenter with a spherical tip. *Russ. Metall.* **2021**, *2021*, 563–569. [CrossRef]
106. Kumaraswamy, A.; Rao, V. High strain-rate plastic flow behavior of Ti-6Al-4V from dynamic indentation experiments. *Mater. Sci. Eng. A* **2011**, *528*, 1238–1241. [CrossRef]
107. Ito, K.; Arai, M. Simple estimation method for strain rate sensitivity based on the difference between the indentation sizes formed by spherical-shaped impactors. *Int. J. Mech. Sci.* **2021**, *189*, 106007. [CrossRef]
108. Kashfi, M.; Goodarzi, S.; Rastgou, M. Plastic properties determination using virtual dynamic spherical indentation test and machine learning algorithms. *J. Mech. Sci. Technol.* **2022**, *36*, 325–331. [CrossRef]

109. Yew, C.H.; Goldsmith, W. Stress distributions in soft metals due to static and dynamic loading by a steel sphere. *ASME J. Appl. Mech.* **1964**, *31*, 635–646. [CrossRef]
110. Rudnitsky, V.; Djakovicht, V. Material testing by the method of dynamic indentation. *Nondestruct. Test. Eval.* **1996**, *12*, 253–261. [CrossRef]
111. Nobre, J.; Dias, A.; Gras, R. Resistance of a ductile steel surface to spherical normal impact indentation: Use of a pendulum machine. *Wear* **1997**, *211*, 226–236. [CrossRef]
112. Clough, R.; Webb, S.; Armstrong, R. Dynamic hardness measurements using a dropped ball: With application to 1018 steel. *Mater. Sci. Eng. A* **2003**, *360*, 396–407. [CrossRef]
113. Liu, B.; Soares, C. Effect of strain rate on dynamic responses of laterally impacted steel plates. *Int. J. Mech. Sci.* **2019**, *160*, 307–317. [CrossRef]
114. Meguid, S.; Shagal, G.; Stranart, J.; Daly, J. Three-dimensional dynamic finite element analysis of shot-peening induced residual stresses. *Finite Elem. Anal. Des.* **1999**, *31*, 179–191. [CrossRef]
115. Almasri, A.; Voyiadjis, G. Effect of strain rate on the dynamic hardness in metals. *ASME J. Eng. Mater. Technol.* **2007**, *129*, 505–512. [CrossRef]
116. dos Santos, T.; Srivastava, A.; Rodríguez-Martínez, J. The combined effect of size, inertia and porosity on the indentation response of ductile materials. *Mech. Mater.* **2021**, *153*, 103674. [CrossRef]
117. Nguyen, N.V.; Kim, J.; Kim, S.E. Methodology to extract constitutive equation at a strain rate level from indentation curves. *Int. J. Mech. Sci.* **2019**, *152*, 363–377. [CrossRef]
118. Atkins, A.; Tabor, D. Plastic indentation in metals with cones. *J. Mech. Phys. Solids* **1965**, *13*, 149–164. [CrossRef]
119. Casem, D.; Pittari, J.; Swab, J. High-rate indentation using miniature Kolsky bar methods. In *Dynamic Behavior of Materials*; Mates, S., Eliasson, V., Eds.; Springer: Cham, Switzerland, 2022; Volume 1, pp. 63–65.
120. Casem, D.; Grunschel, S.; Schuster, B. Normal and transverse displacement interferometers applied to small diameter Kolsky bars. *Exp. Mech.* **2012**, *52*, 173–184. [CrossRef]
121. Wu, B.; Cao, Y.; Zhao, J.; Ding, W. The effect of superimposed ultrasonic vibration on tensile behavior of 6061-T6 aluminum alloy. *Int. J. Adv. Manuf. Technol.* **2021**, *116*, 1843–1854. [CrossRef]
122. Zhu, D.; Mobasher, B.; Rajan, S.; Peralta, P. Characterization of dynamic tensile testing using aluminum alloy 6061-T6 at intermediate strain rates. *ASCE J. Eng. Mech.* **2011**, *137*, 669–679. [CrossRef]
123. Lesuer, D.; Kay, G.; LeBlanc, M. *Modeling Large-Strain, High-Rate Deformation in Metals;* Technical Report UCRL-JC-134118; Lawrence Livermore National Laboratory: Livermore, CA, USA, 2001.
124. Clayton, J.; Casem, D.; Lloyd, J.; Retzlaff, E. *Toward Material Property Extraction from Dynamic Spherical Indentation Experiments*; Technical Report ARL-TR-9520; US Army Research Laboratory, Aberdeen Proving Ground: Adelphi, MD, USA, 2022.

MDPI AG
Grosspeteranlage 5
4052 Basel
Switzerland
Tel.: +41 61 683 77 34

Metals Editorial Office
E-mail: metals@mdpi.com
www.mdpi.com/journal/metals

www.ingramcontent.com/pod-product-compliance
Lightning Source LLC
Chambersburg PA
CBHW061606210926
51850CB00009B/2233